'This insightful volume is a testament to the resilience and innovation of Meghalaya's communities in environmental stewardship. It serves as an inspiring blueprint for others worldwide, emphasizing the importance of local engagement in sustainable development.'

**Mr Aibanshngain Swer,** *Director, Meghalaya Institute of Governance, Shillong, India*

'The book showcases how local communities in the North East address environmental challenges of the region. The work presents narratives of grassroots initiatives and demonstrates the potential of community-driven conservation approaches for sustainable development.'

**Dr M. Dinesh Kumar,** *Executive Director, Institute of Resource Analysis and Policy, Hyderabad, India*

# MEGHALAYA MIRRORS

*Meghalaya Mirrors* offers an insightful exploration of innovative, community-led initiatives that are shaping environmental conservation, biodiversity preservation, and socio-economic development in Meghalaya, India.

Between the state's rich natural heritage and pressing environmental challenges such as deforestation, mining, and climate change, this book emphasizes the vital role of communities in protecting their resources and promoting sustainable development. Through case studies, grassroots perspectives, and examples of indigenous knowledge systems, the book showcases the resilience and creativity of local communities.

Part of the Innovations, Practice and the Future of Public Policy in India series, this open access volume will be an essential resource for policymakers, researchers, development practitioners, and stakeholders committed to understanding and advancing community-driven environmental action in ecologically sensitive regions. It will also be of interest to researchers working on climate change, climate action, social sciences, environmental science, geography, and South Asia studies.

**Anjal Prakash** is Professor - Public Policy at Flame University, Pune, India. Earlier he was the Clinical Associate Professor (Research) and Research Director at the Bharti Institute of Public Policy, ISB. With over 25 years of experience in water, climate, and policy research across South Asia, he has served as Coordinating Lead Author for the IPCC's Special Report on Oceans and Cryosphere and contributed to the IPCC AR6. He specialises in urban resilience, gender, and social inclusion, and has authored numerous publications, including a number of edited books.

**Sampath Kumar**, IAS, serves as Development Commissioner and Principal Secretary, Government of Meghalaya. With 26 years of public administration experience, he pioneered the State Capability Enhancement Project (SCEP), promoting adaptive governance, convergence, and data-driven decision-making. Under his leadership, Meghalaya achieved a significant reduction in acute malnutrition and institutionalised its Early Childhood Development Mission. A Harvard Kennedy School alumnus, he continues to champion systemic governance reforms rooted in community participation and iterative learning.

**Gunanka D. B.**, an Indian Forest Service officer, is the Additional Secretary to the Government of Meghalaya. With over 16 years spanning corporate IT and public administration, he works across institutional reform, natural resource management, and data-driven governance. He has led major externally aided projects financed by the World Bank, JICA, KfW, and ADB, strengthened geospatial systems, and advanced community-led conservation models. His work blends systems thinking with grounded implementation, contributing to applied research and policy innovation in ecologically sensitive regions. Currently, he is on study leave to pursue an MPA-DP at SIPA, Columbia University, through the Fulbright-Nehru Scholarship and Columbia Dean's Distinguished Scholarship.

## Innovations, Practice and the Future of Public Policy in India

Series Editors: **Ashwini Chhatre**, Bharti Institute of Public Policy, ISB, India; **Anjal Prakash**, Bharti Institute of Public Policy, ISB, India; and **Aarushi Jain**, Bharti Institute of Public Policy, ISB, India

The field of public policy is broad, interdisciplinary, and multifaceted. It requires constant communication between stakeholders – including academics, practitioners, and the general public – to address important public policy design and concerns around implementation.

This open access series from the Bharti Institute of Public Policy, Indian School of Business, India, brings together contributions and diverse perspectives of academics and practitioners in the field of public policy in India and records innovative ideas and best practices to facilitate knowledge transfer, replication, or scaling. Public policy must be proactive and focused on the future and work on developing ideas that may be scaled up, customised or applied across jurisdictions. The books in this series focus primarily on the voices of academics and practitioners and offer crucial insights for corrective action and creation of more comprehensive policy designs. By critically analysing cases and policy designs, it offers insights into the feasibility of innovative models where theory meets practice in the pursuit of the public good.

### Editorial Board
1. AK Shiva Kumar, Independent Consultant and Visiting Professor, Indian School of Business, India
2. Rajeev Malhotra, O.P. Jindal Global University, India
3. Purnamita Dasgupta, Institute for Economic Growth, India
4. Ram Sewak Sharma, Former CEO of National Health Authority
5. Sanjaya Baru, Political Commentator and Policy Analyst
6. Rohini Chaturvedi, Climate and Land Use Alliance (CLUA)

**Reimagining Institutions**
*Collaborative Pathways to Social Development in India*
*Edited by Pushpinder Puniha and Aarushi Jain*

**Pathways to Rural Prosperity**
*Livelihood Interventions and Transformation in India*
*Edited by Ved Arya, Ayushii Balutia, Ashwini Chhatre, and Nivedita Narain*

For more information about this series, please visit:
www.routledge.com/Public-Policy-in-India/book-series/PPI

# MEGHALAYA MIRRORS

## Reflections of Community-Led Environmental Action

*Edited by Anjal Prakash, Sampath Kumar and Gunanka D. B.*

Routledge
Taylor & Francis Group

LONDON AND NEW YORK

Designed cover image: Meghalaya Basin Development Authority (MBDA), Government of Meghalaya. Photograph by MBDA staff.

First published 2027
by Routledge
4 Park Square, Milton Park, Abingdon, Oxon OX14 4RN

and by Routledge
605 Third Avenue, New York, NY 10158

*Routledge is an imprint of the Taylor & Francis Group, an informa business*

© 2027 selection and editorial matter, Anjal Prakash, Sampath Kumar and Gunanka D.B; individual chapters, the contributors



For Product Safety Concerns and Information please contact our EU representative GPSR@taylorandfrancis.com. Taylor & Francis Verlag GmbH, Kaufingerstraße 24, 80331 München, Germany.



*British Library Cataloguing-in-Publication Data*
A catalogue record for this book is available from the British Library

ISBN: 978-1-041-22883-7 (hbk)
ISBN: 978-1-041-23107-3 (pbk)
ISBN: 978-1-003-73538-0 (ebk)

DOI: 10.4324/9781003735380

Typeset in Sabon
by Deanta Global Publishing Services, Chennai, India

# CONTENTS

**PART III**
**Green Financing: Innovations in Funding Climate Action**   213



**PART IV**
**Governance and Policy**   289

# TABLES

# LIST OF CONTRIBUTORS

**Subhash Ashutosh**, IFS (Retd.), is Co-Chairman and Director, Centre of Excellence (NRM & Sustainable Livelihoods), Meghalaya Basin Development Agency, Government of Meghalaya. A retired Indian Forest Service Officer and former Director General, Forest Survey of India, he is recognised for expertise in natural resource management, agroforestry, and remote sensing/GIS applications for forest monitoring and carbon inventory. He leads initiatives linking scientific forestry with community livelihoods, including a recent bamboo forest resource project aimed at generating carbon credits and community benefits through climate-resilient, innovative forest management.

**Van Shanborlang Buhphang** is Assistant General Manager at the State Project Management Unit of MegLIFE under MBDA. With over 18 years in rural development, he oversees planning, coordination, and monitoring across plantation, soil and water conservation, and livelihood sectors. He previously worked with Nehru Yuva Kendra, MIE, and MBMA, contributing to youth development, enterprise promotion, and financial inclusion. He holds a Master's in Rural Development and emphasises community-driven progress.

**T. Niang Suan Ching** is Assistant General Manager (HR) at MBDA, with qualifications in hospital management and an MBA in HR. She leads strategic HR initiatives for externally funded programmes and played a key role in launching the Apprenticeship Training Program to address workforce skill gaps. Her work strengthens institutional capacity, talent development, and human capital systems across the organisation.

**Lavinia Mary Dkhar** is Technical Specialist–NRM & Sustainable Livelihoods at MBDA's Centre of Excellence. She works on ecosystem restoration, PES, bamboo and agroforestry systems, and climate adaptation. With experience in biodiversity studies, monitoring frameworks, and policy-linked education projects, she has contributed to medicinal plant research, environmental baseline studies, and academic teaching. Her work integrates scientific assessment with community-focused development.

**Fettleman Dohling** is Assistant General Manager (State GIS & UAV) at MBDA and in-charge of the Meghalaya State GIS & UAV Centre. With postgraduate training in IT and Geoinformatics, he has over a decade of experience applying GIS, remote sensing, and UAV analytics to natural resource management. He leads enterprise GIS platform development, spatial infrastructure creation, inter-departmental coordination, field mapping, and capacity building across Meghalaya.

**Thomas Iangjuh** holds a Master's degree in Biochemistry from North-Eastern Hill University (NEHU), Meghalaya, India. He has worked extensively on projects related to strawberry cultivation, floriculture, biofertilisers, and community-based green technologies. His work emphasises the integration of biotechnology with rural development, aimed at empowering local communities through sustainable innovations. He has contributed immensely towards the implementation of key projects aimed at the development of sustainable agricultural practices and biotechnology in Meghalaya.

**Rajaul Karim** has over 10 years of experience in NRM and currently works with MBMA–MBDA on capacity building, data analysis, and technical reporting. He has contributed to major state initiatives, including Forest Management Plans, PES, bamboo assessments, MegARISE baseline studies, and statewide horticultural inventories. Previously, he worked on aromatic plants, essential oil extraction, and NRM boundary demarcation across Meghalaya.

**Jillianda Kharjana** is Water Resource Manager with expertise in springshed development, soil and water conservation, and natural resource management. With a B.Tech in Water Resource Engineering and a diploma in Geoinformatics, she contributed to rejuvenating critical springs during her tenure at MBDA. She now supports plantation and conservation interventions at Earthtree Enviro Pvt. Ltd., combining scientific knowledge with community-based approaches.

**Effie Gwyneth L. Kharjana** is Technical Assistant at MBDA's Centre for Water Resources, with six years of experience in water resource projects.

She previously worked on wind energy engineering and GIS-based forest planning. At MBDA, she leads research, project management, and technical decision-making across spring mapping, wastewater treatment, and stream rejuvenation initiatives, contributing to water security and innovation across Meghalaya.

**James T. Kharkongor** is Deputy Project Director at MBMA and General Manager of GREEN Meghalaya+ and has over 23 years in NRM and community-led conservation. He leads large-scale programmes on PES, forest restoration, and springshed management. Formerly with NERCORMP and IIFM, he has advanced participatory NRM, certification systems, and sustainable livelihoods in rural Meghalaya. He holds a PG Diploma in Forest Management.

**Mr. Paka I Yo Suja** is a Technical Assistant at the Institute of Natural Resources, MBDA, holding a BSc in Forestry and MSc in Environment, currently working under Megh Aroma Mission in East and West Jaintia Hills on essential oil cultivation, distillation, and reclaiming mine-degraded land for sustainable community development. Mr. Baniaraplang Syiemiong is a Technical Assistant at the Institute of Natural Resources, MBDA, with an MSc in Forestry from Mizoram University, specializing in agroforestry, conservation, and ecosystem restoration, and actively contributing to Megh Aroma Mission through medicinal plant cultivation, oil extraction, and landscape management initiatives. Ms. Rikermeka Kharsahnoh is an Assistant Manager at the Institute of Natural Resources, MBDA, engaged in natural resource management and leading Megh Aroma Mission activities focused on rehabilitating mining-affected areas, promoting sustainable aromatic crops, and supporting livelihoods, with training in forest management and organic farming certification. Ms. Diadema Irini Nongrum is a Field Assistant at the Institute of Natural Resources, MBDA, a graduate in Horticulture and Forestry, actively involved in Megh Aroma Mission by promoting medicinal plants for land reclamation, conducting awareness programs, site inspections, training, and monitoring plantation and essential oil activities. Mr. Markshield Warjri is a Technical Assistant at the Institute of Natural Resources, MBDA, working since 2020 on the Aroma Mission to restore mine-degraded lands across large areas using aromatic crops, improving soil conditions, and creating sustainable livelihood opportunities through essential oil production. Ms. Mourdome B Marak is a Field Assistant at the Institute of Natural Resources, MBDA, holding an MSc in Rural Development, currently working on Megh Aroma Mission and land reclamation by mobilizing farmers, conducting awareness, verifying plantation sites, and providing training in aromatic crop cultivation and essential oil extraction.

**Rishotskhem Kharkongor** is Field Assistant at the Institute of Natural Resources, Shillong, India, and works on restoring mining-affected lands through medicinal and aromatic plant cultivation under the Megh Aroma Mission. With an MSc in Applied Genetics, he supports cultivation monitoring, essential oil distillation, and quality assurance. His work advances ecological restoration and livelihood improvement by applying scientific techniques to degraded landscapes.

**Bandarihun Kharlor** is Assistant Manager at the Meghalaya Basin Development Authority (MBDA), Shillong, with over seven years of experience (joined 6 April 2018). Promoted from Programme Associate to Assistant Manager in 2024, she specialises in programme management, field coordination, data collection and analysis, training, and awareness. Her project work includes acid mine drainage neutralisation, stream restoration using Green Bridge Technology, springshed management under the Jal Abhyaranya Campaign and MegARISE, and monitoring soil and water conservation progress. She also contributes to documentation, reporting, and stakeholder capacity building.

**Smt. Rikmenlang Khongsngi** formerly served as Programme Associate at MBMA's Centre of Excellence for NRM & Sustainable Livelihoods. With 3+ years of experience, she contributed to the Seed Ball Initiative, Forest Management Plans, Bamboo Resource Assessments, and baseline studies under MegARISE. Her work advanced community-led ecological restoration and supported state-wide assessments of horticultural and arecanut crops, promoting sustainable landscape planning.

**Catherine Derica Kyndiah** is Senior Manager of the Village Data Volunteer Programme at MBMA. Trained in social work and community development, she has worked across government and nonprofit sectors in policy research, training, and capacity building. Her work integrates data with local knowledge to strengthen public services, heritage preservation, and rural development. She has also taught as a lecturer and consulted on grassroots social development initiatives.

**H. Liansuanmung** serves as Data Management Executive at the Centre of Excellence (Natural Resource Management & Sustainable Livelihoods) under the Meghalaya Basin Management Agency (MBMA–MBDA). Over the past three years, he has led key functions in data management, validation, processing, analysis, and report compilation, while also designing report templates and managing graphic design for official publications. He has contributed to major initiatives such as Forest Management Plans, Bamboo Resource Assessments, Seed Ball Initiative, and MegARISE baseline survey. He is currently involved in state-level activities, including Arecanut and Horticultural Crop Survey, Bamboo Certification Pilot, and Bamboo Growth Study.

**Jyswill Lyngdoh Nongpiur** is Manager for Environment Management at the Meghalaya Basin Management Agency Centre of Excellence (NRM & SL). With over a decade of experience in Natural Resource Management (NRM), he leads planning, training, field implementation, monitoring, and report writing for environmental and forestry initiatives across Meghalaya. He has contributed to Forest Management Plans, bamboo resource assessments, bamboo certification initiatives, bamboo carbon credit initiatives, Bamboo Growth Studies, seedball initiatives, Payment for Ecosystem Services (PES) design, the MegARISE baseline survey, and statewide horticultural and arecanut surveys. He is also actively involved in pioneering community-based Biochar initiatives aimed at promoting sustainable agriculture, carbon sequestration, waste-to-resource management, and climate resilience. His work strengthens community-driven conservation, sustainable land-use planning, ecosystem restoration, and evidence-based environmental management across Meghalaya.

**Junie P. Lyngdoh** holds a PhD in Allied Health Sciences and a Master's in Biochemistry. With over 20 years of experience, she has led significant work in orchid and banana micropropagation, biofertilisers, and sustainable agriculture. She has pioneered ARC-based potato seed production and supported indigenous knowledge systems. Her contributions include promoting women-led bio-entrepreneurship and publishing research on traditional healthcare and community-based biotechnological initiatives in Meghalaya.

**Reccica D. Lyngkhoi** holds a Master's in Microbiology and has worked at the Bio-Resources Development Centre since 2017. Her work focuses on microbial bioinoculants, biofertiliser field trials, and establishing community seed-saving systems. She contributes to sustainable agriculture by supporting grassroots interventions and developing proposals for state and central agencies. Her research-driven approach strengthens biotechnological development and food security initiatives in Meghalaya.

**Dokatchi K. Marak** is Senior Manager at MBDA, brings over a decade of experience in development practice in Meghalaya. She has led major plantation and forest conservation initiatives, delivered capacity-building programmes, and supported ecosystem services projects. With a background in Sociology and Tourism, she is also an accomplished trainer, communicator, and event facilitator. Her work spans stakeholder engagement, project management, and community mobilisation towards ecological restoration and sustainable livelihoods.

**Raynold Marwein,** with over eight years of experience, has been serving as a Field Engineer now been promoted to Assistant Manager. His expertise includes the design, estimation, and supervision of structures for water quality improvement and management – such as Acid Mine Drainage neutralisation, stream restoration using Green Bridge Technology, and springshed management under the Jal Abhyaranya Campaign and MegARISE Project.

**Tilaris Marwein** is Watershed Landscape Conservator with over three years of experience in heritage preservation, community development, and stakeholder engagement. She works closely with local institutions on participatory planning and conservation. She played a key role in preparing the UNESCO nomination dossier for the Jingkieng Jri/Living Root Bridge Cultural Landscape and supports collaborative management approaches for natural and cultural heritage sites in Meghalaya.

**Delanmi Massar** is Deputy Manager for M&E and Knowledge Management under MegLIFE and began his career interning with *The Guardian* before working in teaching and communication roles. He holds degrees in English and Mass Communication and has contributed to knowledge management and livelihood initiatives at MBMA since 2020. His work spans documentation, field coordination, and supporting income-generating activities across sustainable development programmes.

**Sucielia Mylliemngap** is General Manager (HR) at MBDA, and has over a decade of experience in workforce planning, employee engagement, and organisational development. With a PG Diploma in Industrial Relations, she has led major HR reforms and capacity-building initiatives across the organisation. She also volunteers with community-focused NGOs and is committed to fostering inclusive growth and empowering marginalised groups.

**B. Padmapriya** is research fellow at the Bharti Institute of Public Policy, ISB, Punjab, India. She recently submitted her PhD in Economics, focusing on service-sector output and productivity in India. With prior experience at ISB's Centre for Analytics and Finance, she works on development studies, labour economics, and public policy. An ICSSR fellow, she has presented internationally and published in reputed journals, with technical expertise in STATA, SPSS, R, and E-Views.

**Justin Pathaw** is Field Engineer at MBDA with expertise in water quality structures, acid mine drainage treatment, and springshed management. Since 2021, he has contributed to water conservation structures under the MegLIFE Project, conducted spring mapping, trained field facilitators, and supported data analysis and documentation. His work strengthens community-focused water management and environmental restoration efforts.

**Gordon Steven Pde** is Programme Associate–Knowledge Management under the MegARISE Project. With a Master's in Political Science and three years of experience, he specialises in documentation, storytelling, and community engagement. His work highlights grassroots voices, crafts success stories, and supports proposal development recognised at state and national levels. He is committed to strengthening community-driven development through communication and field insights.

**Cheryl Bernice Pohrmen** is Assistant Environmental Engineer with the Meghalaya State Pollution Control Board (joined 30 June 2025) and has six years of experience in environmental and water-related work. Earlier, she served as Junior Project Fellow at the Institute of Natural Resources, Shillong, India, MBDA (October 2022–October 2023) on the "Springshed Rejuvenation for Water Security in Himalaya" project under the Jal Abhyaranya Campaign (NMHS). Her work covered spring mapping management, training GFAs/VCFs on water-monitoring data, and support for wastewater and acid mine drainage treatment projects across Meghalaya.

**Bratati Purkayastha** has over a decade of experience in livelihoods, monitoring, and capacity development, particularly through JICA-funded projects. She previously worked as a special educator, Research Associate, and Additional Director under the Tripura Forest Department. She has supported international bidding processes and currently serves as Technical Specialist for JICA-funded MegLIFE and KfW-supported MegARISE projects. Her work spans marketing, NRM, post-harvest systems, and project management across Northeast India.

**Wansah Pyrbot** is Botanist specialising in ecology with a PhD from the Department of Botany, NEHU, Shillong, India. Her research focuses on water quality assessment, urban wastewater impacts, and coal mine seepage effects on aquatic plant diversity in Jaintia Hills, Meghalaya. She has published work on acid mine drainage neutralisation, stream ecorestoration, and toxic element analysis in major international journals. Dr Pyrbot is committed to advancing environmental research and ecosystem restoration.

**M Wanlambok Sanglyne** is Botanist and researcher with expertise in plant biotechnology, conservation biology, seed physiology, algal studies, and ethnobotany. He has completed his PhD from the Department of Botany, North-Eastern Hill University (NEHU), Meghalaya, India, his work focused on saving endangered Citrus species using advanced biotechnological approaches. He is currently working as a Knowledge Management Associate at BRDC. He has published numerous research articles and received recognition from the

Botanical Survey of India (BSI) for his efforts made in the conservation of endangered Citrus species. His work combines traditional knowledge with modern science for plant conservation and development.

**Tremie M. Sangma** is Technical Specialist at MBDA, leading NRM and sustainable livelihood initiatives in the Garo Hills. With expertise in forest ecology, biodiversity, and community-based conservation, she supports ecological restoration in challenging landscapes. She works closely with indigenous communities to strengthen livelihoods and environmental resilience, driving evidence-based and participatory approaches to natural resource management.

**Veveane Sayo** is Senior Manager–Knowledge Management at MBMA, leading communications for World Bank-, JICA-, and KfW-supported projects. She has produced documentaries, talk shows, animations, and NRM anthems, and curates success stories for NITI Aayog. She has organised major events, including the G20 Meghalaya meet and JICA workshops. A former journalist with The Quint, she also teaches media in tourism as a guest lecturer.

**Nangteibor Shabong** is Development Management Professional specialising in strategy, M&E, and government liaison. He has contributed to major initiatives under World Bank–MBDA programmes and NRLM, supporting field-level transformation across Meghalaya. With a strong technology-driven approach, he has designed scalable training platforms, M&E systems, and collaborative frameworks. He effectively bridges grassroots implementation with high-level planning, ensuring operational clarity and long-term developmental impact.

**Hygina Siangbood** is Project Scientist and Team Lead at MBDA's Institute of Natural Resources, Shillong, India, and specialises in medicinal and aromatic plant promotion. She conceptualised the Megh Aroma Mission and leads ecological restoration in mining-affected areas. With a PhD in Plant Ecology, she trains farmers, conducts action research, and promotes NRM-based livelihoods. Her work integrates scientific innovation with conservation-driven income generation.

**Gaurav Singh** is Remote Sensing and GIS Specialist leading the RS & GIS component of the JICA-assisted MegLIFE project. With academic training from IIT Roorkee, IIRS–ISRO, and Aligarh Muslim University, Uttar Pradesh, India, he has worked on climate–health research and supported national e-learning initiatives. His expertise spans natural resource management, GIS-based planning, and capacity building across departments in Northeast India.

**Fidellia Sun** (BTech Agricultural Engineering; PG Diploma in Sports Coaching) is Athletic Coach with the Department of Sports and Youth Affairs, Meghalaya (joined 4 April 2023), with eight years of overall experience. She previously worked at the Institute of Natural Resources, MBDA, Shillong, India, as a Technical Assistant and Team Head for the Centre for Water Resources, supporting spring mapping, water quality testing, and springshed management across Meghalaya. Her expertise spans water and wastewater treatment, rejuvenation of polluted streams, and research-led project management, alongside sports coaching and youth development.

**Wankit K. Swer** is Development Practitioner with over a decade of experience in sustainability, renewable energy, NRM, and climate governance. At MBDA, he has led community-centric planning, institutional innovation, and climate resilience initiatives. His work integrates technology, policy design, and grassroots implementation, contributing to state-level strategies for green growth. He is committed to enabling community-led sustainable development.

**Goldenstar Thongni** is Visual Communicator with expertise in documenting natural and cultural heritage. His work significantly contributed to the nomination dossier for the Jingkieng Jri/Living Root Bridge Cultural Landscape. Skilled in community mobilisation, he builds trust with local stakeholders while using visual tools for heritage monitoring. He advocates for technology-enabled conservation and combines storytelling with community engagement for sustainable landscape protection.

**Raymond Wahlang** is Environmental Scientist specialising in UAV and GIS applications. Currently, Vertical Head of UAV operations, he leads state-wide surveys in sensitive landscapes, ensuring high-precision spatial data. With research experience in carbon stock assessment and sequestration, he has contributed to R&D, training, and geospatial capacity-building. He represents his department at national forums and co-authors research on UAV–GIS integration.

**John Kearney Wanniang** is Senior Manager at the State Project Management Unit of MegLIFE under MBDA. With over 12 years of rural development experience, he leads planning, coordination, and monitoring of the SALT component, supporting field beneficiaries in sustainable agriculture. He specialises in participatory approaches, enterprise promotion, and community engagement, contributing significantly to institution building and inclusive development across Meghalaya.

# PREFACE

## Meghalaya Mirrors: Reflections of Community-Led Environmental Action

Meghalaya, a region defined by its cascading waterfalls, steep terrain, and rich biodiversity, stands at a critical juncture. While the state possesses a profound natural heritage, it faces complex environmental challenges, from deforestation and unsustainable mining to the shifting patterns of climate change. In a landscape where over 90% of forest land is owned and managed by communities rather than the state, the conventional top-down approach to conservation is not merely inefficient; it is impossible. This volume, *Meghalaya Mirrors*, was born from the realisation that the solution to these ecological crises lies in the resilience and ingenuity of the people themselves.

This book is a compilation of practice-based scholarship that documents a paradigm shift in governance – one that moves from centralised planning to community-led landscape management. As editors coming from diverse backgrounds – public policy, forestry, technology, and administration – we sought to capture the "state capability" approach, which emphasises building local leadership, convergence, and adaptive governance.

To understand the philosophy behind this book, one might look to the *Jingkieng Jri* – the living root bridges of Meghalaya described in Chapter 5. Just as these bridges are not built in a day but are patiently guided by generations of community members to connect opposite banks over turbulent waters, the initiatives detailed in this book represent a patient, intergenerational effort to bridge the divide between modern government policy and traditional grassroots wisdom. Both the bridges and these policies rely on the strength of living roots – the community – to withstand the changing climate.

The chapters within are organised into four thematic reflections, each mirroring a different facet of Meghalaya's environmental journey. First, we begin by exploring how local voices are empowered to drive change. Readers will discover the "Village Community Facilitators" (VCFs) – a cadre of local youth trained as professionals in natural resource management – and the revival of community nurseries and seed banks that safeguard food security and biodiversity.

Second, we challenge the notion that traditional communities resist modernity. The case studies here demonstrate the seamless integration of Indigenous knowledge with cutting-edge technology, from the use of GIS and UAVs for mapping community forests to the adoption of Sloping Agriculture Land Technology (SALT) and the neutralisation of acid mine drainage.

The third part focuses on Green Financing that details pioneering financial mechanisms, including India's first large-scale Payment for Ecosystem Services (PES) scheme – Green Meghalaya – and the development of carbon finance projects that monetise conservation efforts for the benefit of farmers.

Finally, we examine the institutional frameworks, such as the Meghalaya Basin Development Authority (MBDA)/Meghalaya Basin Management Agency (MBMA), that bridge the gap between policy design and on-ground implementation, ensuring that successful experiments can be scaled and institutionalised.

This volume is intended for policymakers, researchers, and development practitioners who seek to understand how decentralised planning can function in ecologically sensitive regions. It serves as a blueprint, illustrating that when communities are trusted with data, technology, and financial agency, they become the most effective guardians of their environment.

We acknowledge the support of the Platform for the Development Research and Communications Project, funded by the Bill & Melinda Gates Foundation, in bringing this research to fruition. It is our hope that the reflections in *Meghalaya Mirrors* will resonate far beyond the hills of the Northeast, offering lessons for a world increasingly in need of sustainable, community-driven solutions.

**Anjal Prakash, Gunanka D. B. and Sampath Kumar**
2 December 2025

# ACKNOWLEDGEMENTS

This volume, *Meghalaya Mirrors: Reflections of Community-Led Environmental Action*, is the culmination of a collective journey involving policymakers, practitioners, researchers, and, most importantly, the communities of Meghalaya.

We extend our deepest gratitude to the Bill & Melinda Gates Foundation for their support through the Platform for Development Research and Communications Project, which made the research and compilation of this work possible.

We express our sincere gratitude to the political leadership of the State, particularly the Hon'ble Chief Minister, whose vision and guidance have been pivotal in advancing Meghalaya's commitment to community-led natural resource management and in sustaining the momentum of these transformative efforts. We are profoundly indebted to the Government of Meghalaya, particularly the Meghalaya Basin Development Authority (MBDA) and the Meghalaya Basin Management Agency (MBMA). Their willingness to reimagine governance through a bottom-up approach provided the fertile ground for the initiatives documented in these chapters. We specifically acknowledge the support of the World Bank, whose financing for the Community-Led Landscape Management Project (CLLMP) was instrumental in launching many of the innovations discussed here, such as the Village Community Facilitator (VCF) model and the creation of Village Natural Resource Management Committees (VNRMCs). We also acknowledge the contributions of JICA, KfW, and IFAD for their support in projects like MegLIFE, MegARISE, and MLAMP, which have further scaled these efforts.

We also recognise the dedication, hard work, and passion of the staff of the State Project Management Unit (SPMU) and the District Project Management Units (DPMUs) of CLLMP. Their commitment to working closely with communities and ensuring that processes remain true to the spirit of participatory governance has been central to the success of the project.

The heart of this book rests with the Indigenous communities of Meghalaya. We dedicate this work to the thousands of Village Community Facilitators (VCFs), members of the Village Natural Resource Management Committees (VNRMCs), and Village Employment Councils (VECs) who have not only participated in these initiatives but *truly owned them*. Their leadership, grounded in deep community trust, has transformed this project from a government programme into a people-led movement.

We applaud the VCFs and VNRMCs for their remarkable commitment to mastering geospatial technology, building robust datasets, and developing the skills needed to plan, manage, and protect their landscapes. Their willingness to learn, adapt, and lead from the front has restored forests, revived springs, strengthened governance, and inspired thousands across the state.

We express our gratitude to the traditional institutions, the Dorbar Shnongs and the Nokmas, whose partnership has been central in uniting ancestral wisdom with modern conservation tools. Their guidance ensured that every intervention remained rooted in community values, cultural continuity, and collective stewardship.

This book is, above all, a tribute to the people of Meghalaya, whose courage, curiosity, and leadership continue to shape a future where communities themselves are the custodians of their forests, waters, and knowledge.

We are grateful to our technical partners who helped bring policy and practice together. IORA Ecological Solutions and Rabobank contributed to advancements in carbon finance. CSIR-CIMAP provided key technical support on aromatic grasses and land reclamation. Arghyam supported the creation of learner-oriented content. Chirag, ACWADAM, PSI, and Parsari worked together on Springshed Management. NESFAS strengthened the work through its Agroecology Learning Circles. FES supported the development of technical tools for NRM planning. We also acknowledge the Soil and Water Conservation Department (SWCD) for its support in building VCF capacity, and the Bharti Institute of Public Policy (BIPP) at the Indian School of Business (ISB) for its academic guidance. Special thanks to Padma Priya Balabhadrapatruni, Project Specialist at BIPP, ISB, for coordination and support.

Finally, we thank our diverse group of contributors, ranging from senior government officials and scientists to programme managers and technical specialists, who took the time to document their experiences and insights. We also thank the editorial and production team at Routledge for their guidance in bringing this manuscript to a global audience.

The Editors

# ABBREVIATIONS

| | |
|---|---|
| ACA | Agroforestry Carbon Associate (trained facilitator in community agroforestry projects) |
| AI | Artificial Intelligence |
| AMD | Acid Mine Drainage |
| ANR | Assisted Natural Regeneration (reforestation technique) |
| ASI | Archaeological Survey of India |
| ATP | Apprenticeship Training Programme (skills development initiative) |
| BCM | Billion Cubic Meters (unit of volume, e.g. for water resources) |
| BRDC | Bio-Resources Development Centre (Meghalaya) |
| C&RD | Community and Rural Development (as in C&RD Block administration) |
| CBD | Convention on Biological Diversity |
| CLLMP | Community-Led Landscape Management Project (World Bank-funded natural resource project in Meghalaya) |
| CoE | Centre of Excellence (advanced institution or hub for specialized training) |
| CRU | Carbon Removal Unit (represents one metric ton of $CO_2$ sequestered) |
| CSB | Central Silk Board (India) |
| CSIR | Council of Scientific and Industrial Research (India) |
| DAC | Development Assistance Committee (OECD) |

| | |
|---|---|
| DIF | District Innovation Fund |
| DPMU | District Project Management Unit |
| EAP | Externally Aided Project (development project funded by external agencies) |
| ECJ | Eastern Clouded Leopard (scientific shorthand or project code) |
| EGH | East Garo Hills (district in Meghalaya) |
| EJH | East Jaintia Hills (district in Meghalaya) |
| EKH | East Khasi Hills (district in Meghalaya) |
| ENVIS | Environmental Information System (India's environment data center network) |
| FAO | Food and Agriculture Organization of the United Nations |
| FMP | Forest Management Plan |
| FOCUS | Farmers' Collectivization for Upscaling Production and Marketing Systems (Meghalaya's FOCUS farmers' support scheme) |
| FRA | Forest Rights Act |
| GA | Geographical Area (often in context of forest area statistics) |
| GFA | Green Field Associate (block-level community forest monitor in PES scheme) |
| GFW | Global Forest Watch |
| GHG | Greenhouse Gas |
| GIF | Grassroot Innovation Fund |
| GIS | Geographic Information System |
| GLAD | Global Land Analysis & Discovery (satellite forest monitoring alerts) |
| GM | GREEN Meghalaya (state Payment for Ecosystem Services scheme) mbma.org.in |
| GM+ | GREEN Meghalaya Plus (expanded phase of GREEN Meghalaya scheme) |
| GOI | Government of India |
| GP | Producer Group (farmers' collective) |
| GPS | Global Positioning System |
| GREEN | Grassroots-level Response towards Ecosystem Enhancement and Nurturing (acronym in GREEN Meghalaya PES scheme) mbma.org.in |
| HDI | Human Development Index |
| IACM | Integrated Adaptive Co-Management (strategy for resource management) |

| | |
|---|---|
| ICAR | Indian Council of Agricultural Research |
| ICIMOD | International Centre for Integrated Mountain Development |
| ICOMOS | International Council on Monuments and Sites |
| IFAD | International Fund for Agricultural Development |
| IISc | Indian Institute of Science (Bangalore) |
| ILO | International Labour Organization |
| INBAR | International Network for Bamboo and Rattan |
| INR | Indian Rupee (₹) |
| IRS | Indian Remote Sensing (satellite) |
| ISFR | India State of Forest Report |
| IUCN | International Union for Conservation of Nature |
| IVCS | Integrated Village Cooperative Society (community co-op model) |
| IWMI | International Water Management Institute |
| JFM | Joint Forest Management |
| JICA | Japan International Cooperation Agency |
| JJLCCL | Jingkieng Jri/Lyu Chrai Cultural Landscape (living root bridges cultural landscape of Meghalaya) |
| JSYS | Jala Samvardhane Yojana Sangha (Watershed & tank management agency, Karnataka) |
| KML | Keyhole Markup Language (GIS file format for maps) |
| KfW | Kreditanstalt für Wiederaufbau (German Development Bank) x.com |
| LISS | Linear Imaging Self Scanner (satellite sensor on IRS) |
| LPC | Liters per Capita (often in context of water use; also Land Possession Certificate in Meghalaya) |
| LULC | Land Use and Land Cover |
| MBDA | Meghalaya Basin Development Authority |
| MBMA | Meghalaya Basin Management Agency |
| MCLLMP | Meghalaya Community-Led Landscape Management Project |
| MGNREGA | Mahatma Gandhi National Rural Employment Guarantee Act (India's rural employment guarantee law) |
| MGNREGS | Mahatma Gandhi National Rural Employment Guarantee Scheme (the program under MGNREGA) |
| MIS | Management Information System |
| MoA | Ministry of Agriculture (Government of India) |
| MoEFCC | Ministry of Environment, Forest and Climate Change (Government of India) |
| MoU (pl. MoUs) | Memorandum of Understanding |

| | |
|---|---|
| MRV | Measurement, Reporting and Verification (protocol) |
| MSCCC | Meghalaya State Council for Climate Change |
| MegARISE | Meghalaya's project for Adaptation/Resilience in Special Ecosystems (Protection of Vulnerable Catchment Areas project in partnership with KfW) x.com |
| MegLIFE | Meghalaya's Community-Based Forest Management and Livelihoods Improvement project (JICA-supported forestry/livelihood project) |
| NBPGR | National Bureau of Plant Genetic Resources (India) |
| NESAC | North Eastern Space Applications Centre (Shillong) |
| NFP | National Forest Policy (Government of India) |
| NGH | North Garo Hills (district in Meghalaya) |
| NGT | National Green Tribunal (India) |
| NOC | No Objection Certificate |
| NRLM | National Rural Livelihoods Mission (Government of India) |
| NRM | Natural Resource Management |
| NTFP | Non-Timber Forest Product |
| OBIA | Object-Based Image Analysis (remote sensing technique) |
| PES | Payment for Ecosystem Services |
| PG | Producer Group (farmers' collective) |
| PGS | Participatory Guarantee System (community-based organic certification) |
| PPP | Public-Private Partnership |
| PRA | Participatory Rural Appraisal |
| PRASARI | Professional Assistance for Sustainable Agricultural and Rural Innovation (NGO named PRASARI) |
| QPM | Quality Planting Material |
| REDD | Reducing Emissions from Deforestation and Forest Degradation (UN climate mechanism) |
| RFA | Recorded Forest Area |
| RUSLE | Revised Universal Soil Loss Equation (soil erosion model) |
| SALT | Sloping Agricultural Land Technology |
| SDC | Swiss Agency for Development and Cooperation |
| SDG | Sustainable Development Goal |
| SEWA | Self-Employed Women's Association |
| SGH | South Garo Hills (district in Meghalaya) |
| SHC | Soil Health Card |
| SLIC | State-Level Implementation Committee |

| | |
|---|---|
| SOC | Soil Organic Carbon |
| SPMU | State Project Management Unit |
| SWGH | South West Garo Hills (district in Meghalaya) |
| SWKH | South West Khasi Hills (district in Meghalaya) |
| TDS | Total Dissolved Solids (water quality metric) |
| TEK | Traditional Ecological Knowledge |
| UAS | Unmanned Aerial System (drone system) |
| UAV | Unmanned Aerial Vehicle (drone) |
| UC | Utilization Certificate (financial reporting document) |
| UNDP | United Nations Development Programme |
| UNEP | United Nations Environment Programme |
| UNESCO | United Nations Educational, Scientific and Cultural Organization |
| USLE | Universal Soil Loss Equation |
| VCF | Village Community Facilitator (local youth para-professional for NRM) |
| VEC | Village Employment Council (local governance body for rural employment schemes) |
| VET | Vocational Education and Training |
| VNRMC | Village Natural Resource Management Committee |
| VO | Village Organization |
| VPIC | Village Planning and Implementation Committee |
| WGH | West Garo Hills (district in Meghalaya) |
| WJH | West Jaintia Hills (district in Meghalaya) |
| WKH | West Khasi Hills (district in Meghalaya) |
| WWAP | World Water Assessment Programme (UNESCO) |

# 1

## EDITORIAL

## Meghalaya Mirrors – A Blueprint for Community-Led Environmental Resilience

*Anjal Prakash, Sampath Kumar, and Gunanka D. B.*

Meghalaya, the north-eastern state of India, faces a multitude of challenges, spanning environmental, socio-economic, and governance domains (Bhattacharjee et al., 2025). Environmental degradation is a predominant concern, primarily caused by deforestation, unsustainable shifting cultivation (Jhum) (Tamuli & Bora, 2022), extensive coal mining (Nomani et al., 2021), and monoculture practices (Lyngkhoi et al., 2022). These activities contribute to severe soil erosion, nutrient depletion, and reduced agricultural productivity (Badavath, Sahoo & Samal, 2024). Specifically, coal mining has led to acid mine drainage (AMD), contaminating water bodies and posing a significant environmental threat (Kshetriya et al., 2021).

Water security is severely threatened, with springs – the primary water source for 80% of villages – experiencing declining discharge and drying up during dry seasons (Majaw, 2023). This is attributed to land use changes, increased water diversion, pollution, and shifting rainfall patterns (Chakraborty & Saikia, 2022). The region also faces a decline in indigenous crop varieties and overall biodiversity loss (Balusamy et al., 2024).

From a governance perspective, Meghalaya grapples with a lack of comprehensive spatial data and reliance on outdated, manual survey methods, which hinder effective natural resource management, conservation efforts and contribute to land tenure conflicts (Lyngdoh et al., 2023). Institutional gaps include fragmented climate action, poor coordination, and low awareness at the village level (Kharmylliem & Kipgen, 2025). Additionally, there is a significant skill gap in advanced technologies such as Geographic Information Systems (GIS) and Unmanned Aerial Vehicles (UAVs), crucial for large-scale development projects. Traditional conservation efforts have

DOI: 10.4324/9781003735380-1

historically overlooked community leadership, impacting successful local engagement.

This volume addresses the critical challenge of environmental degradation and climate vulnerability faced by Meghalaya, a fragile and ecologically rich region in Northeast India. Rapid deforestation, unsustainable agricultural practices, mining, and infrastructure development have severely impacted the region's forests, soil, water sources, and biodiversity. The region's heavy dependence on community-owned lands exacerbates these pressures, often leading to uncoordinated or ineffective conservation efforts. Moreover, a lack of access to modern technology, insufficient community participation, and limited financial incentives have hindered the sustainable management of natural resources.

Another pressing issue is the vulnerability of Meghalaya's water resources. The state's reliance on springs, which are declining due to land use changes, climate variability, and anthropogenic impacts, threatens water security for rural communities (Sharma & Laskar, 2021). Similarly, land degradation from practices like shifting cultivation and mining has led to erosion, loss of soil fertility, and ecological imbalance, posing significant threats to local livelihoods and food security (Singh & Chaudhary, 2023).

This book recognises that isolated, top-down approaches are insufficient to meet these complex challenges. Instead, it advocates for community-led, participatory strategies that harness indigenous knowledge, innovative technology, financial mechanisms, and effective governance. Its overarching goal is to create resilient ecosystems, bolster local livelihoods, and promote sustainable development through locally driven action, integrated policies, and science-based solutions, thereby addressing the region's multifaceted environmental crises comprehensively and inclusively.

## 1.1 Book Structure and Key Themes

The book is thoughtfully structured into four distinct yet interconnected parts, each addressing a critical facet of environmental action in Meghalaya, emphasising the pivotal role of local communities. This volume is organised into four thematic sections, each exploring essential facets of community-led environmental action in Meghalaya.

Part I on community engagement emphasises empowering local voices, showcasing initiatives like building community professionals, establishing nurseries for forest restoration, seed saving for food security, preserving cultural and natural heritage, and supporting grassroots innovations. These chapters demonstrate how community participation serves as the backbone of sustainable environmental change. Part II of the book introduces innovative solutions. It focuses on harnessing technology to bolster climate resilience. It highlights practical approaches such as sloping land technology, neutralisation of acid mine drainage, spring mapping, biodiversity-friendly seedball initiatives, and integrating advanced tools into natural resource management, illustrating how science and community efforts can synergise for ecological preservation.

Part III is on green financing, which explores financial mechanisms vital for sustainability. It includes chapters on payments for ecosystem services, sustainable forest management, land reclamation with aromatic grasses, agroforestry projects for carbon sequestration, and bamboo resource assessment, showing how innovative funding can incentivise and scale conservation efforts. Part IV focuses on governance and policy issues and reviews strategic policy impacts, including case studies on Meghalaya's governance institutions like Meghalaya Basin Development Authority (MBDA), and addresses skill development programmes, emphasising the importance of institutional support and capacity building for long-term sustainable development.

### 1.1.1 Community Engagement: Empowering Local Voices for Environmental Change

This part underscores the vital role of local communities in driving environmental change in Meghalaya. It highlights how empowering residents through capacity-building initiatives – such as training village professionals – creates a strong foundation for sustainable resource management. The establishment of community nurseries for forest restoration exemplifies collective efforts to revive degraded lands, while seed saving programmes enhance food security and protect indigenous crop varieties. Preserving cultural and natural heritage demonstrates the importance of traditional knowledge and indigenous practices in conservation, fostering a sense of ownership and responsibility among local populations. Additionally, supporting grassroots innovations showcases how community-driven solutions can address unique environmental challenges with locally tailored approaches. Overall, this section affirms that genuine environmental transformation depends on amplifying local voices, nurturing community leadership, and integrating traditional wisdom with modern conservation strategies, thereby fostering sustainable development rooted in community participation. The five chapters in this section deal with some of these issues.

Dokatchi K. Marak and Nangteibor Shabong, in their chapter, "Building a Cadre of Community Professionals" (Chapter 3), highlight a pioneering model for natural resource management (NRM) through the development and deployment of Village Community Facilitators (VCFs). These trained local youths, initially introduced through the World Bank-funded Community-Led Landscape Management Project, undertake critical field-level responsibilities, fostering community participation and supporting project goals. The research indicates the effectiveness of training rural youth for specialised roles such as GIS-based natural resource mapping, seed ball training in over 1,840 schools, spring mapping, and the preparation of Forest Management Plans (FMPs). This model offers unconventional employment, bridging skill development with community service, and providing valuable livelihood opportunities that enhance socio-economic standing. The mutual benefits are clear: Implementing agencies gain from the local knowledge, trust, and commitment of VCFs, while youth acquire hands-on experience, skills, and dignified employment, demonstrating potential for scalability and sustainability, particularly in remote regions. This cadre is presented as a promising strategy for advancing participatory NRM, fostering local ownership and enhancing grassroots environmental initiatives.

Bratati Purkhayastha and Van Shanborlang Buhphang, in their chapter focusing on "Community Nurseries for Forest Restoration" (Chapter 4), address Meghalaya's significant ecological degradation due to deforestation, shifting cultivation, mining, and monoculture practices. This chapter details a decentralised, community-based forest restoration programme. Recognising the limitations of centralised strategies in a region where over 90% of the land is under community ownership, the Meghalaya Basin Development Authority (MBDA) and Meghalaya Basin Management Agency (MBMA) initiated this programme. Through participatory planning, indigenous species were cultivated in 739 community nurseries, producing over 3.5 million seedlings and enabling the restoration of more than 37,137 ha of degraded land. This initiative not only improved seedling survival rates, reaching up to 90%, but also significantly reduced transportation costs and logistical challenges, especially during monsoons. Innovative monitoring using GIS and MIS enhances decision-making and accountability. Beyond ecological benefits, the project yielded substantial socio-economic gains through financial incentives, micro-enterprise development, buy-back arrangements, and profits for self-help groups. This model exemplifies a scalable and cost-effective approach, blending traditional knowledge with modern tools and institutional support, influencing similar interventions across Meghalaya.

Junie P. Lyngdoh and Reccica D. Lyngkhoi, in their chapter, "Community Seed Saving for Food Security" (Chapter 5), focus on a community-driven initiative for conserving indigenous and heirloom crop varieties in Meghalaya. In response to declining traditional agricultural practices and biodiversity loss, the Bio-Resources Development Centre (BRDC) established model seed saving units. These units integrated traditional knowledge with modern scientific techniques like biofertiliser application and sustainable farming, resulting in the conservation of over 90 crop varieties, the training of 300 farmers, and enhanced seed viability, soil health, and crop yields. Notably, women-led self-help groups played a pivotal role in crop selection, storage, and management. The project improved food security, reduced reliance on commercial seeds, and generated additional income, with the model successfully replicated in multiple villages.

In Chapter 6, "Preserving Cultural and Natural Heritage," Catherine Derica Kyndiah, Goldenstar Thongni, and Tilaris Marwein highlight indigenous communities' active role in conserving the Jingkieng Jri/Lyu Chrai Cultural Landscape (JJLCCL). This landscape, shaped by the Khasi and Jaintia people, encompasses sacred forests, monoliths, agricultural fields, and living root bridges, reflecting generations of shared knowledge and a deep bond with nature. The research emphasises how traditional governance systems like the Dorbar collaborate with state-supported initiatives to sustain conservation, a departure from past efforts that often ignored community leadership. With support from MBMA and the Community-Led Landscape Management Project (CLLMP), communities restored 74 living root bridges, built 25 nurseries for *Ficus Elastica*, and revived traditional homes (Iing Mariang). They also mapped territories using GPS, recorded oral histories, and engaged in research collaborations. The chapter reveals how grassroots leadership, traditional wisdom, and scientific partnerships reinforce one another, presenting the JJLCCL as a powerful model of cultural landscape conservation.

Chapter 7, titled "Supporting Grassroot Innovations" written by Veveane Sayo, Delanmi Massar and Gordon Pde, explores the role of grassroots innovations in advancing sustainable development and climate resilience within Meghalaya's unique socio-ecological context. Through the CLLMP and its Innovations fund, community-driven initiatives in areas like waste management, sustainable agriculture, biodiversity conservation, and traditional knowledge preservation have been identified and supported. A comprehensive review highlights their economic, social, and environmental impacts, emphasising their replicability and contribution to policy discourse. Key findings show significant local income generation, environmental restoration, and reinforcement of indigenous knowledge systems.

### 1.1.2 Innovative Solutions: Harnessing Technology for Climate Resilience

This part highlights how technology is transforming climate resilience in Meghalaya by providing practical, scalable tools for environmental management. It features initiatives like Sloping Agricultural Land Technology (SALT), which promotes sustainable farming on hill slopes, reducing soil erosion and enhancing productivity. The neutralisation of acid mine drainage showcases how community-based, low-cost geochemical methods can restore water quality affected by mining activities, safeguarding local ecosystems. Hydrological spring mapping and springshed development demonstrate how scientific tools help communities understand and manage vital water sources, ensuring water security amid changing climate patterns. The seedball initiative exemplifies biodiversity conservation through simple yet effective dispersal techniques that engage local youth and schoolchildren. Integration of advanced technologies such as GIS, remote sensing, and UAVs enables precise mapping of natural resources, supports better land use planning, and enhances monitoring of ecological changes.

These solutions exemplify how blending traditional knowledge with modern science empowers communities to address environmental challenges proactively, improve resource management, and build resilience against climate impacts. Overall, this part underscores the importance of technological innovation as a catalyst for sustainable, community-driven climate action. It showcases how various technological and scientific solutions, often integrated with traditional practices, are being employed to address pressing environmental challenges in Meghalaya, focusing on climate resilience. Five chapters address these issues.

John Kearey Wanniang, in the Chapter 8, "Sloping Agriculture Land Technology (SALT)," examines the implementation and outcomes of SALT in Meghalaya, a hilly state facing severe land degradation from traditional farming methods like Jhum cultivation and monoculture. These practices cause extensive soil erosion, nutrient depletion, and declining agricultural productivity. SALT is presented as a sustainable agroforestry-based solution tailored for sloping terrains, aiming to restore soil health and improve rural livelihoods. The research investigates whether SALT can be a scalable and sustainable method to counteract environmental degradation while enhancing food security and economic resilience. Key interventions included the establishment of nitrogen-fixing contour hedgerows, integrated crop cultivation, and farmer capacity-building. Findings reveal that SALT significantly reduces soil erosion and improves soil parameters (pH, nitrogen, phosphorus, potassium, organic carbon), leading to enhanced crop yields and farm diversification. Despite initial barriers like labour intensity and financial constraints, SALT proved adaptable and replicable with institutional support. The study highlights SALT as a replicable model for climate-resilient agriculture in hilly regions, demonstrating how community-led interventions combining indigenous knowledge and scientific methods can reverse land degradation.

Chapter 9, "Neutralisation of Acid Mine Drainage Contaminated Water,", focuses on the issue of *Neutralisation of Acid Mine Drainage* (AMD). AMD from coal mining poses a significant environmental threat in Meghalaya, contaminating water bodies with highly acidic conditions (pH 2–4). This study evaluates Open Limestone Channels (OLCs) as a cost-effective and community-driven solution for neutralising AMD-contaminated water. Thirty-one OLCs were constructed using locally sourced limestone and vetiver grass for phyto-remediation, with active community participation. Results showed substantial improvements in water quality, with pH levels increasing by 3 to 4 units, meeting Indian standards. Chambered OLCs achieved 90% effectiveness compared to 60% for drainage ditches, also reducing maintenance and enhancing heavy metal removal. Despite challenges like limestone armouring and site-specific constraints, community participation was pivotal for sustainability and water security. OLCs, integrated with natural remediation, are presented as a scalable and sustainable approach to AMD treatment.

Jillianda Kharjana and Gaurav Singh, in their Chapter 10 titled, "Hydrological Spring Mapping and Springshed Development," show how springs are the primary water source for 80% of Meghalaya's villages and how crucial it is for drinking and irrigation. The chapter shows that anthropogenic activities (mining, unsustainable agriculture, forest extraction) and shifting rainfall patterns threaten this reliance, leading to reduced spring discharge and dry springs, forcing women to travel farther for water. The "Springshed Management" approach, introduced by the MBMA, addresses the lack of baseline data by providing crucial information on springs, enabling sustainable management plans. Monthly monitoring and recharge interventions aim to enhance groundwater recharge, restore spring flow, and build long-term water security for vulnerable communities.

S. Ashutosh, Tremie M. Sangma, and Rikmenlang Khongsng, in their chapter "Seedball Initiatives for Biodiversity Conservation" (Chapter 11), discuss an eco-friendly seed ball programme launched by the State government to combat widespread forest loss and environmental degradation from shifting cultivation and coal mining. Under the CLLMP, in collaboration with MBMA, this initiative engages local communities, especially schoolchildren, in restoring degraded ecosystems. Seed balls, made from a mixture of soil, compost, biochar, and native tree seeds, are prepared and dispersed by students, promoting environmental education and responsibility. Between October 2022 and June 2023, 3 million seed balls were prepared and dispersed by 75,000 students from 1,840 schools, engaging about 282 VCFs. A post-dispersal survey showed a 55% average germination and survival rate. This inexpensive and simple approach is an adaptable and scalable model for ecological restoration, recognised as a Best Practice by NITI Aayog in 2024.

Chapter 12 by Raymond Wahlang and Fettleman Dohling, focuses on "Integrating Technology for Natural Resource Management." This chapter highlights a transformative initiative that integrates modern geospatial technologies – geographic information systems (GIS), remote sensing (RS), and unmanned aerial vehicles (UAVs) – with deep community engagement to overcome challenges like ecological fragility, lack of spatial data, and outdated manual survey methods. More than 1,100 VCFs and Village Data

Volunteers (VDVs) were trained in GPS data collection, and youth were empowered through apprenticeship programmes in GIS analysis and UAV-assisted mapping. These community actors led boundary delineation, Land Use Land Cover (LULC) classification, inventory for forest management plans (FMP), and Measurement, Reporting, and Verification (MRV) of forest ecosystems. This participatory model generated granular, real-time, and verifiable datasets, integrating Traditional Ecological Knowledge (TEK). UAVs enabled high-resolution mapping in remote, hilly terrains, filling historic data gaps and resolving territorial ambiguities. Outcomes included improved watershed protection, guided afforestation, support for Payment for Ecosystem Services (PES) schemes, and strengthened village-level governance through precise planning tools and dispute resolution. This initiative demonstrates a scalable, policy-relevant model for decentralised natural resource management.

### 1.1.3 Green Financing: Innovations in Funding Climate Action

The part on *Green Financing* explores innovative financial mechanisms essential for scaling climate action and sustainable development in Meghalaya. It emphasises the importance of Payment for Ecosystem Services (PES), which provides direct financial incentives to local communities and landowners for conserving forests and maintaining ecological health. Such schemes motivate community participation and help fund conservation initiatives that would otherwise lack sustained resources. The section also covers sustainable forest management plans that balance ecological preservation with livelihood needs, ensuring long-term resource sustainability. Additionally, projects like land reclamation using aromatic grasses demonstrate how eco-friendly techniques can generate income while restoring degraded landscapes. The agroforestry-carbon project exemplifies integrating climate mitigation with livelihood enhancement, enabling farmers to earn carbon credits through sustainable land use. Further, community-based assessments of bamboo resources reveal how local inventories can unlock investment and policy support for economic growth tied to ecological conservation. Overall, these innovative financial models highlight the crucial role of funding mechanisms in translating community efforts into scalable, impactful climate resilience actions.

Five chapters in this section delve into innovative financial mechanisms that incentivise and support conservation efforts, recognising the economic value of ecosystem services and sustainable practices.

The chapter by James T. Kharkongor and Lavinia Dkhar, "Conservation of Forests and Payment for Ecosystem Services" (Chapter 13), describes the Payment for Ecosystem Services (PES) approach under GREEN Meghalaya and its expanded version, GREEN Meghalaya+, as India's first large-scale initiative to compensate forest custodians for maintaining natural forest cover. Given that over 90% of Meghalaya's forest land is community or individually owned, the scheme offers direct financial incentives for forest conservation, biodiversity protection, and climate resilience. Since its inception in 2022, it has enrolled over 3,300 beneficiaries, protecting over 51,000 ha, and disbursing approximately $5.78 million USD. The upgraded GREEN Meghalaya+ aims to expand coverage to over 100,000 ha, with enhanced incentives for ecologically significant sites. Implementation is guided by a robust Measurement, Reporting, and Verification (MRV) protocol and GIS-based monitoring. Despite challenges related to land tenure, verification in remote terrains, institutional capacity, and cultural diversity, the programme sets a pioneering precedent for community-driven, performance-based conservation.

Chapter 14, on "Forest Management Plans for Sustainable Forestry," is written by Lavinia M. Dkhar, Jyswill Nongpiur and Rajaul Karim. It presents the Forest Management Plan (FMP) in Meghalaya in an integrated, participatory framework for sustainable forest governance, balancing ecological preservation with socio-economic advancement. Spanning 400 villages and managing 1,102 km$^2$ of forest, the initiative combined traditional knowledge with modern tools such as remote sensing and GIS, RS and mobile apps to support biodiversity assessments, forest inventories, growing stock estimation, carbon stock estimations, and sustainable harvesting. Central to the plan was the engagement of over 1,200 VCFs, ensuring grassroots participation and ownership. Despite challenges like difficult terrain, climate impacts, and stakeholder conflicts, the project achieved significant outcomes, including assessment of timber and carbon volumes, while reinforcing decentralised forest governance. The FMP serves as a scalable, community-driven model for climate-resilient and legally compliant forest management, delivering both environmental and livelihood benefits.

Chapter 15, "Reclamation of Mining Affected Land Using Aromatic Grass," by Hygina Siangbood and Rishot Skhem Kharkongor, talks about *Aromatic Grasses for Land Reclamation*. This study examines a community-driven land reclamation initiative using aromatic grasses, particularly Cymbopogon winterianus (Citronella), across 328.5 ha in 46 villages to restore mining-affected soils and improve socio-economic conditions. Coal mining in Meghalaya has left vast tracts of land barren and degraded. The research investigated the effectiveness of phytoremediation using aromatic grasses to enhance soil fertility and water retention. Post-intervention analyses showed notable improvements in soil pH, organic carbon, and moisture retention, and significant reductions in heavy metal levels. Spring water discharge also increased. These ecological gains were coupled with socio-economic benefits: Over 2,000 farmers (62% women) were trained, enabling income diversification and generating substantial returns and employment. The project combined GIS-based site selection, community mobilisation, and partnerships with national institutions, also demonstrating strong carbon sequestration potential. This integrative approach bridges ecological restoration with livelihood improvement and gender empowerment, offering a replicable and scalable solution for post-mining restoration.

Chapter 16 on "Agroforestry-Carbon Project in Meghalaya" by Lavinia M. Dkhar describes a pioneering effort to combat climate change and enhance rural livelihoods through agroforestry practices and carbon finance mechanisms. Initiated by MBDA in collaboration with Iora Ecological Solutions and supported by Rabobank's Acorn platform, the project targets 2,000 villages, promoting reforestation, conservation, and community-led agroforestry to generate Carbon Removal Units (CRUs), improve soil health, support biodiversity, and establish green livelihoods. As of 2024, over 9,100 farmers have enrolled, managing over 17,500 ha. A participatory approach with trained Agroforestry-Carbon Associates facilitates community mobilisation and capacity building. Despite challenges like financial constraints, the initiative shows strong community engagement and a scalable model for integrating climate mitigation with socio-economic development.

Nangteibor Shabong and Pyndapbor Marbaniang write in Chapter ,17 "Leveraging Bamboo Resource" about the transformative impact of community-led bamboo resource assessment in Meghalaya under the CLLMP. Despite rich bamboo biodiversity, Meghalaya historically lacked policy coherence and baseline data for effective bamboo sector development. This initiative used a participatory methodology, mapping over 2,100 million culms using a stratified grid-based approach, engaging VCFs, and applying geospatial tools. The assessment provided a species-wise inventory of bamboo distribution, leading to investor proposals exceeding ₹800 crore. Integrating traditional knowledge with modern tools built community capacity and localised data cadres, positioning Meghalaya for policy advancement and industrial collaboration. This model offers a replicable framework for resource governance, enabling sustainable economic growth through community engagement and shifting to a holistic, data-driven NRM strategy.

### 1.1.4 Governance and Policy

This part underscores how effective institutional frameworks and strategic policies are vital for advancing sustainable environmental management in Meghalaya. It features a comprehensive case study on the Meghalaya Basin Development Authority (MBDA), illustrating how bottom-up, community-driven governance models can facilitate large-scale natural resource management, data collection, and climate resilience. The chapter highlights the creation of innovative institutions like Village Natural Resource Management

Committees and the development of geospatial platforms, which enhance transparency, coordination, and local participation. These policy initiatives have strengthened governance, fostered multi-stakeholder collaborations, and embedded climate considerations into regional planning. Additionally, efforts to bridge skill gaps through targeted training programmes ensure institutional capacity building, enabling communities and officials to implement environmental strategies effectively. This part highlights the importance of integrating traditional governance structures with modern policy tools to foster inclusive, sustainable development. Ultimately, evidence from Meghalaya demonstrates that strategic policy formulation and strong governance are fundamental to institutionalising environmental resilience and ensuring long-term climate adaptation.

Two papers in this final section contribute to specific initiatives within broader institutional frameworks, demonstrating how successful community-led actions can inform and shape state-level policy and governance structures, ensuring long-term sustainability and scalability.

Wankit K. Swer examines the "Case Study on the Policy and Governance Impact of the MBDA" (Chapter 18). He looks at the policy and governance impacts of the MBDA, a unique institutional innovation by the Government of Meghalaya. By adopting a bottom-up, community-driven approach, MBDA has successfully facilitated scalable models of NRM, data-driven governance, and climate resilience. Key policy outcomes include the launch of a state policy for Village Natural Resource Management Committees (VNRMCs), the establishment of a State GIS and UAV Centre, creation of the Meghalaya State Geo Portal, strengthening of the Meghalaya State Council for Climate Change and Sustainable Development, and drafting of a pioneering Climate Emergency and Green Growth Framework. These developments show how practical implementation and real-time data directly inform policy. The study identifies prior institutional gaps like fragmented climate action and a lack of coordination. MBDA's innovations – community-led governance, spatial data use, and cross-departmental convergence – are deemed scalable and replicable.

T. Niang Suan Ching and Sucielia Mylliemngap, in their Chapter 19, "Bridging Skill Gaps," evaluate the Apprenticeship Training Program launched by the Meghalaya Basin Management Agency (MBMA) in December 2021 to address a significant skills gap in Meghalaya, particularly for Externally Aided Projects (EAPs) requiring advanced technologies like GIS, UAVs, and drone-assisted remote sensing. The programme aims to strengthen local technical capacity and support large-scale development initiatives. Apprentices gained skills in GIS, drone operations, finance, IT, and environmental management. High employment absorption rates reflect strong market demand and directly contributed to project success, such as the

Meghalaya Community-Led Landscape Management Project (MCLLMP). The programme enhances the employability and long-term sustainability of EAPs, demonstrating its potential for replication.

## 1.2 Unique Contributions of *Meghalaya Mirrors*

*Meghalaya Mirrors* stands out due to several unique contributions that collectively paint a comprehensive picture of effective environmental action.

First, there is an emphasis on Community-Led Models and Ownership. The overarching theme is the prioritisation of local communities as key actors in NRM. This is not merely a theoretical assertion but is demonstrated through concrete examples like the VCFs, community nurseries, seed saving initiatives led by women's groups, and traditional governance systems like the Dorbar. The book consistently showcases how empowering local voices leads to more sustainable and impactful outcomes.

Second, the integration of Traditional Knowledge and Modern Technology is a striking feature, which is the seamless blend of indigenous wisdom and contemporary scientific tools. Chapters detail the use of GIS, remote sensing, UAVs, and mobile apps for mapping, monitoring, and data collection, while simultaneously valuing and integrating Traditional Ecological Knowledge (TEK). This fusion is shown to resolve historic data gaps, enhance precision, and foster local ownership, as seen in spring mapping, forest management plans, and bamboo resource assessment.

Third, the focus on Innovative Green Financing Mechanisms offers pioneering insights into innovative financial models like the Payment for Ecosystem Services (PES) under GREEN Meghalaya, which directly compensates forest custodians. Similarly, the agroforestry-carbon project connects local farmers to global carbon finance mechanisms, demonstrating how economic incentives can drive conservation and sustainable livelihoods. These approaches provide tangible economic benefits, making conservation attractive and sustainable for local communities.

Many initiatives described in this volume, such as the VCF model, community nurseries, seed saving units, SALT practices, OLCs for AMD treatment, seedball programmes, and the MBDA's governance innovations, are explicitly highlighted as scalable, cost-effective, and replicable models. This suggests that the lessons learned in Meghalaya are not unique to its context but offer practical frameworks for other regions facing similar environmental and socio-economic challenges. Beyond ecological restoration, the book consistently demonstrates the socio-economic benefits derived from these initiatives. Chapters detail income generation through micro-enterprises, employment opportunities (e.g., VCFs, aromatic grass cultivation for women), and improved food security. The active and prominent role of women, particularly in seed saving and aromatic grass cultivation, underscores a commitment to gender empowerment within environmental action.

## 1.3 Critical Evaluation: Strengths and Weaknesses

This book presents a compelling case for the effectiveness of community-led environmental action, grounded in practical outcomes and empirical evidence. There is evident impact and tangible results: The chapters in this book consistently report measurable successes. For instance, community nurseries achieved up to 90% seedling survival rates, SALT improved soil parameters and crop yields, OLCs increased water pH by 3–4 units, and the seedball programme saw a 55% germination rate. The PES scheme enrolled over 3,300 beneficiaries, protecting over 51,000 a, disbursing ₹48 crores. These quantifiable results lend significant credibility to the approaches advocated.

Second, the book does not focus on isolated interventions but portrays an integrated approach that addresses multiple facets of NRM – from soil health and water conservation to forest restoration, biodiversity, and sustainable agriculture. The recognition that ecological systems are interconnected and that solutions must be multi-pronged is a significant strength.

Second, the chapters in this book show strong policy relevance and influence. Several chapters demonstrate the direct influence of grassroots implementation on state-level policy formulation. The MBDA's success, for example, directly led to the launch of policies for Village Natural Resource Management Committees and the establishment of state-level geospatial centres. This highlights the importance of ground-up data and experiences in shaping effective governance frameworks. The continuous emphasis on training and capacity building for VCFs, VDVs, farmers, and youth underscores a genuine commitment to empowering local communities. This builds in-house technical capacity and reduces reliance on external experts, ensuring the long-term sustainability of interventions.

Third, the initiatives are designed to be context-specific, acknowledging the unique socio-ecological challenges of Meghalaya. The adaptation of SALT to sloping terrains, the use of locally sourced limestone for AMD neutralisation, and the integration of traditional knowledge in seed saving are examples of this flexibility and cultural sensitivity.

## 1.4 Significance and Future Research Avenues

*Meghalaya Mirrors* hold profound significance for both regional and global discourse on sustainable development and climate action. The book presents Meghalaya as a living laboratory for integrated, climate-sensitive development, especially in challenging terrains with weak local institutions and high vulnerability to climate change. It offers a holistic model that weaves together ecological restoration, socio-economic upliftment, technological innovation, and robust governance.

It empirically validates the critical importance of community participation and local ownership in achieving environmental and developmental goals. This serves as a powerful counter-narrative to top-down approaches, reinforcing the idea that sustainable solutions are best co-created with those on the ground. The success of PES and agroforestry-carbon projects provides concrete proof-of-concept for green financing mechanisms that link environmental conservation directly to economic benefits for local communities. This can inspire similar incentive-based conservation models elsewhere, moving beyond traditional funding paradigms.

The various initiatives contribute directly to building climate resilience by enhancing soil health, restoring water sources, promoting sustainable agriculture, and conserving forests – all critical buffers against the impacts of climate change.

Building on the identified challenges and the forward-looking perspectives within the chapters of the book, several crucial avenues for future research emerge:

- *Long-Term Monitoring and Impact Assessment*: While current results are promising, there is an explicit need for long-term monitoring to optimise the effectiveness and adaptability of technologies like OLCs. Similarly, future research should investigate the long-term career outcomes of apprentices and assess the programme's scalability to truly understand sustained impact and replicability. This would provide robust evidence of durability beyond initial project phases.
- *Optimisation of Innovative Solutions*: Further research is advocated for alternative neutralisers, integrated treatment techniques, and ecosystem restoration related to AMD treatment. This suggests a continuous improvement cycle for these nascent technologies, aiming for greater efficiency and broader applicability.
- *Strengthening Grassroots Innovation Ecosystems*: The chapter proposes future research directions to strengthen grassroots innovation ecosystems and to inform policy formulation at the regional and national levels. This would involve understanding the dynamics of local innovation, identifying pathways for nurturing and diffusing these innovations, and developing mechanisms for their effective integration into broader development frameworks.
- *Integration of Emerging Technologies*: The apprenticeship programme points towards exploring the integration of emerging tools such as AI-driven GIS and advanced climate resilience strategies. This highlights the need to constantly evolve technological approaches to remain

at the forefront of NRM and climate action. Similarly, the potential for AI and advanced analytics to further refine data collection, monitoring, and decision-making for FMPs and spatial data initiatives remains a ripe area for exploration.

- *Policy Development for Resource Governance*: The success of bamboo resource assessment leads to calls for policy advancement and industrial collaboration. Future research could focus on developing specific policy frameworks and regulatory mechanisms that effectively integrate scientific data, traditional knowledge, and economic incentives to ensure sustainable resource governance and attract investment.

- *Addressing Sociocultural Complexities in Implementation*: Research could delve deeper into effectively navigating challenges related to cultural diversity and stakeholder conflicts. Understanding nuanced local governance structures and developing culturally sensitive conflict resolution mechanisms would be critical for sustained community engagement and project success in diverse regions.

## 1.5 Conclusion: A Sustainable Path Forward

This volume paints a compelling picture of a state at the forefront of integrated, community-led climate action. It stands as a testament to the fact that effective environmental stewardship in a complex and vulnerable nation like India requires a nuanced understanding of local contexts, a deep respect for traditional knowledge, and a proactive embrace of innovative solutions and enabling policy frameworks.

The initiatives detailed, from empowering local youth as VCFs and establishing community nurseries to harnessing advanced geospatial technologies and pioneering green financing mechanisms like Payment for Ecosystem Services, collectively illustrate a dynamic and replicable blueprint. Meghalaya's success lies in its ability to foster genuine community engagement, which ensures the relevance and sustainability of interventions. This is seamlessly coupled with the strategic adoption of innovative solutions, often bridging traditional wisdom with scientific methods, leading to measurable improvements in ecological health and socio-economic well-being. Crucially, these efforts are underpinned by forward-thinking policy and governance structures, notably those championed by the MBDA, which facilitate decentralised decision-making, integrate data, and foster multistakeholder collaboration.

The lessons from Meghalaya, as reflected in these chapters, transcend regional boundaries. They offer invaluable insights for other regions grappling with similar environmental and developmental challenges, demonstrating how a truly holistic and locally driven approach can pave the way for a more resilient, equitable, and sustainable future for communities across India and beyond.

This book is not merely a collection of cases from the Community-Led Landscape Management Project (CLLMP), but it is a powerful declaration of what is possible when environmental stewardship is rooted in the hands of local communities. Notably, the Independent Evaluation Group (IEG) of the World Bank has rated the Implementation Completion and Results Report of the project as "High" – a rare distinction within the Bank's ecosystem, underscoring the project's impact, credibility, and replicability across regions (World Bank, 2024).

Through the diverse case studies presented in its chapters, Meghalaya serves as a beacon of participatory natural resource management, where ancient wisdom and modern technology converge to forge a path towards a sustainable and resilient future. It offers a replicable framework for inclusive environmental conservation and climate change mitigation, serving as an invaluable resource for policymakers, practitioners, and researchers glob-ally. By meticulously documenting both successes and challenges, the volume provides critical insights, not just into the "what," but crucially, the "how" of effective environmental action. It underscores that true progress lies in empowering those closest to the land, ensuring that development is both ecologically sound and socially just. The echoes from Meghalaya, reflected in this significant volume, resonate with universal lessons for a planet striv-ing for sustainability.

## References


Badavath, N., Sahoo, S., & Samal, R. (2024). Assessing soil erosion risk in Meghalaya, India: Integrating geospatial data with RUSLE model. *Environment, Development and Sustainability, 2424*, 1–36.

Balusamy, A., Ramesh, T., Moirangthem, P., Talang, H. D., Chanu, L. J., Gopalakrishnan, B., & Choudhury, B. U. (2024). Indigenous wisdom for sustainable mountain agriculture in the eastern Himalayas (Meghalaya). *Indian Journal of Soil Conservation, 52*(1), 46–57.

Bhattacharjee, K., Ganguly, A., Mitra, K., & Mitra, S. (2025). Investigating quality of life and resilience: A case study on tourism service providers in Meghalaya. In U. Chatterjee, A. Bhunia, J. Gupta, & K. Gupta (Eds.), *Sustainability and urban quality of life* (pp. 197–220). Routledge.

Chakraborty, T., & Saikia, U. S. (2022). Climate change implications on groundwater resources in East Khasi Hills, with special reference to greater Shillong, Meghalaya, India. In *Current directions in water scarcity research* (Vol. 5, pp. 3–17). Elsevier.

Kharmylliem, B., & Kipgen, N. (2025). Village councils, social capital and sustainability: A study of urban water management of Shillong in Meghalaya, India. *Water Policy, 27*(3), 301–316.

Kshetriya, D., Warjri, C. D., Chakrabarty, T. K., & Ghosh, S. (2021). Assessment of heavy metals in some natural water bodies in Meghalaya, India. *Environmental Nanotechnology, Monitoring & Management, 16*, 100512.

Lyngdoh, A. W., Kumara, H. N., Babu, S., & Karunakaran, P. V. (2023). Community reserves: Their significance for the conservation of mammals in a mosaic of community-managed lands in Meghalaya, Northeast India. *PLoS One, 18*(1), e0280994.

Lyngkhoi, D. R., Singh, S. B., Singh, R., & Baruah, B. (2022). Shifting cultivation to settled agriculture: Land ownership rights and cropping pattern in Meghalaya, Northeast India. *Indian Research Journal of Extension Education, 22*(5), 150–154. https://doi.org/10.54986/irjee/2022/dec_spl/150-154

Majaw, B. (2023). Complex and worrying questions to Meghalaya's water crisis. *Sustainable Water Resources Management, 9*(2), 49.

Nomani, M. Z. M., Osmani, A. R., Salahuddin, G., Tahreem, M., Khan, S. A., & Jasim, A. H. (2021). Environmental impact of rat-hole coal mines on the biodiversity of Meghalaya, India. *Asian Journal of Water, Environment and Pollution, 18*(1), 77–84.

Sharma, K., & Laskar, N. (2021). Effect of climate change on spring discharge management system of the himalayan region in India. In *Advanced modelling and innovations in water resources engineering: Select proceedings of AMIWRE 2021* (pp. 105–117). Springer.

Singh, V., & Chaudhary, N. (2023). Land degradation, desertification, and food security in North-East India: Present and future scenarios. In *Sustainable development goals in Northeast India: Challenges and achievements* (pp. 153–166). Springer Nature.

Tamuli, T. M., & Bora, A. K. (2022). Impact of Jhum cultivation on forest ecosystem and environment management policies in Meghalaya, India. *Ecology, Environment, and Conservation*, 344–350. https://doi.org/10.53550/eec.2022.v28i02s.056

World Bank (2024). *Meghalaya - Community-Led Landscape Management Project (P157836): Implementation Completion and Results Report (ICR)*. Independent Evaluation Group (IEG), Report No: ICR00006597, World Bank Group.

# Community Engagement: Empowering Local Institutions for Environmental Change

# 2

# ECOLOGICAL TRANSITIONS AND ENVIRONMENTAL CHANGE IN THE NORTH-EASTERN HILLS

*B. Padmapriya*

## 2.1 Introduction

The north-eastern region of India plays a pivotal ecological role as one of the world's biodiversity hotspots and a major source of water, climate regulation, and ecological balance for the Indian subcontinent. Its dense forest cover, intricate river networks, and unique topography support vital ecosystem services such as carbon sequestration, groundwater recharge, and protection against natural disasters. Extensive aquatic and terrestrial habitats from lowland marshes to alpine meadows nurture thousands of endemic plant and animal species, underpinning both ecological stability and socio-economic well-being in the region.

However, these benefits are under increasing threat. Deforestation, much of it driven by shifting cultivation (Jhum), illicit logging, uncoordinated infrastructure development, and mining, is fragmenting habitats and accelerating biodiversity loss. Soil erosion and land degradation are exacerbated by crop cultivation on steep slopes and excessive grazing, while frequent landslides and floods, recently intensified by altered rainfall and climate variability, pose persistent risks to both environments and communities. Scientific studies reveal significant declines in forest cover between 2001 and 2020, ranging from 5% to 17% across multiple states, with Assam, Nagaland, and Tripura among the worst affected (Lotha et al., 2024; Das et al., 2023; Jain and Kasaudhan, 2025).

The urgency of research into ecological transitions and environmental impacts in the North-East is reflected in several recent national and regional initiatives. Secondary data, encompassing land use mapping, climate records, biodiversity assessments, and hydrological models, enable researchers to

DOI: 10.4324/9781003735380-3

systematically characterise deforestation trends, track changes in rainfall, and assess ecological degradation at multiple scales. These approaches are indispensable for revealing long-term transformations in landscape structure, quantifying ecosystem vulnerability, and informing conservation priorities for sustainable development and climate adaptation.

Moreover, the region's rich traditions of community-based resource management and indigenous knowledge contribute to resilient adaptation strategies; however, these mechanisms themselves are threatened by population pressure and rapid urbanisation. As observed, the influx of outsiders and rapid socio-economic shifts frequently undermine established conservation norms, amplifying ecological instability and resource use conflicts (Meenu, 2021).

Against this mounting backdrop of ecological pressure, the chapter analysis on climate change, deforestation, and land-use intensification emerges as both timely and crucial. By examining precipitation, soils, topography, land use, forest cover, hydrology, and natural disasters like floods, this chapter addresses the complex interplay of ecological transitions that shape the future sustainability and resilience of the north-eastern region.

## 2.2 Geographic and Environmental Context

The North-East India consists of seven sister states (Arunachal Pradesh, Assam, Manipur, Meghalaya, Mizoram, Nagaland, and Tripura) and it spans longitudinally from 89°41′57″ E to 97°26′16″ E and latitudinally from 21°56′15″ N to 29°29′46″ N, covering approximately 2,55,088 km² or 7.8% of India's total area.

### 2.2.1 Topography

The north-eastern region is predominantly mountainous, with about two-thirds of its total land area falling under hilly terrain. In contrast, less than one-tenth of the region is made up of plateaus, while roughly one-third consists of plains. The major terrain features of the North-East states are as follows (Table 2.1).

The mountainous terrain is the dominant feature of the region, covering an area of 165,241 km², which constitutes 63.30% of the total area of North-East India. This terrain includes much of the states of Arunachal Pradesh, Nagaland, Manipur, Mizoram, parts of Tripura, and Sikkim. These areas are characterised by rugged, hilly landscapes that define the physical geography of the region. On the contrary, the plateau region spans 21,532 km², accounting for only 8.12% of the total area. Meghalaya is the state primarily represented by this terrain, known for its elevated and relatively flat landforms, which contribute to the region's distinctive geography. The

**TABLE 2.1** Terrain features of the north-east states

| Name of the terrain | Area in km² | Percentage of the total area of North-East India |
| --- | --- | --- |
| Mountain Area | 165,241 | 63.30 |
| Plateau | 21,532 | 8.12 |
| Plains | 75,406 | 28.74 |
| (a) Brahmaputra plain | 56,195 | 21.43 |
| (b) Barak plain | 6,922 | 2.64 |
| (c) Pasighat, Roeing and Namsai plain | 3,250 | 1.24 |
| (d) Tripura plain | 6,300 | 2.40 |
| (e) Imphal valley plain | 1,843 | 0.70 |
| (f) West Garo-Brahmaputra plain | 879 | 0.33 |
| Total | 262,179 | 100 |

*Source:* Dikshit and Dikshit, 2014.

remaining 28.74% of the region consists of plains. These plains are divided into several distinct regions, including the Brahmaputra Plain in Assam, which spans 56,195 km² (21.43%); the Barak Plain, also in Assam, covering 6,922 km² (2.64%); the Pasighat, Roing, and Namsai Plain in Arunachal Pradesh, which extends over 3,250 km² (1.24%); the Tripura Plain, covering 6,300 km² (2.40%); the Imphal Valley Plain in Manipur, which spans 1,843 km² (0.70%); and the West Garo-Brahmaputra Plain in Meghalaya, covering 879 km² (0.33%). Together, these plains form an integral part of the region's fertile landscape, supporting agriculture and settlements across the area.

### 2.2.2 Weather and Precipitation

The weather conditions in the North-East region, expressed by temperature and precipitation over a period of time, are observed as follows. Across most parts of the world, weather data are systematically recorded, often daily or even hourly, through networks of meteorological stations equipped with specialised instruments. The density of these stations largely determines the accuracy of weather forecasting and the reliability of regional climate assessments. However, in North-East India, the network of such stations remains relatively sparse, limiting the precision of localised climate monitoring and prediction. Recognising this gap, the India Meteorological Department (IMD) has, in recent years, installed a greater number of rain gauges and automatic weather stations throughout the region. The temperature and rainfall data are calculated with the respective mean values of the four seasons.

### 2.2.3 Temperature

The temperature in the North-East region varies with height. In the Brahmaputra and Barak plains, temperature conditions are generally moderate, though slightly warmer than those observed in the adjoining hills and plateaus. Given their lower elevation, typically between 50 and 100 m above sea level, the plains experience temperatures that are about 8°C to 10°C higher than the upland regions. Beyond elevation, continentality, or distance from the sea, plays a significant role in shaping winter temperatures. Areas located farther inland tend to record cooler winters, while the valleys and western parts of the region, being closer to the Bay of Bengal, experience a moderating effect from sea breezes.

During summer, temperature contrasts between the plains and the hills remain evident, though the overall variation is relatively small across the region. Both zones exhibit closely clustered maximum temperatures, with the West Tripura plains recording the highest summer temperatures in the entire north-eastern region. This pattern highlights the combined influence of altitude, location, and maritime proximity in determining the temperature distribution across the diverse landscapes of North-East India.

The temperature data from 1950 to 2020 for five selected cities in North-East India, namely Cherapunji, Dibrugarh, Gangtok, Guwahati, and Imphal, indicate a consistent warming trend across the region. Cherapunji, known for its heavy rainfall, has experienced a temperature rise from 21.0°C in 1950 to 23.0°C in 2020, suggesting a shift in local climate patterns that may affect its hydrological cycle and ecosystems. Similarly, Dibrugarh and Guwahati, which are located at lower altitudes, show temperature increases from 24.0°C and 22.8°C in 1950 to 26.0°C and 24.5°C by 2020, respectively.

**TABLE 2.2** Decadal series of annual mean temperature

| Year | Cherapunji | Dibrugarh | Gangtok | Guwahati | Imphal |
| --- | --- | --- | --- | --- | --- |
| 1950 | 21 | 24 | 16.2 | 22.8 | 23.4 |
| 1960 | 21.2 | 24.3 | 16.5 | 23 | 23.6 |
| 1970 | 21.5 | 24.6 | 16.7 | 23.2 | 23.9 |
| 1980 | 21.8 | 24.9 | 17 | 23.5 | 24.2 |
| 1990 | 22.1 | 25.2 | 17.3 | 23.7 | 24.5 |
| 2000 | 22.4 | 25.5 | 17.5 | 24 | 24.7 |
| 2010 | 22.7 | 25.8 | 17.8 | 24.2 | 25 |
| 2020 | 23 | 26 | 18.1 | 24.5 | 25.3 |

*Source:* Author's own computation.

This warming could be linked to urban heat island effects, urbanisation, and industrialisation. Gangtok, situated at higher altitudes, exhibits a slower but steady increase from 16.2°C in 1950 to 18.1°C in 2020, indicating that even high-altitude areas are not immune to global temperature shifts. Imphal follows a similar trend, rising from 23.4°C in 1950 to 25.3°C in 2020. This regional warming suggests significant implications for local ecosystems, agriculture, and water resources, with potential challenges such as heat stress, changes in precipitation patterns, and impacts on health, particularly in urbanised areas. The data highlights the need for further exploration of the environmental and socio-economic impacts of temperature rise in North-East India, especially given the region's sensitivity to climate change.

### 2.2.4 Rainfall

Rainfall in North-East India is overwhelmingly a summer phenomenon, beginning in late April and continuing through October, with June–July as the peak. Along the southern and western margins, the monsoon typically arrives earlier, often in May or early June, and reaches its maximum in June; places such as Agartala, Cherrapunji, and Dhubri follow this pattern. Across the region, more than 90% of annual precipitation comes from the southwest monsoon, whether in the Brahmaputra valley (e.g., Dhubri, Guwahati, Sibsagar, Pasighat) or on the Meghalaya Plateau (e.g., Cherrapunji, Shillong). Even in comparatively drier locations like Lumding and Imphal, winter contributions are minimal (about 8% and 11%, respectively). While totals vary due to rainfall intensity and localised rain-shadow effects, summer monsoon rain remains the dominant source of precipitation throughout the north-east.

The rainfall data from 1950 to 2020 reveals a general increase in annual precipitation across the five regions in north-eastern states (Table 2.3). Cherapunji, known for its heavy rainfall, shows the most significant rise, from 535.0 cm in 1950 to 630.0 cm in 2020, reflecting broader climatic shifts potentially linked to changing atmospheric moisture and monsoon patterns. Similarly, Dibrugarh and Gangtok experience gradual increases in rainfall, rising from 277.5 cm to 348.2 cm and from 305.8 cm to 345.0 cm, respectively, indicating changes in local weather patterns that could affect agriculture and water availability. Guwahati and Imphal also show modest increases in rainfall, from 195.0 cm to 218.0 cm and from 145.0 cm to 173.8 cm, suggesting growing precipitation that may influence urban infrastructure and flood management. These trends underscore the need for climate adaptation strategies to address the potential impacts of increased rainfall, including flooding risks and challenges for agriculture and water resources across the region.

**TABLE 2.3** Decadal series of annual mean rainfall

| Year | Cherapunji | Dibrugarh | Gangtok | Guwahati | Imphal |
|---|---|---|---|---|---|
| 1950 | 535 | 277.5 | 305.8 | 195 | 145 |
| 1960 | 547.2 | 287.2 | 312.1 | 198.5 | 149.2 |
| 1970 | 560 | 297.4 | 319 | 202 | 153.5 |
| 1980 | 575 | 308 | 325 | 205.5 | 157.8 |
| 1990 | 590 | 318.2 | 330.5 | 208 | 162 |
| 2000 | 602.3 | 328.5 | 336 | 212.2 | 165.3 |
| 2010 | 615 | 338 | 341.2 | 215.5 | 169.5 |
| 2020 | 630 | 348.2 | 345 | 218 | 173.8 |

*Source:* Author's own computation.

### 2.2.5 Water Resources

Rainfall is distributed unevenly across the planet. Arid deserts receive little to no precipitation, while equatorial belts experience persistently high rainfall throughout the year. Most other regions fall between these extremes, with precipitation varying according to their position relative to large-scale pressure belts, prevailing wind systems, and the shape and orientation of surrounding mountains that force moist air to rise.

According to the Central Water Commission, groundwater availability in the North-East is overwhelmingly shaped by the monsoon, with approximately 71% of the total annual recharge derived from monsoon inputs and roughly 29% from the non-monsoon season. The region's total annual groundwater recharge is 35.61 bcm, of which 29.22 bcm, which is around 82%, is classified as annually extractable, while natural discharges account for about 10% of recharge.

State-wise, Assam dominates the groundwater availability, contributing roughly 75% of total regional recharge and 73% of the extractable resource, followed by Arunachal Pradesh (13% of recharge). Smaller hill states contribute modest volumes but often exhibit high extractable fractions (88–91% in Mizoram, Nagaland, and Meghalaya), reflecting limited natural outflows and relatively good storage potential. An important outlier is Tripura, where the non-monsoon share of recharge is relatively high (34%) and natural discharge is the largest in the region (19% of its recharge), yielding the lowest extractable share (76%) among the states.

In short, most groundwater in the North-East is recharged during the monsoon, and Assam supplies the largest share of the region's total. However, states differ in important ways: Tripura relies more on non-monsoon recharge and loses a larger share of water through natural discharge, making it more vulnerable in the dry season, while the other north-eastern

**TABLE 2.4** Ground water resources in north-east states (BCM)

| States | Ground water recharge Monsoon season | | Non-monsoon season | | Total annual ground water recharge | Total natural discharges | Annual extractable ground water resources |
|---|---|---|---|---|---|---|---|
| | Recharge from rainfall | Recharge from other sources | Recharge from rainfall | Recharge from other sources | | | |
| Arunachal Pradesh | 1.96 | 0.94 | 1.06 | 0.56 | 4.52 | 0.41 | 4.0 |
| Assam | 17.92 | 1.15 | 6.52 | 0.94 | 26.53 | 2.56 | 21.4 |
| Manipur | 0.4 | 0.0 | 0.11 | 0.01 | 0.52 | 0.05 | 0.4 |
| Meghalaya | 1.29 | 0.01 | 0.42 | 0.0 | 1.72 | 0.17 | 1.51 |
| Mizoram | 0.19 | 0.0 | 0.03 | 0.0 | 0.22 | 0.02 | 0.2 |
| Nagaland | 0.36 | 0.33 | 0.08 | 0.02 | 0.79 | 0.08 | 0.71 |
| Tripur | 0.81 | 0.06 | 0.22 | 0.22 | 1.31 | 0.25 | 1.0 |
| Total of NE | 22.93 | 2.49 | 8.44 | 1.75 | 35.61 | 3.54 | 29.22 |

*Source:* Central Water Commission (2023).

states are more clearly monsoon-driven and retain a higher fraction as usable groundwater.

Despite abundant rainfall and runoff, actual water use in the North-East is limited. Steep, mountainous terrain and the scarcity of broad plains (beyond the narrow Brahmaputra and Barak belts) restrict large reservoirs and canals, so much of the rain flows away rather than being stored.

Based on the compiled data of annual extractable groundwater and current withdrawals, North-East India presents a low overall development level of its groundwater resources alongside a strong spatial concentration of use. Region-wide, the annual extractable stock is 29.22 bcm, while current extraction is 2.90 bcm, implying a development ratio of 9.9%. Use is irrigation-led (73.3% of pumping), followed by domestic (26.6%) and industrial (0.8%) demand, indicating a largely agrarian water footprint with limited industrial pressures. Spatially, withdrawals are Assam-centric; the state accounts for 91.4% of total regional pumping and exhibits the highest extraction-to-resource ratio (12.4%), consistent with larger alluvial aquifers and irrigated agriculture. The remaining hill states extract far smaller volumes and show markedly lower development ratios: Arunachal Pradesh (0.75%), Nagaland (2.8%), Meghalaya (3.3%), Mizoram (5.0%), and Manipur (10%), with Tripura also near (10%) (Table 2.5).

These contrasts point to differentiated management needs. In Assam, the policy emphasis should be on irrigation efficiency, conjunctive use, and managed aquifer recharge to moderate growth in demand; in the hill states, drinking-water reliability, spring-shed and catchment recharge, and appropriately scaled irrigation are more salient.

TABLE 2.5 Ground water extraction in north-east states (BCM)

| State | Annual extractable ground water resources | Irrigation | Industrial | Domestic | Total |
|---|---|---|---|---|---|
| Arunachal Pradesh | 4 | 0.02 | 0.01 | 0.01 | 0.03 |
| Assam | 21.4 | 2.06 | 0.01 | 0.58 | 2.65 |
| Manipur | 0.4 | 0.02 | 0.0002 | 0.02 | 0.04 |
| Meghalaya | 1.51 | 0.003 | 0.0007 | 0.05 | 0.05 |
| Mizoram | 0.2 | 0 | 0 | 0.01 | 0.01 |
| Nagaland | 0.71 | 0.002 | 0.00002 | 0.02 | 0.02 |
| Tripur | 1 | 0.02 | 0.0007 | 0.08 | 0.10 |
| Total of NE | 29.22 | 2.12 | 0.02 | 0.77 | 2.9 |

*Source:* Central Water Commission (2023).

### 2.2.6 Forest Coverage

According to the Forest Survey of India (2021), approximately 64.66% of the area in the Seven Sisters region is under forest cover, making it one of the world's twelve recognised biodiversity hotspots. Except for Assam, all the north-eastern states maintain between 50% and 80% of their total area as forested land. The region's warm climate, combined with heavy rainfall, encourages dense forest growth in the lower altitudes. Even in the higher elevations, the abundant summer rainfall and cooler winter temperatures help maintain soil moisture by reducing evapotranspiration, thereby sustaining rich vegetation across both plains and hill ecosystems.

Across the North-East, the distribution of forest area (RFA) varies sharply by state when viewed in both proportional and absolute terms. Manipur has the highest share of forests in its geographical area (GA), with 78.0% of its 22,327 km² under forests (17,418 km²), ranking first. Arunachal Pradesh combines a very high proportion (61.5%) with the largest absolute forest extent in the region (51,540 km² of 83,743 km² GA), placing it second and underscoring its role as the region's primary forest landscape. Tripura ranks third with 60.0% forest cover (6,295 km² out of ~10,486 km² GA as reported), while Nagaland follows with 52.0% (8,632 km² of 16,579 km²). Meghalaya shows a mid-level proportion (42.3%, 9,508 km² of 22,429 km²), and Mizoram a lower share (35.4%, 7,479 km² of 21,081 km²). Assam, despite having a GA comparable to Arunachal (78,438 km²), records the lowest proportional forest cover at 34.2% (26,836 km²), reflecting extensive non-forest land uses in the Brahmaputra plains. Taken together, the data indicate that conservation priorities must consider both metrics: States with high percentages but small base areas (e.g., Tripura) may be highly sensitive to incremental loss, whereas states with large absolute forest stocks (notably Arunachal) are pivotal for regional carbon and biodiversity outcomes; simultaneously, low proportional cover in a large state (Assam) points to opportunities for landscape restoration and agroforestry in non-forest zones (Table 2.6).

## 2.3 Key Environmental Challenges

### 2.3.1 Deforestation

According to the Food and Agriculture Organization (FAO, 2022), the quantity and quality of forest cover have declined considerably. This has led to biodiversity depletion, growth of pollution, global warming, climate change, soil erosion, and disruption of the water cycle (Santos et al., 2021). Although causes were inconsistent among north-eastern regions, some of the factors include shifting cultivation, growth without development, rotational felling in tea gardens, conservation, regeneration of growth in shifting cultivation

**TABLE 2.6** Ranking of north-eastern states as per reserved forest area (RFA)

| States | Total geographical area (GA) (km$^2$) | Total RFA | Percentage of forest under GA (%) | Descending order |
|---|---|---|---|---|
| Arunachal Pradesh | 83,743 | 51,540 | 61.5 | 2 |
| Assam | 78,438 | 26,836 | 34.2 | 7 |
| Manipur | 22,327 | 17,418 | 78.0 | 1 |
| Meghalaya | 22,429 | 9,508 | 42.3 | 5 |
| Mizoram | 21,081 | 7,479 | 35.4 | 6 |
| Nagaland | 16,579 | 8,632 | 52.0 | 4 |
| Tripura | 10.486 | 6,295 | 60.0 | 3 |

*Source:* Forest Survey of India (2023).

areas, bamboo regeneration, rubber plantations harvesting, and expansion of the rubber plantation areas (Forest Survey of India, 2017).

Furthermore, issues of forest management, which include the native people enjoying traditional or customary rights over property and the absence of institutional recognition of local populations, worsen deforestation and forest degradation. In addition to this, the expansion of residential zones, converting forests into permanent agricultural land or pastures, intensive shifting cultivation, development of infrastructure, and degradation of forests due to mining activities, selective logging, and the decline in natural regeneration processes are the primary factors of deforestation in the region (Sastry et al., 2007).

Table 2.7 shows the forest cover change in North-East India, for the period from 1993 to 2003. Arunachal Pradesh, with the largest geographical area (83,743 km$^2$), had 76.01% of its area covered by forests in 1993, but this has slightly declined to 61.5% by 2023, with a significant reduction in absolute forest area from 63,661 km$^2$ to 51,540 km$^2$. In contrast, Manipur maintained a stable forest cover percentage of 78% over the three decades, with no change in absolute forest cover (17,418 km$^2$) between 2003 and 2023. Assam shows a decline in forest cover both in terms of area and percentage: From 31.24% in 1993 to 34.2% in 2023, with the absolute forest cover increasing slightly from 24,508 km$^2$ to 26,836 km$^2$. Meghalaya had a significant decrease in its forest cover, with a drop from 70.31% in 1993 to 42.3% in 2023, reflecting a sharp decline in forest area from 15,769 km$^2$ to 9,508 km$^2$. Mizoram experienced a similar trend, with its forest cover percentage falling drastically from 84.67% in 1993 to 35.4% in 2023, a decline in forest area from 18,697 km$^2$ to 7,479 km$^2$. Nagaland, although showing a decline from 86.54% in 1993 to 52% in 2023, still retains a relatively high forest cover, with forest area remaining stable at around 8,632 km$^2$. Finally,

**TABLE 2.7**  Forest cover in north-eastern states, 1993–2023

| States | Geographical area (km²) | 1993 | | 2003 | | 2023 | |
|---|---|---|---|---|---|---|---|
| | | Forest cover (km²) | % of Forest cover | Forest cover (km²) | % of forest cover | Forest cover (km²) | % of forest cover |
| Arunachal Pradesh | 83,743 | 63,661 | 76.01 | 51,540 | 61.53 | 51,540 | 61.5 |
| Assam | 78,438 | 24,508 | 31.24 | 27,018 | 34.45 | 26,836 | 34.2 |
| Manipur | 22,327 | 17,621 | 78.92 | 17,418 | 78 | 17,418 | 78 |
| Meghalaya | 22,429 | 15,769 | 70.31 | 9,496 | 42.34 | 9,508 | 42.3 |
| Mizoram | 21,081 | 18,697 | 84.67 | 16,717 | 79.3 | 7,479 | 35.4 |
| Nagaland | 16,579 | 14,348 | 86.54 | 8,629 | 52.05 | 8,632 | 52 |
| Tripura | 10,486 | 5,538 | 52.81 | 6,293 | 60.01 | 6,295 | 60 |

*Source:* India State of Forest Report (1993, 2003, and 2023).

Tripura experienced an increase in its forest cover percentage from 52.81% in 1993 to 60% in 2023, with forest area increasing slightly from 5,538 km$^2$ to 6,295 km$^2$. These trends indicate that while some states like Manipur and Tripura have maintained or slightly improved their forest cover, others, particularly Arunachal Pradesh, Meghalaya, and Mizoram, have seen notable declines in both forest area and percentage, pointing to the need for targeted conservation efforts, especially in rapidly deforesting areas.

### 2.3.2  Soil Erosion and Land Degradation

Soil erosion ranks among the most critical environmental challenges facing the world today. It endangers human life, agricultural performance, and even the livelihoods of the marginal farming communities relying on land resources. It also leads to the erosion of fertile topsoil, depletion of nutrients, and siltation of rivers and reservoirs (Pimentel and Burgess, 2013). The region is suffering because of extensive soil erosion, soil fertility reduction, and deforestation in North-East India, leading to widespread ecological deterioration (Ansari et al., 2022). Due to the sharp slopes of the land, high rainfall, and the absence of methods of rainwater management and soil conservation, thousands of millions of tonnes of soil and a considerable quantity of nutrients are believed to be washed away every year (Pandey et al., 2007).

Land-use alterations, especially the transformation of farming (jhum) by exposing the land surfaces and reducing the plant cover, have accelerated the process of soil erosion (Ansari et al., 2022). The disturbances caused by people such as deforestation, agricultural encroachment into forests, road networks, rapid urbanisation, and other developmental processes, poor soil depth, low structural stability, and large rainfall erosivity further contribute to soil loss and land degradation (Mishra et al., 2022). Through this, more soil erosion will cause less crop production, a reduction in soil nutrient levels, serious siltation of rivers and lakes, and ecosystem and water quality deterioration (Pimentel and Burgess, 2013).

The data indicate substantial but state-specific soil degradation across North-East India (Table 2.8). In absolute terms, the largest degraded area is in Arunachal Pradesh (4,503), while the highest intensity occurs in Mizoram (89.2% of land degraded), followed by Nagaland (60.0%), Tripura (59.9%), Meghalaya (53.9%), Arunachal Pradesh (53.8%), Manipur (42.6%), and Assam (28.2%). The dominant degradation processes differ by state: Water erosion is the principal driver in Arunachal (2,372) and also substantial in Assam (688) and Nagaland (390), consistent with steep topography and intense rainfall. Soil acidity is the leading constraint in Meghalaya (1,030) and Mizoram (1,050) and remains high in Arunachal (1,955) and Manipur (481), while water-logging/flooding is comparatively minor across the region

**TABLE 2.8**  Soil degradation status in north-eastern states

| States | Water erosion | Water-logging/ flooding | Soil acidity | Complex problems | Total degraded area | Degradation (%) |
|---|---|---|---|---|---|---|
| Arunachal Pradesh | 2,372 | 176 | 1,955 | – | 4,503 | 53.8 |
| Assam | 688 | 37 | 612 | 876 | 2,213 | 28.2 |
| Manipur | 133 | 111 | 481 | 227 | 952 | 42.6 |
| Meghalaya | 137 | 07 | 1,030 | 34 | 1,208 | 53.9 |
| Mizoram | 137 | – | 1,050 | 694 | 1,881 | 89.2 |
| Nagaland | 390 | – | 127 | 478 | 995 | 60.0 |
| Tripura | 121 | 191 | 203 | 113 | 628 | 59.9 |

*Source:* Gajbhiye (2006).

but notable in Tripura (191) and Manipur (111). "Complex problems" (co-occurring constraints) are particularly high in Assam (876), Mizoram (694) and Nagaland (478), signalling areas that require multiple, coordinated interventions. Policy implications are therefore location-specific: Prioritise erosion control and watershed treatment (contour bunds, vegetative barriers, check-dams) in Arunachal, Assam and Nagaland; deploy acidity management (liming, organic matter additions, nutrient balancing) in Meghalaya, Mizoram, Arunachal and Manipur; and invest in drainage and floodplain management in Tripura and Manipur, while addressing multi-hazard "complex" sites in Assam, Mizoram and Nagaland with integrated soil and water conservation packages.

### 2.3.3 Water Security Challenges

India's water crisis mirrors a global imbalance between population growth and renewable freshwater supply. The country possesses barely 4% of the world's renewable water resources but supports nearly 18% of the global population (Lahiry, 2017). This disproportion creates intense competition for limited water, making it one of the country's most constrained natural resources, affecting both the economy and human well-being (Panwar, 2022). Water scarcity, defined as the lack of sufficient and reliable water for human and ecological needs (UN-Water, 2006), thus stands among India's most critical challenges of the 21st century.

Within this broader national context, the north-eastern region presents a paradox. Despite receiving heavy rainfall, it is poorly endowed with replenishable groundwater. This anomaly arises from a combination of natural factors, intense precipitation, steep slopes, shallow soils, and rapid surface drainage that limit water infiltration and aquifer formation. The region's mountainous terrain, characterised by rugged relief and thin soil profiles, causes rainwater to flow swiftly into rivers, leaving little opportunity for groundwater recharge. Only in the Brahmaputra valley, with its long and narrow alluvial plains, do conditions permit meaningful infiltration and storage. In contrast, the plateaus of Meghalaya and the mountain ranges of Mizoram, Manipur, and Nagaland contain few aquifers, and those that exist are part of dipping geological strata that discharge water through seepage rather than retain it.

Moreover, the interplay of topography, demographic pressures, and human–environment interactions, coupled with the evolving climatic dynamics over the Bay of Bengal, has altered rainfall regimes and intensified hydrological instability across the region. These shifts have affected the timing, distribution, and quantity of streamflow feeding local water systems, aggravating seasonal water scarcity even in areas historically known for abundant rainfall. This underscores the need for adaptive water management

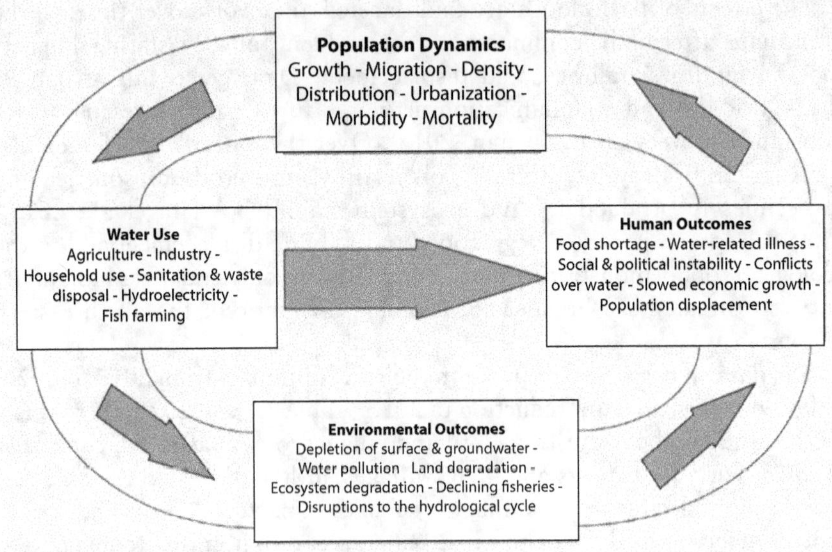


**FIGURE 2.1**    Links between population and fresh water

*Source:* IUCN et al. 1996.

strategies that integrate climate resilience, soil–water conservation, and local hydrological knowledge to safeguard the region's water security (Figure 2.1).

### 2.3.4 Floods

The north-eastern region of India receives abundant rainfall as moist air masses are forced upward by mountainous terrain, a process known as orographic uplift (Rahman & Azim, 2023). The combination of intense monsoon precipitation and steep upstream slopes frequently triggers flash floods. The two major river systems, the Brahmaputra and the Barak, exhibit a recurring flood regime, with the Brahmaputra experiencing floods almost every monsoon season (Dikshit & Dikshit, 2014).

Rapid urban population growth has heightened the region's exposure to flood hazards. This vulnerability is compounded by unequal socio-economic conditions, increased surface runoff, and limited groundwater infiltration capacity. The flood dynamics of the Brahmaputra River are influenced primarily by three factors: The intensity and distribution of rainfall, the longitudinal gradient of the river channel, and the basin ratio, defined as the ratio of the catchment area to river length, which is shaped by basin morphology. In some cases, deeply incised channels may deviate from these general patterns.

The effects of flash floods are multifaceted and evolve over time. In the immediate aftermath, communities face sudden and devastating impacts such as fatalities, injuries, and displacement. Damage to infrastructure, roads, bridges, and communication networks further delays rescue operations and recovery efforts (Chan, 2015). Over the longer term, floods disrupt agricultural cycles, degrade soil fertility, and erode topsoil, leading to declines in productivity and ecosystem stability (Aslam et al., 2023). Livestock losses, uprooted vegetation, and wildlife displacement add to the ecological toll, demonstrating that flash floods not only endanger human life but also threaten the broader socio-ecological fabric of the north-eastern region.

The data shows a clear increase in the scale of impact from 2015 to 2020, followed by a significant reduction in 2023 (Table 2.9). In 2020, there was a peak in the number of districts (30) and villages (5,475) affected, compared to 2015, with 25 districts and 4,763 villages, although the figures for 2023 show a slight decline in the number of villages impacted (4,716). The area of crop damage reached its highest in 2020 (18.28 lakh hectares), compared to 3.388 lakh hectares in 2015, but this dropped significantly in 2023 to 0.21 lakh hectares, indicating an improvement in crop resilience or a reduction in environmental stressors. Similarly, the population affected was highest in 2020 (5.789 million people), but dropped drastically to 1.347 million people by 2023, reflecting a positive trend in mitigation or response measures. While the number of houses damaged peaked in 2020 (8,020 houses), it was notably lower in 2023 (2,984 houses). Tragically, human lives lost were also the highest in 2020 (124 lives), with a significant reduction in 2023 (18 lives). These figures reflect a pattern of growing vulnerability in 2020, but a trend

**TABLE 2.9** Impact of floods in Assam

| Nature of damage | 2015 | 2020 | 2023 |
| --- | --- | --- | --- |
| No. of districts affected | 25 | 30 | 30 |
| No. of villages affected | 4,763 | 5,475 | 4,716 |
| Crop area affected | 3.388 lakh hectares | 18.28 lakh hectares | 0.21 lakh hectares |
| Population affected | 3.667 million | 5.789 million | 1.347 million |
| No. of houses damaged | 6,058 | 8,020 | 2,984 |
| Huan lives affected | 66 | 124 | 18 |

*Source:* Revenue Department, Government of Assam.

towards recovery and improvement in disaster resilience by 2023, likely driven by better preparedness, early warning systems, and disaster response.

## 2.4 Way Forward and Policy Implications

An effective way forward for the North-East is to mesh environmental restoration with livelihood security, rather than treat them as competing objectives. Evidence from the region shows that land degradation, shifting cultivation transitions, and water stress are tightly coupled, so interventions must be bundled at the micro-watershed scale: Soil–water conservation, liming and nutrient management on acid soils, agroforestry and bamboo-based value chains, spring-shed revival in hill catchments, and conjunctive use of surface–groundwater in the plains. Such integrated packages reduce erosion, raise on-farm productivity, and buffer households against rainfall variability, while also rebuilding natural capital (forest cover, soil organic matter) that underpins long-run resilience. Programme design should therefore finance ecosystem-service outcomes (e.g., baseflow augmentation, sediment reduction) alongside income outcomes, and sequence measures to local biophysical conditions and land-use histories.

Strengthening community-based management is central, given the prevalence of customary tenure and community use of forests across the hills. The most durable gains in forest recovery and watershed stability have come where communities control access, enforce fallow cycles, and diversify away from short-fallow jhum into tree crops, NTFPs, and mixed agroforestry. Policy should deepen legal recognition of community institutions, provide technical backstopping (e.g., soil testing, species mix for assisted natural regeneration), and align incentives through payments for ecosystem services and community forestry revenue sharing. Lessons from community forestry and participatory watershed programmes elsewhere, adapted to the North-East's tenure pluralism, suggest that decentralised governance with clear rights and responsibilities improves compliance and lowers enforcement costs, especially when paired with transparent monitoring via remote sensing and simple hydrological metrics.

Policy priorities for forest and watershed conservation should be differentiated by sub-region. In steep, high-rainfall states (e.g., Meghalaya, Mizoram, Nagaland), the emphasis should be on slope stabilisation (contour trenching, vegetative barriers), rehabilitation of degraded jhum fallows through assisted natural regeneration and agroforestry mosaics, liming of acid soils to lift productivity, and protection of headwater springs. In the large alluvial plains (Assam), priorities include floodplain zoning, riparian buffer restoration, managed aquifer recharge, and integration of farm water-saving (micro-irrigation, crop shifts) with catchment-scale sediment management. Across the region, forest cover decline and fragmentation have been documented over recent decades, contrasting with localised stability or gains where institutions are stronger, argue for targeted restoration of critical corridors, stricter control of illegal logging and mining footprints, and impact-screening of roads and hydropower to avoid locking in erosion and landslide risk.

Finally, research and monitoring should close decision-relevant gaps: (1) Long-term, station-level rainfall and temperature analyses linked to flood and landslide incidence; (2) spring-shed hydrology and baseflow response to restoration; (3) high-resolution forest cover and degradation mapping that distinguishes regrowth from primary loss; (4) agronomic trials on acidity management and nutrient balances under hill agroforestry; and (5) socio-economic evaluation of livelihood transitions out of short-fallow jhum. A regional observatory that integrates remote sensing with community-generated data would support adaptive management and transparent accountability, particularly in states already facing the "scarcity amid plenty" paradox in water security.

## 2.5 Conclusion

This chapter provides a comprehensive analysis of the ecological transitions, environmental challenges, and climate-induced vulnerabilities in the northeastern hill region of India. It situates the region as both ecologically vital and environmentally fragile, home to rich biodiversity, dense forests, and complex hydrological systems that sustain livelihoods, regulate climate, and preserve ecological balance across the Indian subcontinent. Through an integration of secondary data on temperature, rainfall, forest cover, land degradation, groundwater availability, and flood impacts, the chapter demonstrates the accelerating pace of environmental change. It identifies deforestation, shifting cultivation, land-use intensification, soil erosion, and water insecurity as the key stressors that undermine the region's ecological resilience. Despite the abundance of rainfall, limited groundwater retention, topographical constraints, and rapid land conversion have created a paradox of "scarcity amid plenty." These trends are compounded by climate variability, which manifests as rising temperatures, erratic precipitation, and frequent floods and landslides. The analysis underscores that while states like Manipur and Tripura have maintained stable forest cover, others such as Arunachal Pradesh, Meghalaya, and Mizoram have experienced alarming declines, threatening both biodiversity and local livelihoods. The chapter concludes that a sustainable pathway for the region must integrate environmental restoration with livelihood security, strengthen community-based management systems, and enhance policy coherence across forestry, water, and land governance to ensure ecological and social resilience.

Placing the north-eastern region within the broader discourse of environment and development is crucial for reimagining India's ecological future. The region embodies a microcosm of the country's most pressing environmental dilemmas, balancing conservation imperatives with livelihood needs in an area of immense ecological value and sociocultural diversity. Its forests and watersheds form the ecohydrological backbone of the Eastern Himalayas, regulating river systems that influence agriculture, energy production, and flood control across eastern India and Bangladesh. Yet, the region's mountainous terrain, intense rainfall, and fragile soils make it particularly sensitive to unsustainable land-use practices and climate variability. Integrating the North-East into national and global debates on sustainable development, therefore, demands recognising its dual identity, as both a provider of ecosystem services and a frontline of environmental vulnerability.

Moreover, the North-East offers unique insights into community-led natural resource management, where customary tenure systems and indigenous ecological knowledge have historically maintained forest and watershed stability. Strengthening and scaling these participatory models can inform broader policy frameworks for climate adaptation and environmental governance across India's other ecologically sensitive zones. In an era marked by deforestation, resource conflicts, and climate migration, the ecological health of the north-eastern hills is not merely a regional concern; it is a cornerstone of India's sustainable development and environmental security. Reaffirming its importance means aligning regional development strategies with global sustainability goals, promoting science-based yet community-driven conservation, and ensuring that environmental governance in the North-East is not peripheral but central to India's environmental policy discourse.

## References


Ansari, M., Choudhury, B. U., Nengzouzam, G., & Islam, A. (2022). Causes and consequences of soil erosion in Northeastern Himalaya, India. *Current Science, 122*(7), 772–789. https://doi.org/10.18520/cs/v122/i7/772-789

Aslam, M,, Hayat, R., Pari, N., Sameen, A., & Ahmed, M. (2023). Climate change, flash floods and its consequences: A case study of Gilgit-Baltistan. In M. Ahmed and S. Ahmad (Eds.), *Disaster risk reduction in agriculture* (pp. 293–310). Springer Nature. https://doi.org/10.1007/978-981-99-1763-1_14

Chan, N. W. (2015). Impacts of disasters and disaster risk management in Malaysia: The case of floods. In D. P. Aldrich, S. Oum, & Y. Sawada (Eds.), *Resilience and recovery in Asian disasters* (pp. 239–265). Springer. https://doi.org/10.1007/978-4-431-55022-8_12

Das, A., Gujre, N., Devi, R. J., Rangan, L., & Mitra, S. (2023). Traditional ecological knowledge towards natural resource management: Perspective and challenges in North East India. In *Sustainable agriculture and the environment* (Chap. 10). Elsevier. https://doi.org/10.1016/B978-0-323-90500-8.00019-1

Dikshit, K. R., & Dikshit, J. K. (2014). North-East India: Land, people and economy. In *Advances in Asian human-environmental research* (pp. 1–25). Springer Science+Business Media. https://doi.org/10.1007/978-94-007-7055-3_7

Food and Agriculture Organization. (2022). *The state of the world's forests 2022: Forest pathways for green recovery and building inclusive, resilient and sustainable economies.* FAO. https://www.fao.org/3/cb9360en/online/cb9360en .html

Forest Survey of India. (2017). *India state of forest report 2017.* Ministry of Environment, Forest and Climate Change, Government of India. https:// ruralindiaonline.org/en/library/resource/india-state-of-forest-report-2017/

Gajbhiye, K. S. (2006). Land utilization. In *Hand book of agriculture* (5th ed., revised and expanded). Indian Council of Agricultural Research.

Jain, R., & Kasaudhan, H. (2025). Impact of deforestation on biodiversity in the Northeastern states of India. *International Journal of Scientific Research & Engineering Trends, 11*(3), 1–3.

Lahiry, S. (2017). *Why India needs to change the way it manages water resources.* Down to Earth.

Lotha, T. N., Ritse, V., Nakro, V., Ketiyala, K., Imkongyanger, I., Rudithongru, L., Hazarika, N., Jamir, L. (2024). Climate change impact and traditional adaptation practices in Northeast India: A review. *Current World Environment, 19*(2), 558–575.

Meenu. (2021). Numerous manifestations of ecology in contemporary poetry from India's Northeast. *Airo Journals.* 240-246

Mishra, P. K., Banerjee, A., & Kumar, P. (2022). Land degradation and soil erosion responses under changing land cover in the Eastern Himalayas, India. *Land, 11*(2), 179. https://doi.org/10.3390/land11020179

Pandey, A., Chowdary, V. M., & Mal, B. C. (2007). Identification of critical erosion-prone areas in agricultural watersheds using USLE, GIS and remote sensing. *Water Resources Management, 21*(5), 729–746.

Panwar, T. S. (2022, March 22). World water day: The case for sustainable solutions. *Deccan Herald.*

Pimentel, D., & Burgess, M. (2013). Soil erosion threatens food production. *Agriculture, Ecosystems & Environment, 185*, 86–90. https://doi.org/10.1016 /j.agee.2013.10.014

Rahman, M. N., & Azim, S. A. (2023). Spatiotemporal evaluation of rainfall trend during 1979–2019 in seven climatic zones of Bangladesh. *Geology, Ecology, and Landscapes, 7*, 340–355.

Sastry, K. L. N., Kandya, A. K., Thakker, P. S., Shangkooptong, L. B., Devi, R. S. Shamungou, N. S., Jagadishwor, K. S., Khaizalian, Thambou K. S., & Singsit, N. (2007). Nationwide forest encroachment mapping using remote sensing and GIS techniques – Manipur state. Manipur Forest Department & Manipur State Remote Sensing Applications. (Report) https://www.isec.ac.in/wp-content/ uploads/2023/07/WP-523-Marchang-Reimeingam-final.pdf isec.ac.in+1

Santos, A. M., et al. (2021). Deforestation drivers in the Brazilian Amazon: Assessing new spatial predictors. *Journal of Environmental Management, 294*, 113020. https://doi.org/10.1016/j.jenvman.2021.113020

UN-Water. (2006). *The United Nations World Water Develop ment Report 2: Water, a shared responsibility.* World Water Assessment Programme (WWAP). Doc. no. UN-WATER/ WWAP/2006/3. UNESCO, Paris, France, and Berghahn Books.

# 3

# BUILDING A CADRE OF COMMUNITY PROFESSIONALS

*Dokatchi K. Marak and Nangteibor Shabong*

## 3.1 Introduction

### 3.1.1 Purpose, Scope, and Significance of the Case

Meghalaya's unique land tenure system and local governance structures often pose challenges to effective implementation of large-scale projects and schemes. However, the key to driving sustainable, climate-adaptive initiatives that benefit both the land and local communities lies in the active participation of people. It is the community members who ensure the long-term sustainability of these initiatives.

Engaging the local population is essential, since it not only offers valuable insights into the real challenges on the ground but also increases the likelihood of project success. When community members take the lead, it fosters a strong sense of ownership and empowerment. In the bargain, it simplifies the process of capacity building and social integration. This approach also facilitates the integration of traditional knowledge, which is crucial in addressing contemporary environmental challenges.

Recognising these factors, the State Government identified the need to train and establish young leaders who understand both the local context and community dynamics, and who can bridge the gap between government institutions and community stakeholders.

These young leaders are known as village community facilitators (VCFs), following the concept of Community Resource Persons (CRPs) under the National Rural Livelihood Mission (NRLM).

CRPs help implement the programme at the grassroots level by mobilising and empowering women through Self Help Groups (SHGs). They focus on

DOI: 10.4324/9781003735380-4

areas like institution building, financial inclusion, livelihood development, and social awareness. Essentially, CRPs act as a bridge between the programme and the community, using their strong understanding of local needs and dynamics to drive programme success.

Also, the Government of Meghalaya drew inspiration from the Civilian Conservation Corps established in the United States in 1933 under President Franklin Roosevelt. Accordingly, it introduced these village community facilitators under the World Bank-funded Community-Led Landscape Management Project (CLLMP).

窗体顶端

Local engagement is key – not only does it provide invaluable insights into the real challenges on the ground, but it also strengthens the likelihood of success. When initiatives are led by people from within the community, there is a natural sense of ownership and empowerment, making the process of capacity building smoother and more integrated. Moreover, such an approach allows for the effective use of traditional knowledge, which remains highly relevant in addressing today's environmental challenges.

### 3.1.2  Review of Literature

Effective local governance is crucial for sustainable natural resource management (NRM), particularly when integrating traditional knowledge and engaging youth. Capacity building initiatives enhance local communities' skills and decision-making abilities, empowering them to manage resources sustainably.

### 3.1.2.1 Capacity Building in Local Governance

Studies have emphasised the need to empower local communities through education, skills development, and collaborative governance to strengthen NRM. They underscore that capacity building strengthens local structures, fostering resilience against environmental challenges (Malahayati & Rasyid, 2020). Similar studies have highlighted the importance of institutional capacity in community-driven NRM, noting that integrating local knowledge into governance systems is key to sustainability (Mohanty, 2009)

### 3.1.2.2 Traditional Knowledge in NRM

Traditional ecological knowledge (TEK), passed down through generations, is a valuable resource in NRM. Research has revealed that integrating TEK with modern scientific approaches leads to more sustainable practices.

Communities' deep-rooted ecological understanding – such as the resource rotation and biodiversity conservation – is often overlooked by external governance systems, yet it is crucial for effective, community-based conservation (Sterling et al., 2022). Researchers have observed that respecting and incorporating such knowledge within governance frameworks improves NRM outcomes.

### 3.1.2.3 Youth Involvement in Conservation

Youth engagement is essential for long-term sustainability. Mohanty (2009) noted that although young people are often passionate about environmental issues, they face barriers to participation in decision-making. The cadre-building initiative of the Meghalaya government plays an important role in involving youth.

The Civilian Conservation Corps in 1933 went about to establish several national parks, notably Yosemite National Park, because the State gave them a platform to engage themselves in conservation efforts (Maher, 2007). These initiatives provided valuable experience and opportunities for young people to contribute to NRM.

The Meghalaya programmes can bridge the gap between generations, combining traditional knowledge with the innovation and energy of younger echelons in the population.

### 3.1.2.4 Policy Frameworks and Community-Based Conservation

Policies play a critical role in supporting local NRM efforts. They should recognise community rights and knowledge, creating an enabling environment for decentralised governance (Angeles & Gurstein, 2000). Policies must balance top-down regulations with bottom-up community initiatives, ensuring that conservation efforts are both equitable and sustainable (Sterling et al., 2022).

### 3.1.3 Identification of Gaps

Meghalaya's **Natural Resource Management (NRM)** policies and various government programmes, introduced over the years to assist in NRM (for forest, water, soil, etc.), had difficulties reaching the most vulnerable areas and the most degraded lands needing these interventions. The Mahatma Gandhi National Rural Employment Guarantee Scheme (MGNREGS), one of India's flagship programmes, which UNDP stated as having the potential to transform rural India into a "more productive, equitable, connected society" (Diengdoh, 2024a, 2024b) could not utilise the 60% allotted to NRM due to a lack of data from the most challenging parts of the State. This lack could be attributed to the absence of a link between the government body and communities.

Additionally, State programmes and schemes did not have a holistic approach, since they did not address conservation efforts and, most importantly, neglected livelihood aspects (Darlong & Barik, 2020)

Clearly, a comprehensive survey covering social, economic, demographic, and environmental factors was required simultaneously in select project villages. This could only be done with the assistance of trained professionals on the ground. Recognising that the government machinery did not have a reach in these villages, the need for training a cadre of such community professionals became apparent.

## 3.2 Theoretical Framework

To ensure effective communication and monitoring across the system, a structured hierarchy was put in place with multiple levels of supervision. At the top, the State Management Unit oversees the entire operation. Then, a district management unit coordinates activities at the district level, followed by a block management unit for operations at the block level. At the grassroots level, village cadres – comprising village community facilitators (VCFs), village data volunteers and "green field associates" (Green field associates are educated youth from the rural areas who are hired by the government and provided training to interact with and support the communities who are interested in the Payment for Ecosystem Services. They serve as the support centre of the State office in the villages and are essential in implementing and overseeing activities within their villages.

This hierarchical structure has facilitated a smooth flow of information and ensured that each level is properly supported and monitored. Various monitoring tools, such as weekly assessments, reporting tools and Geographic Information Systems (GIS)-enabled data collection instruments, ensure an uninterrupted and cohesive flow of the work in accordance with project and State guidelines.

## 3.3 Methodology Used for Capacity Building

To ensure active participation from the ground up, the Community-led Landscape Management (CLLM) Project decided on three village community facilitators (VCFs) in each village; one was to be female. They were to assist in the implementation of project activities locally.

The VCFs received comprehensive training in various areas such as spring and GIS mapping, account keeping, social inclusion, monitoring, and evaluation.

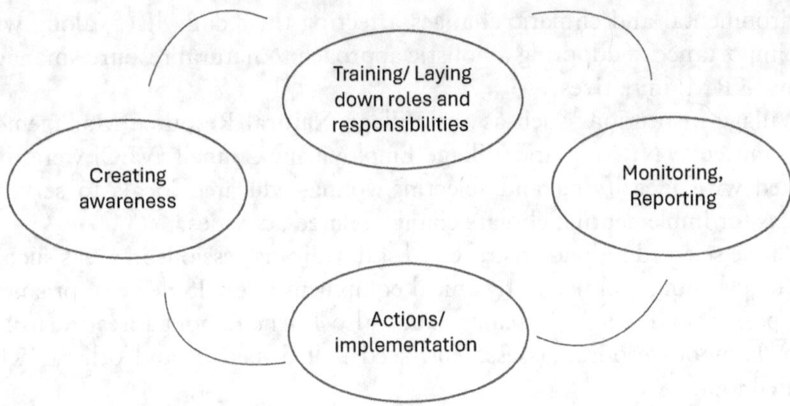


**FIGURE 3.1**    Process flow of capacity building

*Source:* CLLMP project data

A key aspect of this capacity-building initiative was the push to digitise
processes, including data collection and storage, work monitoring, and capturing geo-coordinates. The project also aimed to provide training through
a digital platform, with curriculum access online and electronic participation tracking.

This shift to digital tools streamlined processes like record-keeping of
attendance and distribution of training materials.

To facilitate this transition, the Meghalaya Basin Development Authority
(MBDA) partnered with external organisations for technical support, ensuring the successful implementation of these digital solutions.

Depending on their roles, facilitators are equipped with tools and technology to enable them to perform their duties properly. Here is a short list of the
tools and technology provided:

1.  Tablet preloaded with various mobile applications for data collection.
2.  GPS devices to record location data.
3.  Water tracer for taking readings on various water quality data during
    spring mapping exercises.
4.  Tools for the enumeration of forests.

The process of establishing community professionals began with **raising
awareness** among community members, village elders and leaders, as well
as relevant departments. They engaged in discussions about the various

environmental and climatic changes affecting their daily lives, along with the importance of adopting a holistic approach to natural resource management (NRM) initiatives.

Village institutions such as the Village Natural Resource Management Committee (VNRMC) and Village Employment Council (VEC) were then tasked with **identifying and selecting** young, educated locals to serve as cadets for implementing climate change-related activities.

These selected individuals received skill **training** sessions in areas such as NRM planning and use of technical equipment like GIS devices, prismatic compasses, and weighing scales. They also learned about forest mensuration, bamboo resource assessment, seed ball dispersal, and other NRM-related topics.

The cadre then went on to **implement** activities. These included raising nurseries, initiating afforestation, conducting seed ball training in schools, demarcating village boundaries, measuring forest elements, etc. The VCFs received continuous support throughout the project period. This ensured they gained the knowledge and skills needed for sustainable engagement with and in their communities.

## 3.4 Results

### 3.4.1 Main Case Findings 1

The immediate and tangible effects of integrating a cadre of community professionals into various project activities under the CLLMP were profound. This was particularly so in terms of the seamless bridging of a long-standing gap between State authorities and local communities. Historically, these two groups had often been at odds: the State's centralised governance model sometimes clashed with the lived realities of rural communities that depend directly on the natural environment for their livelihoods.

Yes, the involvement of community professionals embedded within the communities and dealing directly with the people and the landscape marked a pivotal shift. Their active participation in various aspects of project planning and implementation served as a vital bridge. It brought about greater collaboration and mutual understanding between the two sides. This relationship, built on mutual respect and a shared goal of sustainable MNR, represented a significant step towards overcoming the structural and communicative divides that had previously hindered effective environmental governance.

At the heart of this transformation lay the concept of natural resource management. This was not a new or unfamiliar idea: local people had been practising it for generations, in the form of traditional knowledge systems passed down through the ages. Indigenous ways of managing forests, water bodies and land had been honed through centuries of observation and experience. Each community had developed its own set of rules and norms designed to preserve the delicate balance of local ecosystems.

The intricate ways in which these communities interacted with their environment were grounded in a deep cultural understanding and practical know-how. For many, these traditional methods had been the primary means of ensuring the sustainability of the resources they relied on.

Nevertheless, despite the deep wisdom enshrined within these practices, there was often a disconnect between traditional knowledge and the rapidly evolving scientific paradigms that were emerging in modern times. While traditional systems were incredibly effective in their local contexts, they were often limited by their scope. Practices struggled to address the complexities of contemporary environmental challenges such as climate change, population pressure and large-scale deforestation.

Here, science and modern techniques stepped in. By combining traditional knowledge with contemporary scientific methods, the CLLMP offered a unique opportunity to address both the specific needs of local communities and broader, more complex environmental issues at the State level.

The collaboration between community professionals and scientific experts under the CLLMP was designed to maximise the strengths of both worlds. Scientific tools and techniques, such as satellite imagery, GIS, and advanced water management techniques, were used to complement local, traditional knowledge.

These tools allowed for a more nuanced and comprehensive approach to managing natural resources, considering factors that might have been overlooked or inaccessible through traditional methods alone.

For instance, scientific assessments of forest health, soil quality, and water availability could provide valuable insights that help guide community decision-making. At the same time, traditional practices of community-based conservation, such as sacred groves and rotational grazing systems, presented a culturally sensitive and locally appropriate framework for action.

A key outcome of this integration was the way in which it transformed local governance structures and decision-making. By bringing together community members, science-trained experts, and State representatives, the CLLMP created a platform for dialogue and shared decision-making that empowered local people to take ownership of their resources.

This approach was not just about top-down applying scientific solutions; it was about working alongside local people, understanding their concerns and knowledge, and jointly developing solutions that were both scientifically sound and socially acceptable.

The result was a system that benefited both the State and the people, creating a model of resource management that was sustainable, equitable, and adaptable to changing circumstances.

Another important aspect of the CLLMP was its focus on forest, land, and water ownership. These three elements are foundational to the livelihoods of rural communities; they are also central to the concept of NRM. For many communities, land and water are not simply commodities to be exploited for short-term gain; they are integral to their cultural identity and way of life.

In many parts of the world, the legal frameworks surrounding land and water ownership are complex and often do not align with traditional systems of tenure. This has led to tensions and conflicts over resource use because local people have found themselves at odds with State policies or commercial interests. By working with communities to clarify and formalise their rights to these resources, the project helped overcome some of these challenges by giving legal recognition and support for traditional systems of ownership and stewardship.

State involvement in this process was critical, for it helped create an enabling environment for communities to exercise their rights and engage in sustainable resource management. Rather than imposing top-down regulations, the State played a facilitative role, providing technical assistance, policy support and, when necessary, legal structures to protect community land and water rights.

This collaboration ensured that the systems put in place were not only practical but also legally enforceable. It gave communities the confidence to invest in long-term management and made it easier for them to access the resources they needed to thrive. In turn, the State benefited from increased environmental resilience and improved resource management that resulted from these community-led efforts.

Healthy ecosystems, well-managed forests, and sustainable water use not only help safeguard biodiversity but also contribute to climate change mitigation, soil conservation, and water quality improvements. All have far-reaching benefits for the State and its inhabitants.

Further, the project had a profound impact on the livelihoods of local people. Many had been living in poverty, grappling with significant challenges related to environmental degradation and resource depletion. By involving communities in the management of their own resources, the CLLMP helped bring about opportunities for income generation through sustainable practices such as agroforestry, eco-tourism, and the sustainable harvest of non-timber forest products.

These initiatives not only provided economic benefits but also contributed to strengthening intra-community social ties, as people worked increasingly together towards common goals and developed a stronger sense of collective responsibility for their shared environment.

### 3.4.2 Main Case Findings 2

The success of community-driven projects, particularly those aimed at addressing pressing environmental issues, depends largely on the active engagement and support of the very communities they are designed to benefit. The experience gained through various initiatives in Meghalaya underscores this point. It highlights the importance of local involvement and ownership in ensuring the sustainability and effectiveness of such programmes.

A key observation that has emerged from the implementation of the project is the pivotal role village facilitators play in achieving meaningful results. It has also become clear that the success of these facilitators is not solely dependent on their training and technical expertise but also on the degree of endorsement and acceptance from the villages where they are expected to work.

This realisation has led to a crucial shift in the way facilitators are selected. Instead of being appointed by external agencies or project managers, facilitators are now chosen by the communities themselves. This ensures that these professionals are not only well-equipped to address local challenges but also trusted and supported by the people they serve.

The importance of this community-driven approach is evident in the impressive outcomes achieved through the project. By empowering communities to select their facilitators and actively engage in the development and implementation of NRM, the project has succeeded in producing over 400 comprehensive NRM plans for the project villages.

The plans are dealing with a wide range of critical environmental issues, including water scarcity, deforestation, soil degradation, and the consequences of mining. They also contain a roadmap for sustainable resource management, with many proposals directly finding support through funding and technical assistance from the World Bank, the Meghalaya Basin Development Authority (MBDA), and various state departments.

The collaborative efforts between local communities, government agencies, and international organisations have been instrumental in making sure that proposed solutions are both feasible and scientifically sound. This has greatly boosted the prospects for long-term success (Table 3.1).

Encouraged by the success of the core villages, the Meghalaya government expanded the scope of the Project to include over 6,000 villages across the state. This expansion was made possible by incorporating the technical expertise of salient government departments, such as the Soil and Water Conservation Department and the Forest Department, in training and capacity building for NRM.

**TABLE 3.1** Natural resource management plans prepared

| SL. | District | NRM Plans (CLLMP) | NRM Plans (Non-CLLMP) |
| --- | --- | --- | --- |
| 1 | East Garo Hills | 30 | 231 |
| 2 | East Jaintia Hills | 30 | 20 |
| 3 | East Khasi Hills | 93 | 101 |
| 4 | North Garo Hills | 20 | 80 |
| 5 | Ri Bhoi | 30 | 62 |
| 6 | South Garo Hills | 23 | 71 |
| 7 | Southwest Garo Hills | 49 | 91 |
| 8 | Southwest Khasi Hills | 40 | 31 |
| 9 | West Garo Hills | 40 | 110 |
| 10 | West Jaintia Hills | 20 | 178 |
| 11 | West Khasi Hills | 25 | 20 |

*Source:* CLLMP project data.

These departments rendered essential support in equipping community professionals with the knowledge and tools necessary to identify NRM challenges and design appropriate, scientifically based solutions.

The selection of these community professionals, once again taken care of by village institutions like the Dorbars and Nokmas as well as the Village Employment Councils (VECs), has been a critical factor in the success of the project. By involving local leaders and institutions in decision-making, the project has brought about a sense of authority and accountability among the local people, strengthening the effectiveness of the interventions.

The results of these efforts have been nothing short of remarkable. To date, communities themselves have developed over 978 NRM plans. Many of them have become part of the broader development framework of the Mahatma Gandhi National Rural Employment Guarantee Act (MGNREGA).

The plans cover a wide range of natural resource management interventions, from soil and water conservation to forest restoration, and have proved to be a valuable tool for tackling the environmental challenges villages are struggling with.

VECs have already adopted some of these NRM plans. Many more are being actively pursued as part of a broader development agenda for the state. This has created a ripple effect, with other villages and communities expressing interest in replicating the success of the initial project sites.

In addition to the development of NRM plans, a significant achievement of the CLLM Project has been the successful mapping of more than 55,000 springs across Meghalaya. This task involved assessing a variety of properties, such as discharge rates and pH levels. It was a first-of-its-kind exercise in the state, since it involved the collection of comprehensive data under a single coordinating agency.

These data have proven invaluable for understanding the availability and quality of water resources across the state, particularly in areas heavily reliant on spring-fed water sources. The mapping exercise has not only improved the management of water resources but also provided a foundation for future conservation efforts, for it has enabled local communities and government agencies to identify critical water sources that need protection and sustainable management.

Another notable achievement of the project has been the contribution of over 1,200 trained cadre members in the gathering of data for Forest Management Plans (FMPs) through FMP studies. Initially, this data collection focused on the 400 core villages; later, the initiative extended to include 200 non-CLLMP villages.

Involving local cadre members in the data collection ensured that the information gathered was accurate and could be made relevant to community-specific needs. Importantly, the Forest Management Plans have played a crucial role in guiding the sustainable management of forest resources in the state, with a particular emphasis on biodiversity conservation and climate resilience.

Also, trained community professionals have been instrumental in conducting bamboo resource assessments across Meghalaya. Such an assessment involves over 7,000 random sampling plots.

The exercise gave valuable insights into the state's bamboo resources, including their distribution, abundance, and potential for sustainable harvesting. The data collected have been crucial for identifying opportunities to leverage bamboo as a source of livelihood for the local people. As one of the fastest-growing and most versatile plants, it offers a wide range of economic products – from construction materials to handicrafts.

By looking into available bamboo resources and their potential uses, the project has opened new avenues for income generation and poverty alleviation in rural Meghalaya.

An interesting aspect of the project's capacity-building efforts has been the gender distribution of cadre members. Throughout the duration of the project, it became clear that male professionals outnumbered their female counterparts, particularly in technical and physically demanding areas such as finance and GIS mapping.

On the other hand, female participation prevailed in social inclusion and community engagement aspects of the project. Women's involvement proved essential in the proper attention to the needs of marginalised groups – particularly (if not surprisingly) women and children themselves.

This gender imbalance reflects broader educational disparities in the villages, where men are more likely to have access to formal education and training opportunities. In spite of this, the project has made significant strides in promoting gender equality: Increasing numbers of women are participating in technical training and gaining skills in areas traditionally dominated by men.

A closer look at regional variations in gender distribution reveals some interesting trends. In the Khasi and Jaintia regions, female participation in the capacity-building programmes has been higher than in the Garo region. This could have been caused by a variety of factors, including cultural differences, access to education and the role of women in the community.

The higher participation of women in these regions is a promising sign of progress towards gender equity in rural development initiatives, because it suggests that women are increasingly being recognised for their ability to contribute to technical and decision-making processes.

*The* success of the CLLM Project has had far-reaching implications beyond the immediate objectives of natural resource management and community development. One of the key outcomes of the project has been the recognition of trained community professionals as valuable assets by various State Government departments.

Agencies such as DIGI-Agri, FOCUS, the Soil and Water Conservation Department, and the Planning Department have engaged them to carry out a range of surveys and assessments and related activities. These consisted of farmer surveys, agricultural surveys, the installation of automated weather stations, and the identification of water sources.

This continued engagement with government agencies is a testament to the value of the capacity-building efforts under the World Bank-funded project and underscores the potential for community professionals to contribute to broader development objectives.

In conclusion, the experience of the CLLM Project in Meghalaya demonstrates the transformative power of community-driven approaches to environmental management and sustainable development. By empowering local communities to take ownership of their natural resources and actively participate in decision-making, the project has created a model for addressing complex environmental challenges in rural areas.

The involvement of trained community professionals, chosen and endorsed by the communities themselves, has proved to be a key factor in the success of the initiative, ensuring that solutions are both locally relevant and scientifically sound. As the project continues to evolve and expand, it holds great promise for achieving long-term environmental sustainability and improving the livelihoods of rural communities in Meghalaya.

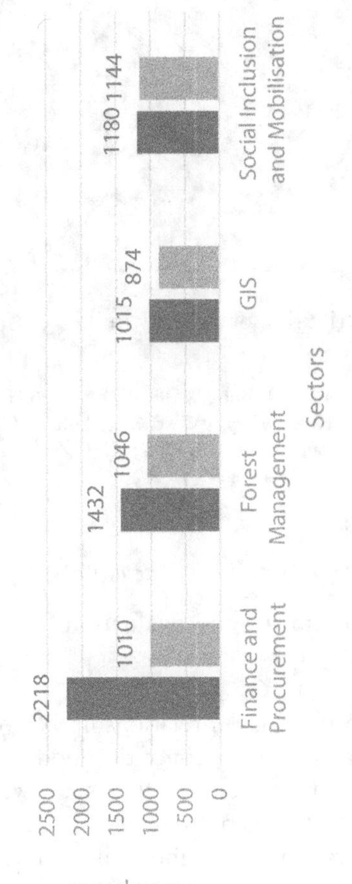


**FIGURE 3.2**   Gender-wise distribution of capacity building

*Source:* CLLMP project data

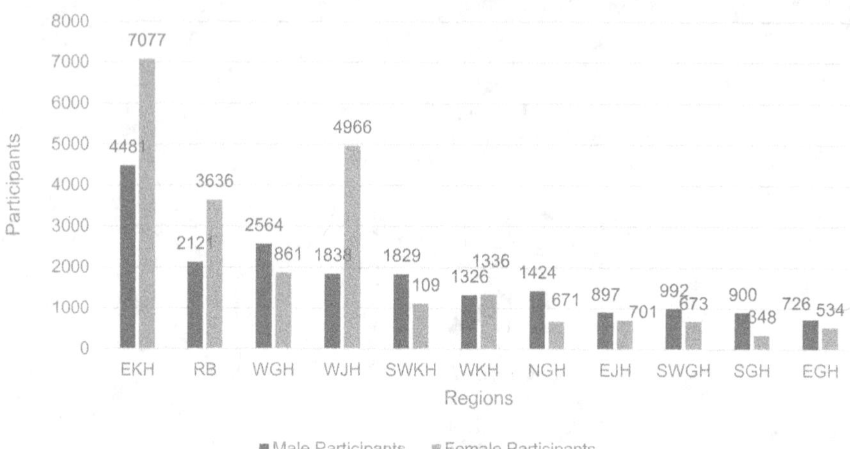


**FIGURE 3.3**  Region-wise distribution of capacity building

*Source:* CLLMP project data

*Notes:* ** EKH – East Khasi Hills, RB – Ri Bhoi, WGH – West Garo Hills, WJH – West Jaintia Hills, SWKH – South West Khasi Hills, WKH – West Khasi Hills, NGH – North Garo Hills, EJH – East Jaintia Hills, SWGH – South West Garo Hills, SGH – South Garo Hills, EGH – East Garo Hills

### 3.4.3 Main Case Findings 3

#### 3.4.3.1 Testimonies from Village Community Facilitators

1.  Karwin K. Marak (Monitoring and Evaluation VCF, Mendal Songma, Bajengdoba)

We, as VCFs, work to bring good relationship between the communities and to encourage and assure them of the good works of the project. Whenever the villages have problems and issues, we work together to find solutions. The CLLMP has been a huge blessing in my life because not only has the project taught me about different kinds of work, but it has also given me immense pride in being identified as a leader in my community. The Village Natural Resource Management Committee (VNRMC) President and Secretary look at me for advice during meetings and trust me to carry the monthly village reports to the main office.

It is because of the position and training CLLMP has given me in my life, that I am recognised as a competent leader to lead my community in decision-making.

Apart from monitoring the physical work carried out in the village every month, I get responsibilities and training such as seed ball training for schoolchildren. I have also received training in forest enumeration, bamboo resource assessment and Monitoring, Reporting and Verification (MRV).

Even though I might not have had the opportunity to learn these in school and pursue further studies, I am grateful that the CLLMP has given me chances.

I have also been taken on several exposure trips by the office to increase our practical learning from other places and institutions such as Chirag in Uttarakhand and SpreadNE in Assam.

2. Banraplang Kharshilot (Environment VCF, Nongdommawria, Mawsynram)

I was scared at first about being selected as a VCF from my village, since I did not have much exposure and did not know much about working in general. But all those fears were put to rest, when I realised that the CLLMP office would train us slowly and handhold us as well as answer our questions about things we did not know. The training and work engagement I received from the CLLMP is more than what I would have learned from school, because not only did it expose me to new subjects but it also sharpened my communication skills and soft skills.

My most memorable moment working under the CLLMP is perhaps the seed ball training programme, which pushed me into a teaching role for students as well as teachers which, I realised, I quite enjoyed. While it involved extensive travelling to schools and exhausted us, it gave me joy like never before, because I knew I was contributing to the knowledge of our children to preserve nature. It also pushed me to learn more about our own plants and indigenous species.

I recently graduated from William Carey University and while I attended my classes and attended my duties as a VCF, I also helped in my brother's barber shop. The additional training and exercise we did for FMP, seed ball training and bamboo resource assessment also help us earn while we conduct the survey. I am proud to say that I am contributing towards the data that our State is collecting.

## 3.5 Discussion

### 3.5.1 Interpretation, Replicability, Practice, and Policy

The cadre of community professionals has been effectively trained and integrated into various roles, contributing significantly to environmental and sustainable development efforts. One of the key roles these professionals have taken up is as "Green Field Associates" under the Payment for Ecosystem Services (PES) programme. In this capacity, they are trained in measurement, reporting, and verification (MRV). This is a specialised process developed by the MBDA to offer performance-based incentives.

In addition, they have taken on extra responsibilities with added incentives, such as serving as trainers for seed ball training in schools, as "Agroforestry Carbon Associates" under the Carbon Credit Programme funded by the Rabo Bank, where they serve as the facilitators for the carbon credit project by visiting the communities, processing their applications and monitoring the progress, and enumerators for forest management planning.

A total of 14,000 community professionals have been trained across the state, with 1,200 receiving ongoing support throughout the CLLM Project period to ensure their continued success and engagement.

After completion of the project, the State Government has continued to involve these trained professionals. It has also expanded the initiative by onboarding additional community facilitators as "Village Data Volunteers" (VDVs). These VDVs, who are voluntarily engaged by the government with added incentives, are responsible for conducting surveys and collecting data that contribute to the State's Sustainable Development Goal targets. To date, Meghalaya has over 1,500 VDVs.

The Community-Led Landscape Management Programme has also played a pivotal role in fostering convergence across various government departments, centralising data under the Centre of Excellence. Several departments have recognised and adopted this convergence model.

Further, to ensure the continuity of natural resource management (NRM) efforts under the MGNREGA and to engage trained professionals as NRM experts at the village level, the State Cabinet passed a resolution in 2023 to form a Natural Resource Management Committee under each Village Environment Committee.

### 3.6 Summary and Conclusion

The Community-Led Landscape Management Programme has proven to be a successful and transformative initiative that has helped bridge the gap between the State and local communities. It has fostered collaboration and shared decision-making and contributed to the more sustainable management of natural resources.

By combining traditional knowledge with scientific expertise and ensuring that communities had the tools, support, and recognition needed to manage their own resources, the CLLMP has created a model for participatory governance and resource management that has the potential to be replicated in other regions and contexts.

Its integrated approach, grounded in respect for local knowledge and guided by a shared commitment to sustainability, offers a promising pathway to achieving more equitable and resilient communities in the face of current environmental challenges.

The VCF model has gained traction. Several departments, such as those for Agriculture, Soil, and Planning, have already adopted it. In these departments, VCFs are serving as data volunteers, collecting vital information related to their villages.

The model proves essential in quickly establishing a cadre capable of carrying out ground-level activities across the state. Recognising their potential, the Government of Meghalaya has appointed 1,200 VCFs as village data volunteers, with a scope for further training and opportunities in sectors like Health, Agriculture, and the Meghalaya Basin Management Agency.

This strategic approach not only supports ongoing initiatives but also strengthens the capacity of local communities to drive development across various sectors.

The success of the CLLMP has brought us important lessons for future development programmes, particularly those that seek to balance the needs of local communities with broader environmental goals. A key takeaway from this experience is the importance of fostering strong partnerships between communities, scientists, and the State.

By recognising and valuing the contributions of all stakeholders, and by creating spaces for dialogue and mutual learning, it is possible to build trust and collaboration that leads to more effective and lasting solutions.

In addition, the programme has shown that traditional knowledge and modern science are not mutually exclusive but can be complementary, with each bringing unique strengths to the table. The integration of both can result in more resilient, sustainable, and culturally appropriate resource management practices that will benefit everyone involved.

## References


Angeles, L., & Gurstein, P. (2000). Planning for participatory capacity development: The challenges of participation and North-South partnership in capacity building projects. *Canadian Journal of Development Studies/Revue canadienne d' études du développement*, 21(Sup.1), 447–478.

Darlong, V., & Barik, S. K. (2020). Community-based approaches to local natural resource management. In KrishnaS. (Ed.), *Agriculture and a changing environment in Northeastern India* (pp. 335–372). Routledge.

Diengdoh, C. V. D. (2024a). 5 Development through participatory governance – MGNREGA in Meghalaya. *India International Centre Quarterly*, 48(3/4), 47–61.

Diengdoh, C. V. D. (2024b) *The working of the Indian constitution*. Taylor & Francis.

Maher, N. M. (2007). *Nature's New Deal: The civilian conservation corps and the roots of the American environmental movement*. Oxford University Press.

Malahayati, M., & Rasyid, L. M. (2020). *The development of vocational village concept based on local wisdom and inclusiveness in North Aceh*, 413, 125–128.

Mohanty, R. K. (2009). Cadre building to strengthen grass root governance. Field actions science reports. *The Journal of Field Actions*, 3, 35–40.

Sterling, E. J., Sigouin, A., Betley, E., Cheek, J. Z., Solomon, J. N., Landrigan, K., & Jones, M. S. (2022). The state of capacity development evaluation in biodiversity conservation and natural resource management. *Oryx*, 56(5), 728–739.

# 4

# COMMUNITY NURSERIES FOR FOREST RESTORATION

*Bratati Purkayastha and Van Shanborlang Buhphang*

## 4.1 Introduction

### 4.1.1 Purpose, Scope and Significance of the Case

Meghalaya, situated in North-East India, is renowned for its rich biodiversity and extensive forests, which span 22,429 km$^2$ at altitudes ranging from 150 to 1,965 m. However, various drivers of deforestation, some of them typical to the state, have been operating since time immemorial in the region. While some of them are age-old practices that are a part of the culture, such as shifting cultivation, others are of recent origin, to take mining, arecanut monoculture, etc., for example.

According to the ISFR 2023 report, Meghalaya experienced a loss of 32.81 km$^2$ of forest cover in 2023 compared to 2022.

This trend shows a significant increase in forest degradation in the state. The State Government, through the Forest & Environment Department and Soil & Water Conservation Department, has made several efforts in massive afforestation programmes to restore the forest areas, but Meghalaya, being the 6th Schedule state (Article 244 of the Indian Constitution), has around 90% of the land under the ownership of the Community and is regulated by the customary laws of the Khasi, Garo and Jaintia tribes, who make up the majority of the state's population. This stands in contrast to many other Indian states, where forest protection and management are under government jurisdiction. Thus, it is very challenging for the State Government to make the large-scale afforestation drive successful.

Therefore, engaging local communities is vital for the success of restoration efforts in the region. The community-based approach has been considered a key strategy for natural resource management (Rout, 2010)

DOI: 10.4324/9781003735380-5

**TABLE 4.1** State/UT wise forest cover inside and outside recorded forest area (RFA)/green wash(GW)

| State /UT | Geographical area | Forest cover inside RFA/GW 2021* | | | | Forest cover inside RFA/GW 2023 | | | | Forest cover change inside RFA/GW | Forest cover outside RFA/GW 2021* | | |
|---|---|---|---|---|---|---|---|---|---|---|---|---|---|
| | | VDF | MDF | OF | Total | VDF | MDF | OF | Total | | VDF | MDF | OF |
| | (A) | | | | (B) | | | | (C) | (C-B) | | | |
| Andhra Pradesh | 162,922.57 | 1,925.22 | 12,588.26 | 9,437.75 | 23,951.23 | 1,925.32 | 12,469.70 | 9,472.74 | 23,867.76 | −83.47 | 72.45 | 1,324.76 | 4,875.18 |
| Arunachal Pradesh | 83,743.22 | 19,633.01 | 26,867.22 | 11,719.58 | 58,219.81 | 19,637.05 | 26,699.94 | 11,836.97 | 58,173.96 | −45.85 | 1,372.87 | 2,971.29 | 3,408.77 |
| Assam | 78,438.00 | 2,789.68 | 8,468.10 | 8,525.61 | 19,783.39 | 2,833.31 | 8,333.78 | 8,529.64 | 19,696.73 | −86.66 | 357.15 | 1,431.29 | 6,753.48 |
| Bihar | 94,163.00 | 319.32 | 2,477.28 | 2,046.80 | 4,843.40 | 375.79 | 2,445.25 | 2,044.70 | 4,865.74 | 22.34 | 11.61 | 819.02 | 1,729.23 |
| Chhattisgarh | 135,192.00 | 5,449.44 | 26,367.76 | 10,616.60 | 42,433.80 | 5,796.08 | 26,070.85 | 10,553.46 | 42,420.39 | −13.41 | 1,605.33 | 5,935.11 | 5,856.64 |
| Delhi | 1,483.00 | 4.06 | 18.75 | 47.65 | 70.46 | 4.02 | 18.24 | 49.94 | 72.20 | 1.74 | 2.66 | 37.74 | 84.50 |
| Goa | 3,702.00 | 518.19 | 325.85 | 380.03 | 1,224.07 | 532.61 | 340.31 | 351.01 | 1,223.93 | −0.14 | 22.56 | 237.61 | 782.98 |
| Gujarat | 196,244.00 | 356.23 | 3,961.48 | 5,133.38 | 9,451.09 | 400.31 | 3,948.17 | 5,041.39 | 9,389.87 | −61.22 | 18.62 | 955.52 | 4,411.34 |
| Haryana | 44,212.00 | 23.99 | 259.08 | 448.63 | 731.70 | 23.99 | 258.77 | 446.17 | 728.93 | −2.77 | 3.18 | 185.98 | 707.36 |
| Himachal Pradesh | 55,673.00 | 2,835.93 | 5,395.79 | 2,440.18 | 10,671.90 | 2,791.82 | 5,431.88 | 2,483.27 | 10,706.97 | 35.07 | 361.75 | 1,828.56 | 2,663.41 |
| Jharkhand | 79,716.00 | 1,444.93 | 5,260.88 | 5,744.00 | 12,449.81 | 1,447.53 | 5,261.05 | 5,793.95 | 12,502.53 | 52.72 | 1,183.38 | 4,380.88 | 5,692.90 |
| Karnataka | 191,791.00 | 3,735.18 | 12,885.56 | 6,300.02 | 22,920.76 | 3,737.90 | 12,876.97 | 6,399.03 | 23,013.90 | 93.14 | 799.82 | 8,294.93 | 7,091.06 |
| Kerala | 38,852.00 | 1,878.93 | 5,429.89 | 2,578.79 | 9,887.61 | 1,877.59 | 5,420.36 | 2,627.89 | 9,925.84 | 38.23 | 164.00 | 3,923.79 | 7,950.54 |
| Madhya Pradesh | 308,252.11 | 6,468.74 | 31,676.75 | 29,651.78 | 67,797.27 | 6,886.21 | 31,366.97 | 29,517.32 | 67,770.50 | −26.77 | 189.37 | 2,462.26 | 6,996.08 |
| Maharashtra | 307,713.00 | 8,497.78 | 15,103.73 | 12,522.82 | 36,124.33 | 9,538.99 | 15,827.39 | 10,744.55 | 36,110.93 | −13.40 | 303.75 | 5,516.60 | 8,968.32 |
| Manipur | 22,327.00 | 893.87 | 5,776.36 | 8,168.05 | 14,838.28 | 893.52 | 5,744.86 | 8,152.43 | 14,790.81 | −47.47 | 10.68 | 484.15 | 1,307.18 |
| Meghalaya | 22,429.00 | 508.98 | 7,599.66 | 6,575.02 | 14,683.66 | 547.93 | 7,502.05 | 6,600.87 | 14,650.85 | −32.81 | 46.47 | 1,537.75 | 728.96 |

*Source:* India State of Forest Report 2023 (Volume 1).

Moreover, in the conventional system, i.e., departmental mode of implementation of afforestation, saplings are raised in central nurseries run by Forest or Soil & Water Conservation Departments. During the monsoon, which is the season for plantations, more than 50% of villages become inaccessible due to the tough terrain of Meghalaya as a result of this.

Transportation of saplings from central nurseries, which are mostly located in urban blocks, is very challenging and less efficient in terms of the survival of planted saplings.

There is less survival of seedlings also due to the planting of non-native species procured from outside the state.

To address this challenge, decentralised village-level low-cost community nurseries have proven to be more effective in motivating the community for large-scale afforestation, inculcating their ownership right from the planning stage and ensuring better survival of planted seedlings by planting native species and reducing transportation shock during carriage to the plantation sites.

This case study highlights the innovative effort undertaken in Meghalaya by MBDA/MBMA, through the training of decentralised village-level community nurseries by engaging local communities right from selecting afforestation models, building their capacity, establishing community nurseries of native species and converting degraded lands into diverse forests. It emphasises the crucial role of community engagement in the restoration of degraded forests, raising 37,137 ha of plantations under mission mode through various schemes and projects implemented by the State Government, under the MBDA/MBMA umbrella.

## 4.2 Organisation of the Chapter

This chapter comprises four main sections:

- Section 4.7.1.1.1 "Establishment of Community Nurseries": Discusses about assessing land, creating awareness, gathering resources, building the capacity of community, establishing nurseries.
- Section 4.7.2.1 "Ensuring Success through Robust Monitoring Using Technology": Showcases ensuring survival of seedlings in nurseries using app-based mobile technology and data analysis through MIS tool.
- Section 4.7.3.1 "Impact and outcome of Community Nurseries": Explores Measurable impacts, Environmental impacts, Outcomes, Unintended gains of Community Nursery initiative.
- Section 4.7.4 "Innovative Approaches Adopted for Implementation": Highlights about the use of innovations like GIS and MIS technology for implementation and monitoring, implementation through Community-Based Organisations like Village Project Implementation Committees

(VPICs) and Village Community Facilitators (VCFs), and the use of innovative Information, Education and Communication (IEC) materials like flipcharts in local languages for generating awareness.

## 4.3 Review of Literature

- The state has experienced a gradual decline in its forest cover, primarily due to human activities such as agriculture, shifting cultivation, wood collection, mining and the expansion of settlements driven by population growth (Report on Identification of Drivers of Degradation, 2028–19, MBMA). Between 2021 and 2023, the open forest in the state increased by 25.85 km$^2$ (ISFR, 2023)
- These have led to a dwindling of water at key sources due to degradation of catchment areas, thereby causing severe stress on the state's population, both rural and urban. This is reflected in the priority accorded by project villages during planning.
- Further, changing weather patterns have led to an overall reduction in the amount of rainfall with a disproportionate reduction in the number of months, leading to higher rainfall but over a reduced period (rain intensity). This has led to other issues such as soil erosion and flash floods, which were otherwise uncommon in the state. Data also indicates an increase in the amount of surface water despite the reduced rainfall, which points to a reduction in ground penetration of water (internal study).
- To achieve the goals of large-scale restoration programmes, several improvements in the restoration productive chain are needed.
- Large-scale restoration programmes may also contribute to the conservation of areas used to collect native seeds, as well as improve local livelihoods. For that, there is a clear need for flexible legislation, local community participation, technical assistance and environmental awareness (Guariguata & Brancalion, 2014; Meli & Brancalion, 2017).
- Studies have also emphasized the importance of community-led ecological restoration (Blaikie, 1985; Mekonnen, 2000).

## 4.4 Identification of Gaps

Meghalaya's unique geography, characterised by its hilly terrain, heavy rainfall and diverse ecosystems, presents several hurdles for making a large-scale plantation drive successful. There is limited availability of quality planting materials suited to the region's microclimates, availability of suitable land for raising nurseries and a lack of the technical knowledge required for sustainable nursery practices, which results in low seedling survival and poor growth rates.

## 4.5 Methodology of the Case

### 4.5.1 Description of Methodology

This case study showcases the community-led initiatives for restoring degraded lands by establishing village-level community nurseries, undertaken by MBDA/MBMA in Meghalaya. The case study focused on the community nurseries established under Externally Aided projects implemented by MBDA/MBMA and attempts to summarise the key experiences, observations, outcomes, impacts and lessons learnt to aid in adoption, replication and scaling up this initiative and adoption of this practice by other government and non-government programmes. A narrative is built using the above data and evidence to present the case study in a structured manner.

### 4.5.2 Justification of Methodology

Based on the hypothesis that socio-economic development in communities that are dependent on resources in biodiversity-rich areas can encourage local stakeholders to actively participate in biodiversity conservation, the approach of involving the community for the production of planting materials was initiated.

This is a collective venture of community members who all play a part, from planning to running it, and all benefit from it. As a larger unit at the community level than would be feasible at the individual level, the approach has more leverage in accessing inputs and selling its outputs.

The community nursery approach fulfils two important roles. Firstly, it enhances the livelihoods of poor, rural people by drawing upon their own inherent plant cultivation skills. The community nature of the unit empowers each worker with a voice and a role in its running. Secondly, it promotes environmental protection.

Community nurseries are managed by resourceful villagers who are trained for the raising of high-quality seedlings in temporary structures made with local materials like bamboo. These nurseries serve as regular and contingent sources of seedlings of desired quality and quantity for large-scale forest restoration initiatives.

## 4.6 Limitations and Challenges

The hilly landscape of Meghalaya makes it difficult to establish nurseries, as land is often fragmented and inaccessible, limiting the availability of space with a gentle slope. Private ownership of land restricts the ability to utilise the land for long-term projects. Additionally, there is a lack of scientific knowledge among local communities regarding seed collection seasons for various species, leading to poor timing in seed sowing, which affects germination. Due to poor understanding of nursery management practices, there is a risk of failure of the nursery. Disease management also poses a significant challenge, as local communities often lack the expertise to identify and manage pests and diseases in the nursery.

## 4.7 Results

### 4.7.1 Main Case Findings (1)

#### 4.7.1.1 Establishment of Community Nurseries

##### 4.7.1.1.1 Creation of Awareness

The creation of awareness about community nurseries is the prime step toward the establishment of village-level community nurseries. Once the villagers were ready to take up the venture, they were trained for various activities of nursery management. The project resource persons, along with the Forest Department and Soil & Water Conservation Department, provide technical support. Project staff were trained to guide villagers on the selection of land, construction of low-cost temporary nurseries, potting media, sources of quality seeds, pest and disease management, etc.

##### 4.7.1.1.2 Training and Capacity Building of the Participating Villagers

The participating villagers were oriented on the basic tenets of quality seedling production under protected conditions. They were trained on improved nursery management technologies like soil mixing, sowing, fertilisation, raised beds, pest and disease management activities, etc. They were made to understand the concept of quality planting materials.

##### 4.7.1.1.3 Establishment of Low-Cost Temporary Community Nursery

After selection of a suitable area taking into consideration the important criteria such as the proximity of a local water source, soil quality, accessibility and environmental and social impact assessments to construct the community plant nursery, our field teams had signed

Memoranda of Agreement (MoA) with the landowner in case of private lands and "No Objection Certificate" (NoC) in case of community lands for utilising land for raising a community nursery. Through a participatory approach, community members selected which plants to raise at the nursery with high commercial value.

The nursery bed preparation included preparing the land, removing rocks and adding soil amendments. Ten raised beds were prepared for accommodating 1,000 seedlings per community nursery. All nurseries are situated near perennial water sources, thus ensuring year-round water availability. Road accessibility was also given due priority for easy transportation of seedlings.

Retired foresters were recruited to facilitate the plantation works in villages. These foresters helped the community to prepare village-level seed calendars based on the availability of native seeds according to the season. These calendars were distributed to villages for the collection of seeds.

Collected seeds were first sown in the mother beds for germination, and after one month, the seedlings were transplanted to the poly bags with dimensions 6"-8" and sufficiently perforated.

One of the members of the community was given responsibility for day-to-day care of the nursery, such as weeding, watering, pest and disease management, etc.

### 4.7.2 Main Case Findings (2)

#### 4.7.2.1 Ensuring Success through Robust Monitoring using Technology

The MBDA/MBMA has developed its own monitoring and evaluation system that tracks both physical and financial activities in detail. This system provides intricate insights into the project's operations and serves as a reliable source of data for this case study. Additionally, Geographic Information Systems (GIS) have been extensively used as an extra layer to validate project interventions and their impacts. The progress of project activities was closely monitored through a Management Information System (MIS), which tracked the completion of activities, financial progress, staff visits, meetings held with various stakeholders and more. Additionally, there is an in-house

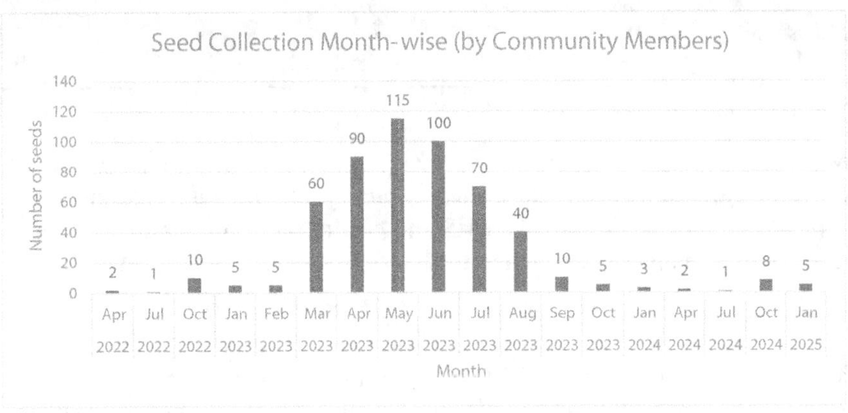


**FIGURE 4.1**  Graph showing month-wise collection of seeds by community members

*Source:* MegLIFE GIS Dashboard

**TABLE 4.2** Block wise total quantity of native seeds collected for raising community nurseries

| Block name | Nursery area (Ha) | Qty of seeds collected (Kgs) |
| --- | --- | --- |
| Rongram | 16.85 | 148 |
| Zikzak | 1.5 | 950 |
| Mairang | 1.32 | 68 |
| Songsak | 3.02 | 885 |
| Betasing | 9.06 | 1,261 |
| Gasuapara | 6.2 | 457 |
| Dambo Rongjeng | 12.39 | 142 |
| Mawkynrew | 9.64 | 176 |
| Mawkyrwat | 7.9 | 305 |
| Tikrikilla | 85,007.36 | 522 |
| Thadlaskein | 11,901.22 | 117 |
| Dalu | 2.65 | 759 |
| Resubelpara | 2.87 | 887 |
| Saipung | 8.7 | 113 |
| Baghmara | 9 | 94 |
| Samanda | 3.93 | 576 |
| Umsning | 7.5 | 56 |
| Umling | 0.31 | 200 |
| Kharkutta | 17.5 | 172 |
| Rongara | 5.19 | 1,781 |
| Gambegre | 5.66 | 613 |
| Mawryngkneng | 3.02 | 29 |
| Chokpot | 4.5 | 1,33,910 |
| Laskein | 0.47 | 28 |
| Salsella | 0.5 | 17 |
| Dadenggiri | 0.4 | 6 |
| Mawshynrut | 0.07 | 6 |
| Total | 160.55 | 18,22.22 |

*Source:* MegLIFE MIS.

Monitoring and Evaluation Team in MBDA/MBMA for intensive technology-based monitoring of the project interventions through site visits and capturing real-time data in the MIS and evaluation of the data at the State Office through analytical dashboards.

Implementation:

4.7.2.1.1 Data Collection: Village Community Facilitators used mobile-based APPs to collect data on seedlings survival, occurrence of diseases, etc. This real-time data was uploaded to the MIS platform.

4.7.2.1.2 Data Analysis: The MIS platform analysed the data to provide insights into seedling survival.

**TABLE 4.3** Block wise total no. of species raised

| Block | No. of species |
| --- | --- |
| tikrikilla | 236 |
| resubelpara | 220 |
| rongara | 215 |
| songsak | 206 |
| dalu | 181 |
| baghmara | 169 |
| zikzak | 160 |
| betasing | 154 |
| dambo rongjeng | 135 |
| gambegre | 132 |
| gasuapara | 121 |
| mawkynrew | 112 |
| samanda | 110 |
| kharkutta | 107 |
| saipung | 105 |
| mawkyrwat | 8 |
| thadlaskein | 94 |
| mairang | 67 |
| rongram | 44 |
| umsning | 54 |
| mawryngkneng | 32 |
| umling | 4 |

*Source:* MegLIFE MIS.

4.7.2.1.3   Reporting: The system generated fortnightly reports that informed project managers on the nursery's progress and highlighted any emerging issues.

4.7.2.1.4   Training and Capacity Building: The local community received training on how to use MIS, ensuring they could independently monitor and manage the nursery in the long term.

Results:

4.7.2.1.5   Improved Monitoring: The real-time data collected through MIS allowed for more informed decision-making and timely interventions.

4.7.2.1.6   Increased Efficiency: The community nursery saw a significant improvement in seedling survival rates due to better resource management and early disease detection.

4.7.2.1.7   Sustainability: With the data-driven approach, the community was able to better plan for future planting seasons and optimise nursery operations.

**TABLE 4.4** Region wise commonly raised species

| Khasi and Jaintia region | | Garo region and Ri-Bhoi district | |
|---|---|---|---|
| Sl. | Species | Sl. | Species |
| 1 | Dieng latyrpad (*Cinnamomum tamala*) | 1 | Chambu (*Eugenia claviflora*) |
| 2 | Dieng sohiong (*Mahomia acanthifolia*) | 2 | Bolbret (*Chickrassia tabularis/Toona ciliata*) |
| 3 | Dieng sohphan (*Artocarpus chaplasha*) | 3 | Chram (*Artocarpus chama*) |
| 4 | Dieng sohphie (*Myrica esculenta*) | 4 | Gamare (*Gmelina arborea*) |
| 5 | Dieng soh ot (*Castanopsis indica*) | 5 | Bolchim (*Duabanga grandiflora*) |
| 6 | *Toona cilliata* (Poma) | 6 | Siso (*Dalbergia sissoo*) |
| 7 | *Michelia champaca* | 7 | Segun (*Tectona grandis*) |
| 8 | *Cryptomeria japonica* | 8 | Bolsal (*Shorea robusta*) |
| 9 | betula alnoides | 9 | Jalpai (*Elaeocarpus floribundus*) |
| 10 | *Schima wallichi* | 10 | Arjun (*Terminalia arjuna*) |
| 11 | *Pinus kesiya* | 11 | Tekring (*Bursera serrata*) |
| 12 | *Exbucklandia populnea* (Dieng Doh) | 12 | Gasampe (*Baccaurea ramiflora*) |
| 13 | Wild apple/Sohphoh (*Docynia indica*) | 13 | Tebrong (*Artocarpus heterophyllus*) |
| 14 | Dieng lieng iong/Alder (*Alnus nepalensis*) | 14 | Titachap (*Michelia champaka*) |
| | | 15 | Amlengga (*Averrhoa carambola*) |
| | | 16 | Jarul Bolassari (*Lagerstroemia speciosa*) |
| | | 17 | Tegachu (*Mangifera indica*) |
| | | 18 | Zam (*Syzygium cumini*) |
| | | 19 | Letchu (*Litchi chinensis*) |
| | | 20 | *Bischopia javanica* |
| | | 21 | *Elacocarpus serratus* |
| | | 22 | *Citrus sinensis/Citrus reticulata* |
| | | 23 | Aritak (*Terminalia chebula*) |
| | | 24 | *Delonix regia* |
| | | 25 | Kimde (*Mesua ferrea*) |
| | | 26 | *Cinnamon tamala* |
| | | 27 | *Albizzia lebbeck* |
| | | 28 | *Phyllantus embilica* |
| | | 29 | *Bauhinia purpurea* |
| | | 30 | *Exbuclandia pupulnea* |
| | | 31 | *Psidium guvajava* |
| | | 32 | *Cassia fistula* |
| | | 33 | *Cassia nodusa* |
| | | 34 | *Dipterocarpus retusus* |
| | | 35 | *Parkia speciosa* |
| | | 36 | *Samanea saman* |
| | | 37 | *Terminalia belirica* |
| | | 38 | *Tetrameles nudiflora* |
| | | 39 | *Eugenia claviflora* |
| | | 40 | *Pongamia pinnata* |
| | | 41 | *Spondias pinnata* |

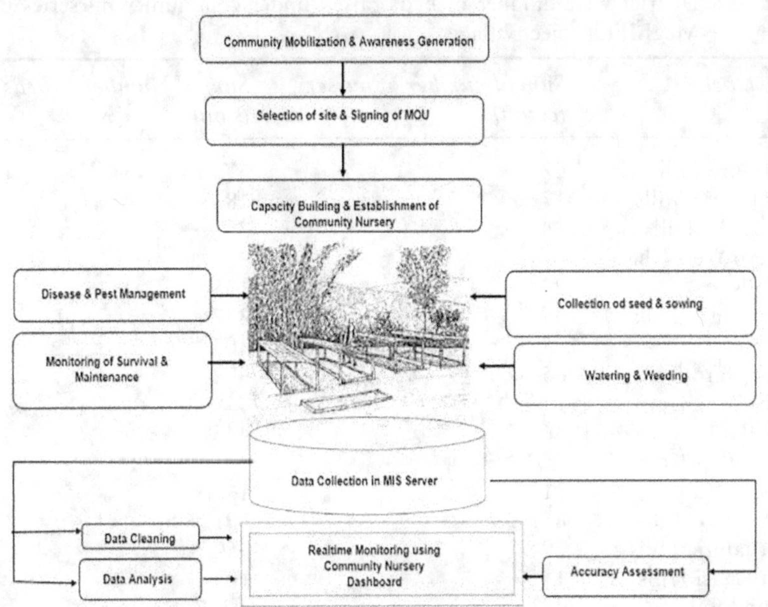


**FIGURE 4.2**  Schematic diagram of approach followed for establishment and monitoring of village level community nurseries

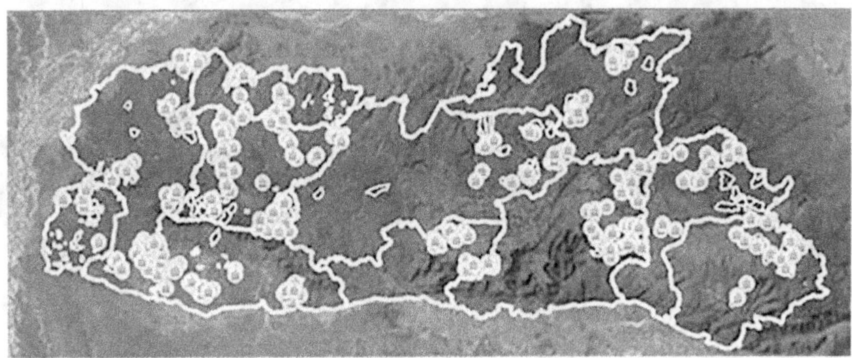

**FIGURE 4.3**  Map showing district wise distribution of community nurseries established under MegLIFE project

*Source:* MegLIFE GIS Dashboard.

### 4.7.3 Main Case Findings (3)

#### 4.7.3.1 Impact and Outcome of Community Nurseries

Empowering local communities to manage tree nurseries and plant trees is a sustainable solution to land degradation, livelihood issues and thus contributing to climate change adaptation and mitigation.

**TABLE 4.5** District wise number of beds raised under community nurseries under MegLIFE project village

| Row labels | Sum of number of nursery (count) | Sum of Number of beds (sum) |
| --- | --- | --- |
| East Garo Hills | 62 | 572 |
| East Jaintia Hills | 17 | 182 |
| East Khasi Hills | 34 | 300 |
| Eastern West Khasi Hills | 23 | 135 |
| North Garo Hills | 35 | 337 |
| Ri Bhoi | 35 | 405 |
| South Garo Hills | 65 | 672 |
| South West Garo Hills | 46 | 443 |
| South West Khasi Hills | 21 | 160 |
| West Garo Hills | 85 | 791 |
| West Jaintia Hills | 29 | 350 |
| West Khasi Hills | 1 | 10 |
| **Grand total** | 453 | 4,357 |

*Source:* https://meglifemis.in

Throughout this journey, the capacity of local community partners in tree nursery development and management, and the awareness and understanding of environmental sustainability has significantly increased.

So far, 739 community-managed Tree Nurseries were established with around 5,000 beds, producing and planting over 3.5 million seedlings in 12 Districts of Meghalaya. Over 37,137 ha of degraded lands were restored through tree planting, soil conservation and agroforestry interventions, utilising over 4.2 million seedlings of which 70% seedlings were supplied by community nurseries.

Community nurseries established under MegLIFE project serve as a model for sustainable practices that contribute to improving livelihoods and environmental resilience.

The project has taken a multi-faceted approach to addressing both environmental and socio-economic challenges in Meghalaya. Some of the impacts are given below:

### 4.7.3.1.1 Measurable Impacts

**Afforestation of 37,137 ha:**
Thirty-seven thousand, one hundred thirty-seven. hectares of degraded or deforested land have already been reforested using seedlings mostly raised

in community-managed nurseries. The goal is not only to restore ecological balance but also to improve the overall forest cover in the region, which has direct implications for biodiversity, climate regulation and soil health.

**Financial Incentives to Communities:**
A core aspect of the project is providing financial incentives to local communities through wages for their active involvement in afforestation activities. These wages act as a supplementary income, empowering families and improving the standard of living while simultaneously motivating community members to participate in the environmental restoration efforts.

**Capacity Building for Micro-enterprises:**
The project is focused on long-term sustainability by building the capacity of local communities to raise and maintain nurseries at the village level. This capacity-building initiative aims to create a model of self-sufficiency where the nurseries function as micro-enterprises. This encourages entrepreneurship within the community, helps in generating employment and ensures that local people can continue to manage and expand afforestation efforts independently.

MBDA/MBMA bought back 221,227 seedlings worth Rs.4.42 lakhs from community nurseries for plantations. Ten thousand, eight hundred four seedlings raised in community nurseries in 22 blocks were also sold to the community members of the same village at subsidised rates for private planting initiatives. The community earned a profit of Rs. 63,975/ by selling seedlings to the public outside the respective villages.

**Profit for Self-Help Groups (SHGs):**
Under the buy-back arrangement with the project, Self-Help Groups (SHGs) that create and manage these nurseries will earn a profit of Rs. 20,000 per

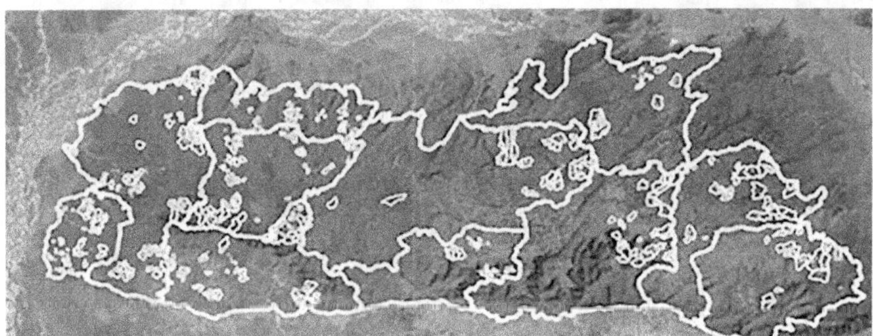

**FIGURE 4.4** Map showing district wise distribution of degraded lands restored through afforestation under MegLIFE project

*Source:* MegLIFE GIS Dashboard.

**TABLE 4.6** Utilisation of saplings for raising plantations

| Row labels | Sum of surviving saplings | Sum of utilised for VPIC plantation |
| --- | --- | --- |
| Baghmara | 118,710 | 74,781 |
| Betasing | 134,502 | 98,757 |
| Dalu | 135,581 | 109,216 |
| Dambo Rongjeng | 323,178 | 132,874 |
| Gambegre | 71,141 | 27,672 |
| Gasuapara | 153,285 | 108,971 |
| Kharkutta | 64,245 | 49,031 |
| Mairang | 139,134 | 120,899 |
| Mawkynrew | 122,938 | 34,748 |
| Mawkyrwat | 130,985 | 92,261 |
| Mawryngkneng | 31,043 | 13,206 |
| Resubelpara | 157,625 | 104,163 |
| Rongara | 111,466 | 59,897 |
| Rongram | 161,577 | 105,317 |
| Saipung | 120,734 | 104,808 |
| Samanda | 86,149 | 1,432 |
| Songsak | 143,892 | 73,101 |
| Thadlaskein | 130,372 | 130,372 |
| Tikrikilla | 120,221 | 51,938 |
| Umling | 125,072 | 54,031 |
| Umsning | 98,103 | 98,103 |
| Zikzak | 98,985 | 58,442 |
| Grand total | 2,778,938 | 1,704,020 |

*Source:* https://meglifemis.in

nursery. This buy-back ensures that SHGs have a steady source of income from the sale of seedlings to the project for afforestation purposes. This incentivises the maintenance and growth of the nurseries while fostering a sense of ownership and responsibility within the community.

**Community Participation (Case study from Miktongjeng Village, Dambo Rongeng Block, East Garo Hills):**
**Cultivating a Greener Future in East Garo Hills**
**Where Community Spirit Meets Environmental Stewardship**
With 760 residents and a deep-rooted agricultural heritage, this small yet vibrant community of Miktongjeng is showing the power of collective action when it comes to restoring nature.

Miktongjeng is playing a pivotal role in forest restoration and afforestation efforts. At the heart of this initiative lies the establishment of community nurseries – vital hubs for nurturing the seedlings required for the village's ambitious forest regeneration efforts.

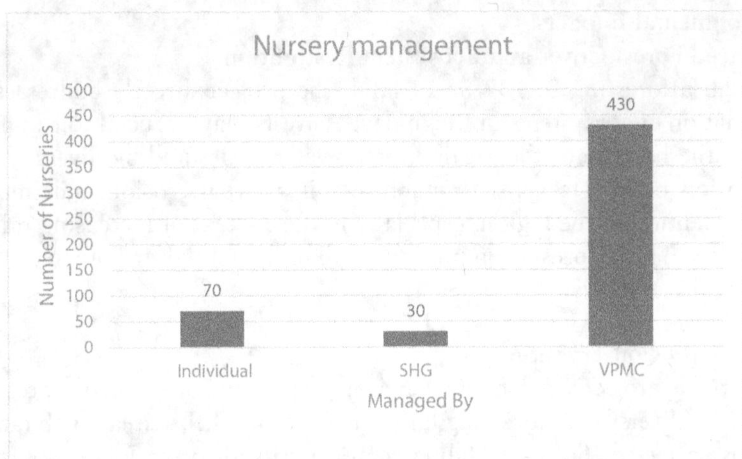


**FIGURE 4.5** Graph showing number of community nurseries managed by individual/SHG/VPIC members of MegLIFE project villages

*Source:* MegLIFE GIS Dashboard.

Starting with an investment of ₹1,10,000 from the MegLIFE project, Miktongjeng took its first steps towards creating a sustainable future with the construction of a nursery featuring 10 beds. The Village Project Implementation Committee (VPIC), along with the Village Community Facilitator (VCF), rolled up their sleeves and worked together to establish a community nursery when the village faced a shortage of saplings. Each villager played an integral role in this labour-intensive task, cultivating a sense of ownership and unity.

The nursery now boasts an impressive stock of 10,000 saplings in polybags, all nurtured and primed for planting. Thanks to the consistent monitoring and collective effort of VPIC members, these saplings have thrived, achieving a remarkable 90% survival rate.

Miktongjeng has proven that incredible things can happen when a community comes together, driven by a shared vision of environmental stewardship and sustainable development. The success of the community nursery has not only addressed the need for seedlings but has also fostered long-term ecological and economic benefits for the village. The villagers, equipped with new skills and knowledge, are now prepared to continue their journey of sustainable forest management for years to come.

The village's collective effort is a shining example of how small actions can lead to monumental changes in the fight against deforestation and climate change. The work done here is laying the groundwork for a greener, more sustainable future – not only for Miktongjeng but also for neighbouring communities looking to follow in their footsteps.

**Environmental Impacts:**
**Enhanced Forest Cover and Ecosystem Restoration:**
Through afforesting 37,137 ha of land, the project directly addresses the degradation of ecosystems in Meghalaya. Forests play a crucial role in maintaining the balance of the local climate, water cycles and biodiversity. The restoration of forests helps to stabilise soil, prevent erosion and improve water retention in the region, especially in the context of increasing rainfall variability. This afforestation initiative will ultimately contribute to a richer, more resilient ecosystem.

**Biodiversity Conservation:**
One of the project's most vital environmental outcomes is the preservation of biodiversity. Afforesting large areas of land helps create habitats for various species of flora and fauna that are critical to the local ecosystem. Moreover, the establishment of Community Reserves, including the protection of sacred groves, ensures that unique and ecologically sensitive areas are preserved. These reserves offer protection for endangered species, help conserve native plants and maintain ecological functions such as pollination, seed dispersal and nutrient cycling.

**Sustainable Forest Management through the Working Scheme:** The Working Scheme, embedded in the project, offers a sustainable model for forest management. By enabling communities to harvest timber while ensuring natural regeneration, this scheme strikes a balance between resource utilisation and conservation. It supports sustainable livelihoods by allowing local populations to benefit from their forests without depleting them. The careful management of forest resources through this model helps ensure the longevity and health of the forest ecosystems.

4.7.3.1.2 Outcomes

**Empowered Communities for Forest Management:**
The project successfully empowers local communities by giving them the tools, resources and knowledge they need to manage their forest resources sustainably. The direct involvement of communities in both afforestation and forest management enhances their sense of ownership and responsibility towards the forest and fosters long-term stewardship.

**Increased Awareness of Forest Conservation:**
Through training, workshops and active participation in forest restoration, community members gain a deeper understanding of the importance of forest conservation. This awareness is vital for building a culture of sustainability, where local populations are interested in preserving their environment for future generations.

### 4.7.3.1.3 Unintended Gains

**Cohesiveness Amongst Communities:**
One of the most powerful and unforeseen outcomes of the project has been the increased **cohesiveness** within the communities. As local people come together to plant trees, maintain nurseries and manage forest resources, they forge stronger bonds. The collective effort required to carry out afforestation and nursery management encourages collaboration, communication and mutual support.

With shared objectives, local groups have created a network of solidarity, which in turn has enhanced their collective resilience to external challenges. This cohesiveness has not only made communities more successful in implementing the project but has also had positive implications for tackling other local issues in a collaborative way, from conflict resolution to enhancing the social fabric of villages.

**Direct Financial Benefits:**
In addition to the wages for afforestation activities, a direct benefit that many community members are experiencing is the **income generated from raising and maintaining nurseries.** Local participants, especially those involved in running and maintaining the community nurseries, receive consistent wages for their work. This economic support is an essential source of income for many families, particularly in rural and underserved areas where other employment opportunities may be scarce.

The wages paid for managing nurseries, as well as the buy-back arrangement for seedlings, ensure that families can rely on this income stream while simultaneously contributing to the project's environmental objectives. This direct financial benefit is crucial in supplementing household incomes, reducing poverty levels and improving overall economic stability in participating communities.

**Additional Income from Carbon Credits:**
One of the most significant unintended benefits of this project is the **generation of additional income through Carbon Credit** trading. As the afforestation efforts lead to an increase in forest cover, the carbon sequestration potential of these areas grows, allowing the project to sell carbon credits. These credits are purchased by organisations or countries looking to offset their carbon emissions, thereby providing the project with additional funding.

For the communities involved, this presents an opportunity to gain supplementary income through the sale of these credits. The financial proceeds from carbon trading can be reinvested into further forest management activities, community development projects, or even used to support local infrastructure improvements. This income source creates long-term financial sustainability for the communities and enhances the project's overall economic impact, while simultaneously contributing to global climate change mitigation efforts.

### Sensitisation and Awareness Regarding Climate Change:

Another unintended but valuable outcome of the project has been the **sensitisation of** communities to the impacts of climate change and the importance of adaptation strategies. Through their involvement in forest restoration, carbon trading and environmental education campaigns, community members have gained a deeper understanding of the role forests play in regulating the local climate, mitigating floods, conserving soil and maintaining water resources.

As communities witness first-hand how their actions can reverse deforestation and improve the environment, this awareness fosters a proactive approach to climate adaptation, where local communities develop strategies to cope with environmental challenges such as erratic rainfall, temperature fluctuations and loss of biodiversity. This sensitisation ensures that climate resilience is built into the very fabric of community development, empowering people to take charge of their environmental future.

### Allied Livelihood Opportunities through Non-timber Forest Products (NTFPs):

In addition to afforestation and timber management, communities have also gained **access to** allied **livelihood opportunities** through the collection, processing and sale of **Non-Timber Forest Products (NTFPs)**. These products, such as medicinal plants, wild fruits, bamboo and mushrooms, are a crucial source of income for many rural communities. The afforestation project indirectly supports the growth and sustainability of these resources by improving forest health and providing a more diverse and abundant ecosystem.

As forests are regenerated and thereby improve biodiversity, the availability of NTFPs increases, providing new avenues for local economic activities. People can harvest and sell these products sustainably, thus contributing to both household incomes and local economies. This diversification of livelihoods reduces dependency on any single income source and helps cushion the community against external shocks, such as crop failures or market volatility.

### Increased Employability through Skill Development:

Another unintended yet highly beneficial outcome has been the **skill development** of individuals involved in the project, particularly Village Community Facilitators (VCFs) employed under the MegLIFE project who have taken on leadership and organisational roles. These facilitators have gained a diverse set of skills through their involvement in training, nursery management, project coordination and outreach activities.

The experience gained from working on the project—ranging from technical forestry skills to community engagement and project management – has made these individuals more **employable**. Many of the VCFs have transitioned to other job opportunities within the government, non-governmental organisations (NGOs) and environmental sectors, capitalising on their new-found expertise. The enhanced skill sets have made them more competitive, helping them secure stable employment and improve their quality of life. In turn, the project has contributed to the personal and professional growth of community members, offering them opportunities for upward mobility and long-term career prospects.

### 4.7.4  Main Case Findings (4)

#### 4.7.4.1  Innovative Approaches Adopted for Implementation

#### 4.7.4.1.1  GIS and MIS Technology

All nurseries established under the project are geo-tagged. These are mapped by location and nursery details can be accessed on the MegLIFE public dashboard of *MegLife Public Portal* and *CLLMP – Centre of Excellence for NRM and Sustainable Livelihood.*

These nurseries are fortnightly monitored through a robust MIS system on indicators like germination status, survival percentage of saplings, occurrence of diseases, availability of plantable seedlings, etc.

#### 4.7.4.1.2  Involving Village Project Implementation Committees (VPICs) and Village Community Facilitator (VCF)

The nurseries are managed by a village-level committee constituted for the implementation of project intervention at the village level. The committee is known as the Village Project Implementation Committee (VPIC). Under the VPIC, a sub-committee has been formed for community-level procurement of materials known as the Local Purchase Committee (LPC), which is responsible for the procurement of nursery equipment like poly bags, water cans, bamboo, shade nets, biopesticides, etc. for the establishment of the nursery.

Further Data Analytic team of MBDA/MBMA converts these into dynamic dashboards for state-level monitoring.

Every Village Community Facilitator (VCF) is given comprehensive training and capacity building and is fully tech-enabled for the management and monitoring of community nurseries. For example, VCF has been provided with tablets and oriented through hands-on training on the basics of data collection methods, templates, processes, frequency of data collection and feeding and basic analytics. His/her services are used for regular monitoring of nurseries by capturing the conditions of the seedlings through a mobile-based app.

### 4.7.4.1.3 The Use of Information, Education and Communication (IEC) Materials, such as Pictorial Community Nursery Flipcharts in Local Languages

The use of flipcharts in local languages plays a significant role in building the capacity of community members to raise successful nurseries. These flipcharts simplify complex information into visually engaging, easy-to-understand formats that are relevant and accessible to diverse audiences, including those with limited literacy. By using illustrations and step-by-step guides, these materials effectively communicate best practices for seed selection, soil preparation, watering, pest management and nursery maintenance. As presented in local languages, they foster better comprehension and active participation, empowering communities to take ownership.

## 4.8 Discussion

### 4.8.1 Interpretation of Findings in Relation to Their Replicability

The five key direct drivers of deforestation in Meghalaya, as a whole, are the following:

- Wood collection (fuelwood and timber).
- Shifting cultivation.
- Settlement expansion.
- Mining.
- Permanent farming (conversion of the forested area into monoculture permanent farming of cashew nut, betel nut, tea and rubber).

The key indirect drivers include poverty, overpopulation, non-availability of alternatives and lack of employment.

Restoration of degraded lands due to these multiple factors mentioned above is one of the primary objectives of the MegLIFE project. Six plantation models are implemented in the project villages for afforestation. These inculcate sustainable land usage practices; one such practice is agroforestry, which enhances the productivity of land, provides sustainable livelihoods to secure and support the livelihoods of the people and converts monoculture lands into diverse land use, thus promoting biodiversity and facilitating soil and water conservation.

Raising plantations on 37,137 ha of degraded lands is a huge task. This requires quality planting materials. Quality planting materials (QPMs) are crucial for successful plantations. QPMs are raised in central nurseries following standard protocol in a controlled environment. Carrying QPMs from central nurseries to the plantation sites is very challenging in villages due to the difficult terrain of Meghalaya and poor accessibility during the monsoon.

Village-level community nurseries raised nearby plantation sites to solve this problem, and this initiative also ensured more than 90% survival of planted seedlings due to less transportation shock during carriage. This is cost-effective also as it reduces transportation costs drastically.

Many government schemes and programmes struggle to achieve 100% success rates in plantations due to poor survival of seedlings in plantations. By adopting the MBDA/MBMA village-level Community Nursery Model, the survival of planted seedlings could be improved and the success rate of plantations could be enhanced. This has already been adopted by the KFW funded project named "Protection of Vulnerable Catchment Areas in Meghalaya (MegARISE)" implemented by Meghalaya Basin Development Authority (MBDA).

## 4.9  Implications for Theory, Practice and Policy

The need for community participation is emphasised by the nature of the 6th Schedule. Conventional rules and laws that are easily implemented in mainland India face challenges in the state. 90% of land is under private and community ownership, thus acquiring land for long-term project interventions is a challenge in Meghalaya. MBDA/MBMA, through a participatory approach and capacity building of the community, has been successful in transforming this scenario by raising 37,137 ha of plantations in private and community land through 3.5 million seedlings raised in village-level community nurseries by the community themselves. The success of this approach has led to the adoption and scaling up of this activity in other projects:

1. Establishment of Village Community Nurseries by Self-Help Groups in MegARISE Project funded by KFW/MGNREGA/National Rural Livelihoods Mission/PM Van Dhan Vikas Scheme/Meghalaya State Bamboo Mission.
2. Decentralised village-level permanent nurseries.
3. Land reclamation using agroforestry & AR ANR plantations.
4. Bamboo resource augmentation.
5. Development of a hub for the production of quality planting material.
6. Linkage to carbon market.

**TABLE 4.7**  Comparison of the Current Study with Existing Literature on Key Factors Influencing Community-Led Forest Restoration

| Factors | Current study | Previous researches |
| --- | --- | --- |
| Community-led restoration of forests | Yes | Yes References: (1), (4), (5), (7), (8), (9) |
| Climate resilience | Yes | Yes References: (10) |
| Community nursery initiative | Yes | Yes Reference: (2), (3) |
| Social development | Yes | Yes Reference: (6) |

## 4.10  Summary and Conclusion

### 4.10.1  Key Findings Summary

This Initiative addresses the challenge of land degradation and biodiversity loss in Meghalaya through the restoration of forests and ecosystems utilising native species raised by local communities in village-level decentralised nurseries managed by communities that promote the planting of native tree species. These nurseries enhance the success of plantation efforts by improving seedling survival rates and reducing costs compared to central nurseries through the engagement of local community members.

This initiative demonstrates how involving the local community in managing the nurseries fosters a sense of ownership and commitment to nurturing the seedlings. As a result, logistical challenges during the monsoon season are minimised, leading to healthier plants and increased biodiversity through the conversion of monoculture into diverse plantation practices.

Overall, this initiative showcases the effectiveness of community involvement in ecological restoration, paving the way for a sustainable future in Meghalaya.

## 4.11  Significance Restatement

The purpose of this study is to better understand the critical role of community participation in restoring degraded lands in Meghalaya through plantation initiatives. This is important because it empowers local communities through capacity building and self-managed village nurseries, successfully establishing 37,137 ha of plantations. This approach not only addresses the challenges of low plantation survival rates but also promotes sustainable

livelihoods. The positive impact of this model has led to its adoption and expansion in other initiatives, such as MegARISE, MGNREGA, the National Rural Livelihoods Mission, PM Van Dhan Vikas Scheme and the Meghalaya State Bamboo Mission. This further reinforces the significance of community-driven, participatory initiatives for sustainable development.

## 4.12  Future Research and Practical Applications

Future research on community participation in land restoration should focus on evaluating long-term ecological outcomes and socio-economic benefits of such community-led initiatives. Investigating methods to enhance the survival rates of plantations and examining indigenous planting techniques can provide valuable insights. Additionally, studies could analyse the impacts of community-driven projects on local biodiversity and climate resilience.

Practical applications of this research can inform policymakers and practitioners about effective strategies for scaling up successful models. This initiative has already been enculturated in another forestry-based project implemented by MBDA and funded by KFW, i.e., Project for Protection of Vulnerable Catchment Areas in Meghalaya (MegARISE Project).

By integrating community engagement into land restoration efforts, similar projects could be implemented in other regions facing degradation. Training programmes could be developed to transfer knowledge and best practices, further empowering local communities. Ultimately, this research could contribute to policy frameworks that prioritise participatory methods in sustainable development, fostering resilience and economic stability in vulnerable communities while ensuring ecological restoration across diverse landscapes.

## References


Agrawal, A., & Gibson, C. (1999). Community and conservation: Beyond enchantment and disenchantment. *Development and Change, 28*, 435–465.

Alpert, P. (1996). Integrated conservation and development projects. *BioScience, 46*, 845–855.

Blaikie, P. (1985). The Political Economy of Soil Erosion in Developing Countries. London: Longman.

Chambers, R. (1983). *Rural development: Putting the last first.* Longman.

Forest Survey of India (2023). India State of Forest Report 2023. Dehradun: FSI.

Gadgil, M. (1992). Conserving biodiversity as if people matter: A case study from India. *Ambio, 21*, 266–270.

Guariguata, M.R. & Brancalion, P.H.S. (2014). Forest restoration governance. Forest Ecology and Management, 329, 292–299.

Identification of Drivers of Deforestation Identification of Drivers of Deforestation in Meghalaya. 2018–19 Report by Rain Forest Research Institute, Indian Institute of Forestry Research & Education, Jorhat-785010, Assam.

Jadhav, A., Saini, P., Ravindra, A., & Singh, S. (2019). Increasing forest or forest cover in India. *Current Science, 116*, 158.

Karnataka Community Based Tank Management Jala Samvardhane Yojana Sangha (JSYS). www.jsysindia.org.

Meli, P. & Brancalion, P.H.S. (2017). Ecological restoration challenges. Restoration Ecology, 25(6), 899–904.

Mekonnen, Z. (2000). Indigenous forest management. Human Ecology, 28(4).

Paul, S. (1987). *Community participation in development projects.* The World Bank experience. World Bank Discussion Paper No. 6. The World Bank.

Rout, S.K. (2010). Community-based natural resource management. Journal of Human Ecology.

Shetty, G. P., Meghana, A., Balakrishna, S. M., Shetty Mahesh, G., Niranjan, H. G., & Narayana Swamy, M. (2024). Importance of quality planting material in Indian agroforestry systems and bio- energy plantations: A review. © *Agronomy, 7(5)*, 177–182.

Srinivas, P., & Singh, H. S. Concept of Community nurseries and seed banks for climate resilient horticulture. https://www.researchgate.net/publication/320957042

# 5

# COMMUNITY SEED SAVING FOR FOOD SECURITY

*Junie P. Lyngdoh, Reccica D. Lyngkhoi,
M Wanlambok Sanglyne, and Thomas Iangjuh*

## 5.1 Introduction

### 5.1.1 Purpose, Scope, and Significance of the Case

Meghalaya boasts a rich diversity of traditional crop varieties. Alarmingly, due to changing land use patterns and the introduction of high-yielding varieties, the indigenous germplasm of various crops is gradually disappearing. Besides, the Public Distribution System has contributed to the decline in the cultivation of traditional food crops such as millet, yams, and maize. For example, there is *Raishan* (*Digitaria cruciata,* var. esculenta Bor), an indigenous cold-tolerant millet crop. It is endemic to the Khasi hills of Meghalaya and cultivated for both food and fodder (Singh & Arora, 1972; Bhat et al., 2019). It is one such crop at risk. Without concerted efforts to conserve these valuable varieties, their loss could significantly affect the region's agricultural heritage, if not food security.

For generations, farmers in Meghalaya have passed down seeds as valuable cultural treasures, adapting crops to their environment. But increasing reliance on commercial seeds, which cannot be replanted, has led to the loss of many traditional varieties. This trend undermines agricultural resilience. A government-sponsored "Community seed saving initiative" in the state promotes the conservation and sharing of locally adapted seeds, supports local entrepreneurship, and strengthens community networks. It aims to empower farmers, preserve agricultural diversity and cultural heritage, and restore endangered varieties, while fostering relationships and promoting biodiversity conservation.

DOI: 10.4324/9781003735380-6

### 5.1.2 *Organisation of the Chapter*

After the introductory outline, a literature review situates the study within existing research, followed by an identification of gaps that the project has paid attention to. The theoretical framework grounds the project in established theory. The methodology for establishing and managing community seed banks follows a participatory, site-specific approach aligned with national guidelines, involving the community in planning, implementation, and monitoring, while integrating sustainable agricultural practices. The main body presents an overview of the results, offering information and a quantitative interpretation of the results. The chapter concludes with a summary of key insights, implications, and future research directions along with reference sources.

### 5.1.3 *Review of Literature*

Seed saving has long been a vital practice for maintaining agricultural biodiversity, allowing farmers to cultivate local varieties well-suited to changing environmental conditions. While global seed banks like the Svalbard Global Seed Vault in Norway help preserve genetic diversity (Da Silva, 2013), they do not always ensure diversity at a local level. Community seed banks (CSBs) have emerged as a solution. They enable farmers to access seeds for future planting seasons and provide emergency supplies when crops are damaged by events like flooding.

In India, organisations like *Navdanya* and the *Annadana* Soil and Seed Savers Network support seed-saving practices. They promote agricultural sustainability, biodiversity, and farmer autonomy. Over 30 years, Navdanya has established 150 seed banks across 22 states. It has trained 750,000 farmers in seed and food sovereignty, thereby preserving genetic resources and encouraging the cultivation of local crops (Navdanya, 2025).

In Meghalaya, where 80% of the tribal population depends on agriculture, women play a central role in seed saving, enhancing food security and community resilience (Meghalaya socio-economic review, 2020). This practice, rooted in traditional knowledge passed down through generations, supports nutritional, medicinal, and cultural needs. But the introduction of high-yielding varieties has eroded this traditional role, particularly as men increasingly engage in commercial farming (Centre for Education and Documentation, 2009).

The prevailing agricultural development model assumed that all farmers would transition to formal seed systems (Douglas, 1980; Frankel & Soulé, 1981). Yet, worldwide, small-scale farmers continue to rely on farm-saved seeds as well as informal systems to access new materials (Byerlee et al., 2007; Louwaars & De Boef, 2012). Climate change further draws attention to the importance of growing locally adapted, genetically diverse varieties that can better withstand environmental stresses. Community-driven models that support women's leadership in agriculture not only foster biodiversity but also strengthen social bonds and community engagement. But barriers such as limited access to resources and lack of decision-making power often hinder women's full impact in sustainable agriculture (Rao & Moharaj, 2023). This brings out the need to tackle these challenges and maximise the role of women in seed saving and conservation.

### 5.1.4 Identification of Gaps

A significant gap was recognised in farmers moving away from traditional agricultural and seed-saving practices. This has led to a decline in local seed varieties and their production and usage, which has undermined key organic farming principles. Also, as part of Meghalaya's "Mission organic" (to improve livelihoods by reducing fertiliser subsidies), a gap was found in bio-input availability. This prompted the Centre to promote sustainable green technologies (SGT). It initiated a seed-saving project at Bio-Resources Development Centre (BRDC) Experimental Farm, later replicated at Cham Cham Village, East Jaintia Hills.

The idea was to restore traditional practices and promote organic seed saving with women Self-help Groups (SHGs). This is generally preferred under various organic farming standards, emphasising sustainability, bio-diversity and a revival of traditional crops. The integration of modern scientific methods into traditional agricultural practices was deemed necessary for ensuring the sustainability of the project.

### 5.1.5 Theoretical Framework

Seed saving is crucial in promoting agricultural sustainability because it helps preserve local plant varieties that are better adapted to the unique environmental conditions of Meghalaya. An agro-ecological approach emphasises ecological balance and biodiversity. It ensures that locally saved seeds contribute to resilient farming systems that can withstand climate change and other environmental stressors. Also, cultural ecology plays a significant role in seed saving in Meghalaya, where agricultural practices are deeply tied to indigenous knowledge and the cultural identity of its peoples. Preserving traditional seeds not only ensures food security but also safeguards cultural heritage passed down through generations.

The project framework envisaged robust community involvement and collaborative efforts in the establishment and functioning of seed banks throughout the State. This way, future generations would always have access to indigenous, local and heirloom seed varieties.

## 5.2 Methodology

### 5.2.1 Description of Methodology

The methodology for establishing and managing community seed banks follows a customised approach. It is in line with the guidelines of the National Bureau of Plant Genetics Resources (NBPGR) for the management of community seed banks in India (Malik et al., 2013). It involves a collaborative approach, engaging SHGs through Village Organisations (VOs). Site selection takes place with a view to accessibility, while local needs guide the determination of crop varieties. The VO's nodal person supervises fund management and implementation, while the BRDC and SHG members take care of monitoring. This process ensures community involvement in design, planning, and sustainable seed-bank operation. Capacity building involves integration with sustainable technologies like composting, bio-inputs including an organic growth promoter (OGP), vermiwash, fishmeal extract, and biofertilisers like phosphate solubilising bacteria (PSB), *Trichoderma viride*, and *Trichoderma harzianum*, as well as integrated pest management. Seed collection, standardisation of agrotechniques, monitoring, and documentation see to the sustainability and effectiveness of seed banks.

### 5.2.2 Justification of Methodology

The methodology for the community seed bank initiative is justified by its adaptability to the specific needs and context of a community. A customised approach helps make the project both relevant and practical for them. Active community engagement is central, since their involvement in all stages determines the success of the initiative and leads to ownership and sustainability. The careful selection of a site fully considers accessibility and alignment with local agricultural needs. Capacity building boosts agricultural resilience. Also, the methodology includes seed collection, which contributes substantively to the conservation and propagation of local germplasm. Robust monitoring, evaluation, and documentation lead to long-term effectiveness and continuous improvement of the seed bank. This holistic approach ensures that a community seed bank not only meets immediate needs but also lays the foundation for long-term sustainability.

### 5.2.3 Limitations and Challenges

A major issue confronting this initiative was the lack of community awareness. Targeted specific programmes, regular meetings, and continuous engagement to encourage active participation were key components of remedial action. Another challenge was the limited source for local seeds. At the same time, the availability of extraneous, high-yielding, hybrid varieties at subsidised rates had made them the preferred choice over local varieties. This required efforts to refocus on indigenous crops and raise awareness about their climate resilience and ecological value along with a spotlight on the negative impacts of hybrid, "imported" varieties.

In addition, lack of baseline data on traditional seeds and traditional knowledge related to seed saving was a problem. Comprehensive documentation and knowledge preservation initiatives attempted to address this defect. Finally, lack of basic amenities such as modern tools, electricity, road connectivity, and local seeds and planting materials hindered progress. These challenges were partially overcome by providing essential resources, mobilising farmers to collect local seeds, providing seeds for trials that were locally unavailable, and arranging for tools, equipment, and infrastructure, where possible, to support community needs. Obviously, addressing these issues requires sustained effort and collaboration at various levels.

## 5.3 Results

### 5.3.1 Model Seed-Saving Unit at the Laitmynsaw Experimental Farm

A model seed saving unit at Laitmynsaw experimental farm was established to conserve local, indigenous and heirloom crop species. To date, 90 different varieties have been conserved under this model unit. These include over 20 indigenous and local varieties with low production rates like millets, sweet potato, pumpkin, and cucumber. These varieties were also revived and promoted in farmers' fields based on their agro-climatic zones by distribution during farmers' training programmes in different villages. Seeds were collected from various parts of Meghalaya through farmer interactions and community awareness programmes, and during field documentation.

To maintain genetic diversity, conservation techniques were employed, so seeds could grow in their natural environment. Some of the seed varieties that have been saved at the experimental farm are listed in Table 5.1.

As part of the initiative, 300 farmers were trained in seed saving techniques and sustainable agricultural practices. This training aimed to reduce farmers' dependency on external seed suppliers and promote self-reliance. Additionally, the project emphasised organic farming methods, encouraging a shift away from chemical inputs to effectuate environmental sustainability. The initiative also explored the effect of biofertiliser and bio-input

**TABLE 5.1** List of some indigenous and local crops saved at Laitmynsaw experimental farm

| Sl no. | Scientific name | Common name | Vernacular name |
| --- | --- | --- | --- |
| 1. | *Eleusine coracana* (L.) Gaertn. | Finger millet | *Rai truh* |
| 2. | *Setaria italica* (L.) P. Beauv. | Foxtail millet | *Krai soh* |
| 3. | *Sorghum bicolor* (L.) Moench | Sorghum millet | *Phñiai* |
| 4. | *Digitaria cruciata* (Nees ex Steud.) E.G. Camus & A.Camus | – | Raishan |
| 5. | *Solanum aethiopicum* (L.) | Bitter tomato | *Sohngang* |
| 6. | *Lagenaria siceraria* (Molina) Standl. | Bottle gourd | *Klong* |
| 7. | *Allium tuberosum* Rottler ex Spreng. | Garlic chive | *Jyllang* |
| 8. | *Cucurbita moschata* Duchesne | Pumpkin | *Pathaw Saw* |
| 9. | *Cucurbita maxima* Duchesne | Pumpkin | *Pathaw Shimon* |
| 10. | *Cucurbita pepo* var. *fastigata* (L.) | Pumpkin | *Pathaw Risang* |
| 11. | *Perilla frutescens* (L.) Britton | Perilla | *Neilieh* |
| 12. | *Cyclanthera pedata* (L.) Schrad. | Slipper gourd | *Soh thliem* |
| 13. | *Vigna umbellata* (Thunb.) Ohwi & H. Ohashi | Rice bean | *Rymbai ja* |
| 14. | *Solanum lycopersicum* L. | Tomato | *Sohsaw Laitkynsew* |
| 15. | *Solanum betaceum* Cav. | Tree tomato | *Sohbaingon dieng* |
| 16. | *Curcuma longa* L. | Turmeric | *Shynrai* |
| 17. | *Brassica oleracea* var. *Botrytis* L. | Cauliflower | *Phulkubi* |
| 18. | *Cucumis sativus* L. | Cucumber | *Sohkhia-Khasi* |
| 19. | *Ipomoea batatas* (L.) Lam. | Sweet potato | *Phankaro saw* |
| 20. | *Ipomoea batatas* (L.) Lam. | Sweet potato | *Phankaro stem* |
| 21. | *Ipomoea batatas* (L.) Lam. | Sweet potato | *Phankaro lieh* |
| 22. | *Flemingia vestita* Benth. ex Baker f. | – | *Sohphlang* |
| 23. | *Solanum melongena* L. | Eggplant | *Sohbaingon pylleng* |

*Source:* BRDC unpublished data and plant of the world online website.

treatments on different crops. Studies revealed that various bio-input treatments improved crop growth, yield, and soil fertility. In the case of lettuce (*Lactuca sativa*), different varieties responded uniquely to biofertilisers, highlighting the need for variety-specific applications. Similarly, experiments on cauliflower (*Brassica oleracea*) demonstrated that *Trichoderma harzianum* significantly enhanced growth and yield. Local pea (*Lathyrus oleraceus*) showed the best results when treated with Pea Rhizobium, which improved plant health and productivity.

**TABLE 5.2** An overview of experimental trials with biofertilisers being conducted at model seed-saving unit at Laitmynsaw Village under BRDC

| Crop | Biofertiliser treatment | Observed benefits |
| --- | --- | --- |
| *Lactuca sativa* L. | *Azotobacter* and *Azospirillum* | Improved plant height, leaf number, and spread |
| *Brassica oleracea* L. | *Trichoderma harzianum* Rifai | Enhanced plant growth, leaf size, and yield |
| *Lathyrus oleraceus* Lam. | Pea *Rhizobium* | Increased plant height, pod length, and seed yield |

*Source:* BRDC unpublished data and plant of the world online website.

Overall, the model seed-saving unit successfully integrated traditional seed conservation with modern, organic farming techniques. By combining community engagement, sustainable agricultural practices, and biofertiliser applications, the initiative contributed to agricultural sustainability and biodiversity conservation. An overview of experimental trials with biofertiliser conducted under this model seed-saving unit is shown in Table 5.2.

### 5.3.2 Community-Driven Seed Conservation Unit in Cham Cham Village

The seed-saving initiative in Cham Cham Village, East Jaintia Hills District, was set up with significant community involvement. Approximately 16% of the village population, 232 individuals out of 1,453 (Population Census, 2011), participated (Figure 5.1). The beneficiaries were members of 29 SHGs who are part of the VOs, a clear collaborative and community-driven approach to sustainable agricultural practices

Before the initiative, Cham Cham Village lacked access to climate-resilient local seeds. This increased their reliance on commercial seeds; as a result, their seed sovereignty decreased. Also, poor seed saving and storage shortened the shelf life of the seeds saved and made them vulnerable to pests and diseases. As a consequence, people limited seed saving to what they could use only. Moreover, limited bio-input availability and conventional farming methods affected soil health and productivity. Community members played an active role in the design, planning, and implementation of the project, and so, the project proved not only beneficial but also relevant to their own agricultural landscape.

The Village Organisations *Nangkiew Shaphrang* and *Iaseiñlang* were instrumental in the selection of crop varieties for the community seed bank. They focused on conserving local crop varieties with an eye on agro-climatic suitability and market value. As a result, these varieties aligned with local needs and preferences. On the whole, the process ensured that the seeds

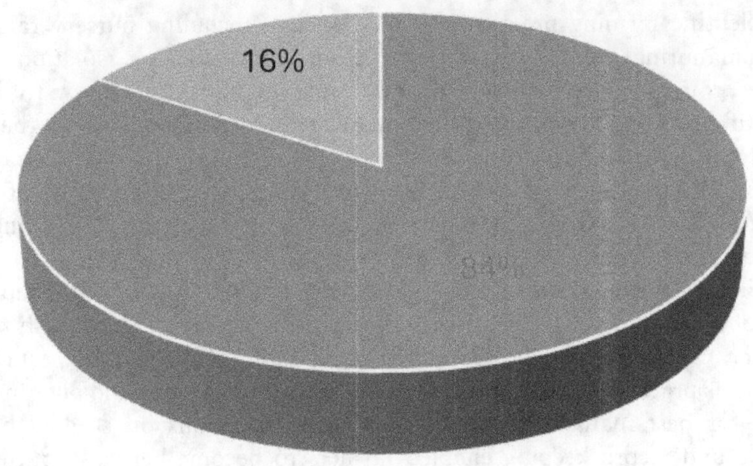


**FIGURE 5.1**   Beneficiary percentage under the seed-saving initiative in Cham Cham Village

*Source:* BRDC unpublished data.

**TABLE 5.3**  List of prioritised crops at Cham Cham for seed conservation

| Sl no | Scientific name | Common name | Vernacular name |
|---|---|---|---|
| 1. | *Capsicum frutescens* L. | Chilli | *Smurit-Pnar* |
| 2. | *Phaseolus vulgaris* L. | Bean | *Toh kper* |
| 3. | *Lathyrus oleraceus* var. *saccharatum* Lam. | Snow pea | *Toh motor* |
| 4. | *Raphanus sativus* L. | Radish | *Muli* |
| 5. | *Rhamphospermum nigrum* L. Al-Shehbaz | Mustard | *Ianem* |
| 6. | *Zingiber officinale* Roscoe | Ginger | *Syin met* |
| 7. | *Zingiber rubens* Roxb. | Bengal ginger | *Syin tro* |

*Source:* BRDC unpublished data and the Plant of the World Online website.

chosen were well-suited to the region's climate, soil, and food requirements, underscoring the relevance and success of the initiative. Seven crops were selected by SHG members for seed saving and multiplication, enabling the preservation of local varieties and contributing to agricultural biodiversity. Table 5.3 lists the seed varieties being saved at Cham Cham Village.

Another critical component of the project was the capacity building of farmers through training on seed-saving techniques, sustainable green technology, and integrated pest management. The training programme focused on scientific farming practices for seed saving, including nursery raising, green manuring, mulching, selection of plants for seed saving, isolation techniques, roughing, seed extraction, and drying methods. In addition to these scientific practices, traditional methods were incorporated, such as the use of ash for pest and disease management.

The "Kolom" method was included in the training programme. It is a traditional seed-saving technique that farmers in Meghalaya are using to produce seeds for Brassicaceae family crops (cauliflower, cabbage, etc.). The training also covered sustainable agricultural practices such as the preparation of bio-inputs like vermiwash, organic growth promoter, and fish meal extract. Farmers were educated on the use of biofertilisers and biopesticides as well as practical applications like yellow-blue sticky traps and pheromone traps for pest management. Hands-on demonstrations on seed viability testing and record-keeping enabled farmers to become better stewards of biodiversity conservation. Farmers could also learn about and take up zero-budget natural farming methods. The BRDC conducted the training for the nodal person of the seed bank. He, in turn, trained the other members.

Additionally, the registration of three local groups under the Participatory Guarantee System (PGS) Organic Certification programme covering 1.73 ha marked a significant milestone. This initiative not only promotes organic farming practices but also strengthens a community's connection to sustainable agriculture. The programme reflects the growing movement towards environmentally friendly farming that enhances food quality and opens market opportunities for the community. VOs set up a customised seed storage (seed bank) to preserve seeds for improved seed viability. It also served as an educational hub for farmers. This facility not only conserved seeds but also provided valuable knowledge on sustainable practices and the importance of preserving agricultural biodiversity. Management of the seed bank remained decentralised and under VO supervision.

The community was able to save 14 crop varieties with a quantity of 1,600 kg seeds (in 2023–2024) at the Cham Cham seed bank, generating an additional income of Rs.1,50,000 through seed sales. This financial gain represents both an economic benefit and a reinforcement of the community's capacity to invest in its agricultural practices. Through this initiative, they were able to promote the cultivation of local varieties that had declined in production, like the smurit-Pnar (chilli), toh kper (bean), sohkhia-Khasi (cucumber), raishan (*Digitaria cruciata*), rai truh (finger millet), krai soh (foxtail millet), and tushop (sunchoke). People have also begun using seed-saving techniques for cabbage and cauliflower because of their high commercial value. These community-led initiatives show the power of local ownership and collaboration in achieving agricultural sustainability. By integrating VOs, SHGs, and local resources, the seed bank initiative not

**TABLE 5.4** List of villages in different districts of Meghalaya where community seed banks is being replicated

| Sl no. | Districts | Villages |
| --- | --- | --- |
| 1. | West Jaintia Hills | *Mulum* |
| 2. | Ri-Bhoi | *Thadnongiaw* |
| | | *Khweng* |
| 3. | East Khasi Hills | *Umsawar* |
| | | *Khapmaw* |
| 4. | East Garo Hills | *Mandalgre* |
| 5. | West Garo Hills | *Darechikre* |
| 6. | Eastern West Khasi Hills | *Mawlum Mawjahksew* |
| | | *Mawkamoit* |

*Source:* BRDC unpublished data.

only helped preserve crop varieties but also empowered the community to take charge of its agricultural future.

The pilot seed conservation unit in Cham Cham Village has proved to be a successful and sustainable model for seed-saving and biodiversity conservation. It also highlights the importance of scientific methods for seed saving to obtain quality seeds.

The success has led to a replication of this model in nine villages covering six districts of the state as shown in Table 5.4. They were initiated in collaboration with other institutions. The comprehensive capacity building programmes and the establishment of a community-managed seed bank have given lasting benefits to the farmers, both in terms of increased yields and economic gains. The positive impact on agricultural biodiversity conservation in the region reflects the initiative's success in promoting environmental sustainability and food security.

### 5.3.3 Effect of Biofertiliser on Zingiber rubens and Zingiber officinale *in Cham Cham Village*

An experimental trial conducted in Cham Cham evaluated the growth and yield responses of *Zingiber rubens* and *Zingiber officinale* to biofertiliser treatments in the village as shown in Table 5.5. *Z. rubens* consistently showed the best growth and yield with PSB treatment, while *Z. officinale* exhibited variable responses, performing best with *Trichoderma viride* at 120 days after sowing (DAS) and PSB at 180 DAS. Overall, PSB proved to be the most effective biofertiliser for both species, significantly increasing yield compared to the control lot. The findings suggest that species-specific biofertiliser applications can optimise *Zingiber* cultivation.

**TABLE 5.5** An overview of growth and yield response of Zingiber species to biofertilisers in Cham Cham Village

| Growth stage (DAS) | Zingiber rubens *Roxb. best treatment* | Zingiber officinale *Roscoe best treatment* |
| --- | --- | --- |
| 60 DAS | PSB | Control |
| 120 DAS | PSB | *Trichoderma viride* Pers. |
| 180 DAS | PSB | PSB |
| Yield increase | PSB showed 10% increase compared to control | PSB showed 12% increase compared to control |

*Source:* BRDC unpublished data.

## 5.4 Discussion

### 5.4.1 Interpretation of Findings in Relation to Model Replicability

The community engagement and seed-saving initiative in Cham Cham Village points to the value of involving local farmers in sustainable agricultural practices. Such community-driven approaches are crucial for long-term success and replication of the initiative because local knowledge and participation ensure better acceptance and adaptation of new techniques (Acosta et al., 2014). The seed-saving initiative involving approximately 16% of the village population in Cham Cham is an example of how sustainable agricultural practices can be scaled up in rural communities. The combination of scientific methods with traditional seed saving and storage practices has improved the quality and shelf life of the seeds saved. This has encouraged farmers to take up seed saving on their fields too. The model, replicated in nine villages covering six districts, would be expanded to cover all 12 districts of the State.

The findings of this study have significant implications for their replicability in different agro-climatic zones of Meghalaya. The positive effects of biofertiliser and bio-input treatments on the growth and yield of various crops, including *Zingiber* species, *Lactuca sativa*, *Brassica oleracea*, and *Lathyrus oleraceus*, suggest that these biofertilisers can be widely applied in different regions with similar agro-ecological conditions, particularly in areas where sustainable farming practices are sought (Sharma et al., 2024). Nevertheless, it is important to consider that the replicability of this model depends on various factors such as soil type, climate, and access to resources. Further research and localised trials in different geographical settings are needed to assess the broader applicability of these bio-input treatments across diverse agricultural landscapes.

### 5.4.2 Implications for Theory, Practice, and Policy

The findings of this study have significant implications across theory, practice, policy, and community engagement, particularly with respect to biofertilisers and seed-saving initiatives.

**Implications for theory:** The study supports the idea that seed saving is essential for maintaining genetic diversity, countering the uniformity of monoculture practices. By preserving a wide range of crop varieties, seed saving contributes to ecological resilience, protecting against pests, diseases, and climate change effects. This aligns with the theory of ecological intensification, which advocates for sustainable productivity using natural inputs. Additionally, the study emphasises the importance of community-driven efforts like seed-saving, because such initiatives preserve indigenous agricultural knowledge and practices vital for sustaining cultural heritage.

**Implications for practice:** Practically, the project provided valuable insights to farmers, particularly smallholders in rural areas. It demonstrated how collective seed-saving efforts can promote biodiversity, reduce dependency on external inputs, and support food sovereignty. The study highlights the effectiveness of biofertilisers such as PSB for *Zingiber* species and *Azotobacter* for *Lactuca sativa* and *Brassica oleracea*, underscoring the need for crop-specific solutions. By incorporating biofertilisers and preserving local seed varieties, farmers can improve soil health, boost crop yields, and develop resilience against climate challenges. Training workshops on seed saving and biofertiliser use can empower farmers to adopt these sustainable practices. The cost for setting up a seed-saving unit at Cham Cham Village was Rs.1.2 million.

**Implications for policy:** Policymakers should address barriers to community seed-saving, ensuring it is integrated into formal agricultural systems without hindering growth. Support for community-driven initiatives could help preserve crop diversity, conserve resources and reduce reliance on commercial seeds. Policies should encourage local food systems, offer subsidies for biofertilisers, and provide incentives for farmers who conserve traditional seed varieties. Ensuring farmers' rights to save and exchange seeds, while protecting against the privatisation of genetic resources, is essential. Additionally, policies that support farmer education and incorporate local knowledge will strengthen agricultural resilience and promote self-sufficiency.

### 5.4.3 Comparison with Existing Literature

The study's emphasis on community engagement in seed-saving initiatives aligns with the work of Duthie-Kannikkatt et al. (2019) in Andhra Pradesh. It underscores the role of local seed-saving programmes in promoting agricultural biodiversity and reducing dependency on commercial seeds. The active participation of farmers in Cham Cham Village, highlighting the significance of community-driven sustainability practices, reinforces findings by Nazeri et al. (2024) that such practices enhance local food security and agricultural resilience.

The role of community-managed seed banks (CSBs) in fostering ownership and responsibility, as observed in this case study, corroborates the findings of Vernooy et al. (2022, 2024). Similarly, Das and Mallick (2024) describe how the *Vrihi* Community Seed Bank in Odisha integrates cultural and social values into its operations, supporting the argument that local management is crucial for the sustainability of CSBs. This aligns with broader literature suggesting that CSBs embedded within local agricultural systems are more likely to endure.

Importantly, the CSB in Cham Cham Village functions not only as a seed repository but also as a hub for knowledge exchange. This is consistent with Das and Mallick (2024), who describe the Vrihi CSB as a platform for farmers to learn about conservation techniques and sustainable agriculture. Likewise, Vansant et al. (2022) observed that CSBs in Malawi reinforce seed-sharing networks and contribute to economic resilience, further demonstrating the multifaceted role of this institution.

Further, this study's findings on the effects of biofertiliser treatments on crop growth and yield align with existing research in sustainable agriculture. Specifically, the observed positive effect of phosphate solubilising bacteria (PSB) on *Zingiber* species and the benefits of *Azotobacter* on *Lactuca sativa* and *Brassica oleracea* are consistent with studies demonstrating the efficacy of these biofertilisers in increasing plant growth and yield (Kalayu, 2019; Benbrik et al., 2020).

In conclusion, the literature converges on the idea that CSBs function beyond seed storage. They are essential institutions for knowledge dissemination, economic support, and cultural preservation. This study builds on existing research by providing further evidence of the role of CSBs in fostering sustainability and resilience in agricultural communities.

## 5.5 Summary and Conclusion

### 5.5.1 Key Findings Summary

The seed-saving unit at Laitmynsaw experimental farm conserved 90 crop varieties, including local, indigenous, and heirloom species, promoting agricultural biodiversity and safeguarding cultural heritage. The initiative used ex-situ conservation to preserve genetic diversity, empowering local farmers with seed-saving skills and sustainable agricultural practices. Around 16% of the village population actively participated in the seed-saving initiative in Cham Cham Village. The community played a crucial role in planning, selecting crop varieties, and managing the seed bank, ensuring the project met local needs.

A community-managed seed bank was established, preserving 14 crop varieties with a quantity of 1,600 kg of seeds. Training in seed-saving, sustainable practices, biofertilisers, and pest management empowered farmers and improved biodiversity conservation. The initiative promoted the preservation of local crop varieties, generating an additional income of Rs.1,50,000 through seed sales. The project also reinforced self-reliance, food security, and environmental sustainability. The community engagement approach has proven more sustainable, conserving biodiversity, improving food security, and increasing economic resilience compared to relying on commercial seeds. The study also drew attention to the significant effect of biofertiliser treatments on growth and yield of various crops in the BRDC experimental farm and Cham Cham Village.

### 5.5.2 Significance Restatement

The success of the seed-saving initiative at Cham Cham Village signifies more than the preservation of crops – it demonstrates how grassroots, community-led action can become a powerful model for ecological sustainability and rural empowerment. By placing ownership in the hands of farmers and integrating traditional knowledge with modern practices, the project has created a replicable framework for biodiversity conservation, food security, and economic resilience. Importantly, the initiative addresses systemic challenges faced by smallholder farmers, such as dependency on commercial seeds and limited access to sustainable inputs. It also reaffirms the critical role of women and community institutions in agro-ecological transformation. The use of biofertilisers and scientific storage techniques further validates the potential for environmentally friendly practices to boost productivity without compromising local biodiversity.

This case thus serves as a model for regions facing similar threats to agricultural diversity, offering insight into how policy, practice, and local leadership can intersect to build more sustainable, self-reliant food systems.

### 5.5.3 Future Research and Practical Applications

Developing a model seed-saving unit in the BRDC Laitmynsaw experimental farm to serve as a field gene bank or in situ collections of indigenous and local varieties of Meghalaya is desirable to further this initiative. Empowering farmers through capacity building programmes regarding various modern techniques and their adoption by the community will strengthen community-led initiatives and build manpower for the management and expansion of such seed banks. Research on integrating biotechnological

methods with traditional practices to improve crop resilience, particularly in the face of climate change, is essential. Also, scaling up community seed-saving initiatives to more regions could empower more farmers, reduce dependence on commercial seed suppliers and support organic farming.

Investigating the economic and market potential of locally conserved crops as well as strengthening policies to protect seed sovereignty will be crucial in ensuring the sustainability of such initiatives. Ultimately, such similar projects can contribute to the broader goal of enhancing food security and effecting environmental sustainability through community-led, organic practices.

## References


Acosta, L. A., Eugenio, E. A., Enano Jr, N. H., Magcale-Macandog, D. B., Vega, B. A., Macandog, P. B. M., & Lucht, W. (2014). Sustainability trade-offs in bioenergy development in the Philippines: An application of conjoint analysis. *Biomass and Bioenergy, 64*, 20–41. https://doi.org/10.1016/j.biombioe.2014.03.015

Benbrik, B., Elabed, A., El Modafar, C., Douira, A., Amir, S., Filali-Maltouf, A., & Koraichi, S. I. (2020). Reusing phosphate sludge enriched by phosphate solubilizing bacteria as biofertilizer: Growth promotion of Zea Mays. *Biocatalysis and Agricultural Biotechnology, 30*, 101825. https://doi.org/10.1016/j.bcab.2020.101825

Bhat, S., Nandini, C., Srinathareddy, S., & Jayarame, G. (2019). Proso millet (*Panicum miliaceum* L.) – a climate resilient crop for food and nutritional security: a review. *Environment Conservation Journal, 20*,113–24. https://doi.org/10.36953/ECJ.2019.20315

Byerlee, D., De Janvry, A., Sadoulet, E., Townsend, R., & Klytchnikova, I. (2007). *World development report, 2008: Agriculture for development*. World Bank Group. United States of America. https://coilink.org/20.500.12592/hbbz4s

Census of India. (2011). *National population register & socio-economic and caste census India, 2011*.

Centre for Education and Documentation. (2009). *Community seed banks in India*. http://base.d-p-h.info/fr/fiches/dph/fiche-dph-8060.html

Da Silva, E. D. (2013). Community seed banks in the semi-arid region of Paraíba, Brazil. In *Community biodiversity management* (pp. 102–108). Routledge.

Das, S. K., & Mallick, S. (2024). Repossession through community participation: A study of Vrihi Community Seed Bank in Odisha. *Social Change, 54*(4), 508–525.

Douglas, J. (1980). *Successful seed programs: A planning and management guide*. Westview Press.

Duthie-Kannikkatt, K., Shukla, S., Rao, M. L. , S., Sakkhari, K., & Pachari, D. (2019). Sowing the seeds of resilience: a case study of community-based indigenous seed conservation from Andhra Pradesh, India. *Local Environment, 24*(9), 843–860. https://doi.org/10.1080/13549839.2019.1652800

Frankel, O. H., & Soulé, M. E. (1981). *Conservation and evolution*. Cambridge University Press.

Kalayu, G. (2019). Phosphate solubilizing microorganisms: Promising approach as bio- fertilizers. *International Journal of Agronomy, 2019*(1), 4917256. https://doi.org/10.1155/2019/4917256

Louwaars, N. P., & De Boef, W. S. (2012). Integrated seed sector development in Africa: A conceptual framework for creating coherence between practices, programs, and policies. *Journal of Crop Improvement, 26*, 39–59.

Malik, S. K., Singh, P. B., Singh, A., Verma, A., Ameta, N., & Bisht, I. S. (2013). *Community seed banks: Operation and scientific management.* National Bureau of Plant Genetics Resources. www.nbpgr.ernet.in

*Meghalaya socio-economic review.* (2020). Directorate Of Economics & Statistics Government of Meghalaya, Shillong. https://des.megplanning.gov.in/documents/Meghalaya-Socio-Economic- Review-2020.pdf

Navdanya website: Community seed bank article. https://www.navdanya.org/living-seed/navdanya-seed-banks

Nazeri, N., Hidayat, R., & El Maza, R. (2024). Encouraging community empowerment and local economic in-dependence in villages through sustainable economic development techniques. *The Es Economics and Entrepreneurship, 3*(2), 239–245.https://doi.org/10.58812/esee.v3i02.371

Rao, R., & Moharaj, P. (2023). Empowering women in climate-resilient farming through sustainable agriculture technologies. *International Journal of Multidisciplinary Research and Growth Evaluation, 4*, 57–265.

Sharma, H., Kagday, M., Poonam, P., Kalia, M., Kalia, S., Pal, A., & Rautela, I. (2024). A sustainable agriculture method using biofertilizers: An eco-friendly approach. *Plant Science Today.* https://doi.org/10.14719/pst

Singh, H. B., & Arora, R. K. (1972). Raishan (*Digitaria* sp.) – a minor millet of the Khasi Hills, India. *Economic Botany, 26*(4), 376–380.

Vansant, E. C., Kerr, R. B., Sørensen, H., Phiri, I. M. G., & Westengen, O. T. (2022). Exchange and experimentation: Community seed banks strengthen farmers' seed systems in Northern Malawi. *International Journal of Agricultural Sustainability, 20*(7), 1415–1436. https://doi.org/10.1080/14735903.2022.2122254

Vernooy, R., Adokorach, J., Gupta, A., Otieno, G., Rana, J. C., Shrestha, P., & Subedi, A. (2024). Promising strategies to enhance the sustainability of community seed banks. *Sustainability, 16*(19), 8665. https://doi.org/10.3390/su16198665

Vernooy, R., Rana, J. C., Otieno, G., Mbozi, H., & Shrestha, P. (2022). Farmer-Led Seed Production: Community Seed Banks Enter the National Seed Market. *Seeds, 1*(3), 164–180. https://doi.org/10.3390/seeds1030015

# 6

# PRESERVING CULTURAL AND NATURAL HERITAGE

## Towards a Sustainable Future: Jingkieng Jri/Lyu Chrai Cultural Landscape, an Exemplar of Community-Led Conservation and Heritage Stewardship

*Tilaris Marwein, Goldenstar Thongni, and Catherine Derica Kyndiah*

### 6.1 Background

*Ko Shnong baieid jong nga pha long,*
*Pha ieng hapdeng ki ranap lum kynjreng,*
*Naduh hyndai hynthai iapha la seng,*
*Hapdeng ki khlaw ki btap ba rben.*

*U Trai nongbuh nongthaw u buh ia ngi,*
*Ba ngin pynneh pynsah ia la riti,*
*Ban ieng ka hok jinglong sotti,*
*Hukum Blei ngin bat ngin ri.*

*Diengjri ba thung longshwa manshwa,*
*Ki im ki sah haduh mynta,*
*Ki thied ki thain katno jingproh,*
*Ban jam lynti sha shilliang ki snoh.*

*Longshwa manshwa ki buh kyntang,*
*Ha shnong ha thaw da law kyntang,*
*Ban bha ban miat ka shnong ka thaw,*
*Duwai phirat h'u Trai nongbuh nongthaw.*

*Khatduh iawai to theh jingshai,*
*Lynti ka hok to ngin beh brai,*
*Ka pop ka sinew to beh sha jngai,*
*Ba lawei jong ngi kan iai phyrnai.*

DOI: 10.4324/9781003735380-7

*Oh, my beloved land!*
***Thou** standest among the hills.*
*From ancient times you have been*
*In the midst of the thick forests.*

*The Lord, our Creator, has created us,*
*For us to keep our customs,*
*To stand up for righteousness.*
*God's law we will keep dear.*

*Diengjri was planted by the forefathers,*
*They are alive today.*
*The roots they weave, how clever they are:*
*They bridge to cross the other path.*

*Forefathers have kept sanctified,*
*In the village with sacred forests,*
*For the good of the community,*
*Praying to the Lord, our Creator.*

*Finally, let us spread the light!*
*The way of righteousness we will follow,*
*Sins and bad omen we will throw afar*
*For a better and brighter future.*

*Take Wahlang*
*52 years old*
*Nongsteng Village*

The essence of *Jingkieng Jri/Lyu Chrai* Cultural Landscape (JJLCCL) is a tapestry that embodies the richness of nature and culture of the Khasi and Jaintia indigenous people

Nestled in the lush green hills of Meghalaya, the JJLCCL is a testament to nature's harmony with human ingenuity. Stepping into the misty forest, one is immediately embraced by nature's tranquillity, the cool breeze carrying the melodies of chirping birds, the steady hum of cicadas, and the very breath of Mother Nature – *Mei Mariang*.

The path descends steeply, revealing breathtaking landscapes of rolling hills draped in soft, cotton-like clouds. Walking barefoot on the uneven stone pathways (*lynti*) offers a deep connection to the land, a therapeutic experience cherished for generations. The challenging trek is eased by the thoughtful placement of resting stones (*mawshongthait*) by ancestors, allowing travellers to pause, contemplate, and appreciate the beauty around them.

As the trail winds through the landscape, cultural heritage is woven into every step. Monoliths (*mawbynna*), memorial stones (*mawsahnam*), and ossuaries (*mawbuhshyieng*) stand as silent storytellers, preserving histories passed down through generations. These sacred markers, along with other culturally significant stones, embody the deep-rooted traditions of indigenous communities who have safeguarded them over time.

The path meanders through lush, revered forests – *lawkyntang* and *law adong* – with water flowing from rocks and crevices nourishing the land and agricultural fields (*bri*) that stretch across the terrain. From a distance, the mighty roar of waterfalls (*kshaid*) and rivers (*wah*) fills the air, blending with the hushed whispers of hidden caves (*krem)*, creating an atmosphere of both serenity and mystery. Together with the amazing living root bridge (*Jingkieng Jri*), meticulously crafted over centuries by Khasi (*Khynriam* and *Pnar*) indigenous communities, these natural wonders serve as reminders of the harmony that exists between humans and their surroundings, influencing the region's ecological and cultural form and making up the *Jingkieng Jri/Lyu Chrai* Cultural Landscape (JJLCCL).

But modernisation and growing infrastructural development have led to a decline in these traditional structures and knowledge sources. With evolving lifestyles and modern ways of doing, it has become of prime importance to preserve and celebrate the cultural and environmental significance of these landscapes.

The indigenous communities of Meghalaya have taken significant steps to safeguard their traditional, cultural, and ecological heritage by continuing practices and having protection guidelines in place through the traditional *Dorbar* system. Community-led efforts, including replanting of *Ficus* trees, reinforcing conservation practices, reviving traditional practices, and raising awareness, have played a vital role in ensuring the survival of these traditions. Government initiatives have also supported the preservation of their age-old knowledge systems. Notably, people made *conscious* decisions to assist and facilitate communities in rediscovering and sustaining these practices.

## 6.2 Journey of Environmental Action

There is now widespread recognition of the need to address the effects of modernisation and development expansion on the deeply interwoven natural and cultural fabric of the JJLCCL, as the decline of traditional structures and knowledge shows.

A Jingkieng Jri/living root bridge team at the Meghalaya Basin Management Agency (MBMA) became actively engaged in community sensitisation, research, and documentation, encouraging and enabling community-led environmental action and promotion of the JJLCCL. The approach involved a dynamic process of planning, experimentation, and innovation, incorporating both traditional wisdom and contemporary thinking and methods.

### 6.2.1 Community Sensitisation

Since 2018, community dialogues have played a pivotal role in strengthening the preservation of the JJLCCL, demonstrating both depth and intensity in engagement. A key aspect of the approach was working closely with those communities that had been nurturing and maintaining these unique living structures and their cultural associations for generations.

Various stakeholders, including village elders, bridge custodians, local authorities, and landowners, participated in community dialogues comprising 367 formal meetings. These dialogues provided a platform to exchange knowledge, document oral histories, and discuss sustainable conservation strategies.

In addition to formal discussions, numerous informal engagements took place, ranging from casual conversations with community members (*Lyngwer dpei*) to interactive field visits and storytelling in a continuous cycle of learning and unlearning.

These interactions helped bridge communication gaps, build trust and encourage active community involvement in preserving the JJLCCL.

Through this multi-faceted and deeply engaging approach, the intensity of the dialogues since 2018 has not only strengthened local stewardship but has also set a precedent for other collaborative and culturally sensitive heritage conservation efforts.

### 6.2.2 Formation of Primary Cooperative Societies and State-Level Federation

Patient and continuous community sensitisation sparked a powerful movement, transforming passive awareness into dynamic action. The urgency to protect cultural and natural heritage drove communities to organise themselves. This led to the formation of 29 primary cooperatives focusing on conservation, research and development. Members include farmers, fishermen, local heads, young people, and women.

These cooperatives came together in a federation at the state level under the *Syrwet U Barim Mariang Jingkieng Jri* – Cooperative Federation Ltd *(SUBMJJCFL)*. It became a collection of communities from across five blocks – Mawsynram, Shella Bholaganj, Laitkroh, Pynursla, and Amlarem – in the East Khasi Hills and West Jaintia Hills District.

The goal was to promote best practices from various communities, fostering collective learning.

### 6.2.3 Conservation and the Community-Led Landscape Management Project (CLLMP) for Cooperatives

Alongside promoting cooperative growth, a Community-Led Landscape Management Project (CLLMP) played a vital role in the facilitation of conservation and preserving traditional knowledge. To safeguard the region's rich cultural and natural heritage, several community-led activities came up, such as the conservation of the *Jingkieng Jri/Lyu Chrai* (living root bridges), the revitalisation of the *Ïing Mariang* (nature homes), and the *Kper Sara* (community nursery) initiative.

### 6.2.3.1 Conservation of Jingkieng Jri/Lyu Chrai

Through conservation activities of the *Jingkieng Jri/Lyu Chrai*/living root bridges, by engaging in their maintenance and restoration, communities not only protected an integral part of their natural heritage but also could pass down essential knowledge and skills to younger generations. Elders in the community would serve as invaluable mentors and sources of traditional knowledge. They would teach youth about generations-long cherished methods of creating and maintaining these living bridges, as well as about their deeper cultural significance.

This intergenerational learning forged a strong bond between the past and future, ensuring that such conservation practices would be sustained for years to come.

These focused efforts resulted in the conservation of 74 living root bridges.

### 6.2.3.2 Ïing Mariang (Nature Home) Revitalisation

While documenting traditional knowledge, it became clear that key indigenous practices, craftsmanship and storytelling traditions were beginning to fade. And fade fast, alarmingly so.

To deal with it decisively, people initiated the *Ïing Mariang*, or nature homes – a space dedicated to sustaining and reviving traditional knowledge, lifestyle embedded in it.

The concept of revitalising nature homes also envisioned a community learning space, where ideas of traditional knowledge and nature conservation could be shared in today's day and age.

Through the CLLMP, 26 *Ïing Mariang* came into being, spread across Shella Bhloaganj, Pynursla, Laitkroh and Amlarem Blocks of the East Khasi Hills and West Jaintia Hills District.

### 6.2.3.3 Kper Sara (Community Nursery) Initiative

The initiative entailed regenerating indigenous plant species, especially those vital for the growth of the *Jingkieng Jri/Lyu Chrai,* so distinctive to Meghalaya. It also aimed to enhance native tree cover and protect natural resources from depletion, while addressing issues like soil erosion, landslides, and biodiversity conservation.

The significance of these nurseries grew in response to climate variability, forest depletion, and an increasing need to improve agricultural and horticultural practices.

Unlike other nurseries, the *Jingkieng Jri* community's nursery was an experiment. Here, people applied traditional methods and innovative ideas from within the communities themselves. They used natural resources only, that is, native plant species such as *tyndong* (bamboo culms), *kriah* (bamboo baskets), *dopwai* (areca leaf sheath), and *sla pashor kait* (banana leaf).

This strategy not only reduced the use of plastic materials in nursery operations but also promoted traditional practices to address plastic waste.

This CLLMP initiative brought about the establishment of 25 community nurseries.

### 6.2.4  Research and Documentation: Safeguarding Traditional Knowledge

Along with practical conservation and ecological restoration efforts, the CLLM Project took up research and documentation to preserve and safeguard traditional knowledge.

As an oral tradition, knowledge used to be transmitted through stories, songs, observations, and other traditional art forms. To grasp the significance and relationship of oral stories and narratives within the cultural landscape, community members and the *Jingkieng Jri* Team made efforts to put together a comprehensive documentation.

A detailed analysis of audio-visual data regarding the JJLCCL enabled learning and recording of profound lessons in resilience, environmental stewardship, and community-led conservation practices. These insights not only contributed to the preservation of the JJLCCL but also served as a model for sustainable development and ecological balance.

The knowledge gained from this process can be applied more broadly to foster a future where cultural heritage and nature coexist in harmony, inspiring sustainable practices across the globe.

### 6.2.5  Strengthening Knowledge through Research Collaboration

Building upon these documentation efforts, collaborative research has enhanced understanding and promoted perpetuation of traditional ecological knowledge. A national *Jingkieng Jri* participatory research conference in 2021 provided a momentous platform, bringing together researchers and experts to assess the ecological significance of Meghalaya's *Jingkieng Jri/ Lyu Chrai*/living root bridges.

These were the key participating institutions:

The Botanical Survey of India – focused on the associated flora of the *Jingkieng Jri/Lyu Chrai*.

The Geological Survey of India – analysed a Jingkieng Jri / Lyu Chrai vulnerability assessment.

The Forest Research Institute – assessed the health status of the living root bridges.

The Indian Institute of Science, Bangalore – studied ecology and genetics of the *Ficus elastica* in relation to the living root bridges.

The Zoological Survey of India – concentrated on the faunal diversity connected to the living root bridge ecosystem.

The **Centre** for Environmental Planning and Technology **University.**

This constructive interaction between community-led conservation and research-driven insights not only strengthened conservation efforts but also deepened scientific understanding, ensuring long-term sustainability of Meghalaya's unique living heritage.

### 6.2.6 Mariang Festival

The momentum of community-led conservation and research collaborations stimulated the organisation of a *Mariang tamasa* (*Mariang* festival) across three blocks in the JJLCCL. It also aimed to encourage community participation and open a platform for innovations and revitalisation of traditional practices through nature-based resources and skills.

An exchange of ideas, practices, knowledge, and skills allowed the younger generation to expand their knowledge through wise elders. Celebrating a close interrelationship between nature, culture, and community, the festival inspired a collective dedication to safeguard traditions and work towards a sustainable future.

### 6.2.7 Nomination for UNESCO World Heritage Site

A nomination process was initiated in recognition of the global importance of preserving this cultural landscape with its outstanding universal value. In March 2022, the *Jingkieng Jri/Lyu Chrai* Cultural Landscape was included in the tentative list for UNESCO World Heritage status.

The nomination was the result of the active participation of the indigenous communities, who had been nurturing and maintaining these bridges along with associated attributes, shaping the overall cultural landscape for generations. They played a vital role in advocating for the recognition of this cultural heritage through documentation, awareness workshops, policy-making, and collaboration with government bodies, as well as researchers. Their involvement has ensured that traditional practices remain central to conservation strategies, simultaneously promoting sustainable practices.

By participating in the nomination process, they had sought to safeguard their ancestral knowledge and secure better protection for their natural and cultural heritage.

### 6.2.8 *Empowering Communities through Capacity Building*

To further strengthen conservation efforts and support the nomination process, a comprehensive capacity-building programme was undertaken. This empowered indigenous communities with the necessary skills and knowledge. Intensive community-led mapping using Geographic Information Systems made a significant contribution in this respect.

As a result of training, people became well-equipped to use GPS equipment, precisely mark the coordinates and construct a substantive cultural landscape map. By integrating traditional knowledge of demarcation with modern mapping techniques, they achieved a more precise and holistic understanding of the landscape's cultural and environmental significance.

### 6.2.9 *Learning Visits for Indigenous Community Members to UNESCO World Heritage Sites*

With sponsorship from the CLLMP, community representatives were able to explore a number of cultures and sites that had achieved UNESCO World Heritage status. This way, they gained valuable knowledge and insights with respect to the best practices for preserving natural and cultural heritage elements and locations. These experiences also deepened their understanding of heritage protection by drawing lessons from globally recognised efforts.

These were the sites of their exposure visits:

- The Kakatiya Rudreshwara (Ramappa) Temple, Telangana, India.
- Rock shelters of Bhimbetka, India.
- The Subak System in the cultural landscape of Bali Province, as a manifestation of the Tri Hita Karana philosophy, Indonesia.
- The Greater Blue Mountains, Australia.
- The Sydney Opera House, Australia.
- The Wadi El Hitan, Egypt.
- Ancient Thebes with its Necropolis, Egypt.
- Nubian Monuments from Abu Simbel to Philae, Egypt.
- Historic Cairo, Egypt.
- Memphis and its Necropolis – the pyramid fields from Giza to Dahshur, Egypt.

### 6.2.10 *Memoranda of Understanding*

Various Memoranda of Understanding (MOUs) created a collaborative framework that integrated traditional knowledge with scientific research for the protection and management of the JJLCCL.

They included MoUs:

- With community representatives from the cultural landscape of Bali Province, Indonesia, with reference to the Subak System.
- With the Geological Survey of India for a vulnerability study of the living root bridges of Meghalaya.
- Between the Sywet U Barim Mariang Jingkieng Jri Co-operative Federation Ltd and the Indian Institute of Science (IIS), Bangalore, for a study on the ecology and genetics of the *Ficus elastica* in relation to the living root bridges.

The agreement with the Bali Province facilitated cultural exchange and shared learning on sustainable heritage practices. The MOU with the Geological Survey of India aimed at a scientific understanding of the structural and environmental challenges of living root bridges. And the partnership between the Cooperative Federation and the IIS, Bangalore, focused on the *Ficus elastica*, considered the cornerstone species of the living root bridges.

These collaborations allowed for an integrated approach, combining indigenous wisdom with scientific expertise to develop informed and sustainable protection and management strategies for the cultural landscape.

### 6.2.11 Cooperation with Different Government Departments

Recognising the need for a collaborative approach to heritage protection, concerned parties made efforts to get various government departments on board to support JJLCCL's conservation and sustainable management activities.

Representatives from the Khasi and Jaintia as well as from the Cooperative Federation, and officials from various state government departments came together in an all-stakeholder meeting on 30 June 2022. The Chief Secretary of Meghalaya held the chair.

The gathering gave attention to the importance of raising awareness about the significance of the Jingkieng Jri/Lyu Chrai Cultural Landscape. It also underscored the essence of its conservation and responsible development in the context of the UNESCO nomination.

### 6.2.12 Strengthening Natural Resource Management Initiatives

Following the June summit, concerted efforts were made to extend sustainable, natural resource management (NRM) practices to more villages within the landscape. Acknowledgement of the crucial role of local communities in conservation was the first step. Then, a series of workshops provided village representatives with the necessary knowledge and skills to adopt eco-friendly NRM practices. Given its significant impact on rural development, activities under the Mahatma Gandhi National Rural Employment Guarantee Act actively encouraged NRM initiatives, promoting sustainable growth and community-driven stewardship.

### 6.2.13 *Shlem Jingtip (Space for Learning)*

Building upon the efforts to strengthen NRM practices, communities also developed initiatives through the *Shlem Jingtip*. This 'space for learning' innovation had sprouted from the creative minds of some indigenous people. It aimed at preserving, studying and protecting intangible knowledge systems present throughout the landscape, so much so that their early traditions and knowledge would continue to thrive across generations.

As these tangible-intangible aspects are preserved and recognised, the community's natural and living heritage will continue to flourish. Emphasising values, traditions and knowledge to the younger generation is crucial, since introducing conservation and cultural pride early in life helps such values take root over time. At the same time, the invaluable insights of knowledge keepers constitute a significant component of indigenous history.

One Shlem Jingtip space area came into being under the CLLMP for the general interest of indigenous communities living across the landscape.

### 6.2.14 Guidelines for the Protection of the Cultural Landscape through the Dorbar System

Various factors affect the integrity of the landscape, such as evolving development needs, environmental degradation, natural disasters, and other anthropogenic activities. In view of these factors, indigenous inhabitants of the state, through traditional institutions known as the *Dorbar*, were already actively occupied with safeguarding the Jingkieng Jri/Lyu Chrai Cultural Landscape. With a deep sense of responsibility and indigenous knowledge systems, they had formulated and implemented guidelines that prioritise conservation.

To ensure sustainable conservation and prevent environmental degradation, they had established strict rules for land use, deforestation, and waste and water resource management. Traditional ecological knowledge was integrated into these regulations, promoting sustainable harvesting of forest resources while maintaining forest conservation to achieve ecological balance.

Further, to deal with tourism pressures, the Dorbar regulated visitor numbers and developed eco-tourism policies that would contribute to maintenance and restoration. Waste disposal and pollution control measures were formulated to prevent environmental harm. Educational initiatives, including awareness programmes, workshops, and knowledge-sharing sessions, in the 'Space for learning' helped in transferring indigenous wisdom to younger generations, ensuring that the cultural and ecological significance of these landscapes would be understood and respected.

### 6.2.15 Indigenous Community Representatives and MBMA Team Members in Global Forums

Indigenous community representatives and MBMA team members had an opportunity to participate in global forums where reps could share the significance of the Jingkieng Jri/Lyu Chrai Cultural Landscape with a global audience. These engagements not only highlighted its unique heritage but also helped in meaningful exchanges around the world. Participants, in turn, brought home fresh knowledge, strengthening local conservation efforts, and reinforcing the importance of protecting traditional wisdom in an evolving world.

The forums included the following:

- ICOMOS General Assembly 2023, Sydney, Australia: Session and Poster on 'Learning's from Jingkieng Jri/Lyu Chrai Cultural Landscape.
- COP 27, Egypt: Presentation on 'Nurturing equal value for all: Complex problem solving through a collective approach.'
- UNESCO Sub-regional Conference on World Heritage, titled 'The next 40 ways forward for South-Asia world heritage, Bhopal, India'; presentation on 'Jingkieng Jri/Lyu Chrai Cultural Landscape.'
- ICOMOS India, North East: Presentation on 'Jingkieng Jri/Lyu Chrai Cultural Landscape.

### 6.2.16 Proposed Amendment of Heritage Act, 2012

The Meghalaya Heritage Act, 2012, brought out to protect heritage sites in the state, was proposed for an amendment. The text was to give a better reflection of joint community and government legal action for the preservation of heritage places in the state for the long-term sustainability of the cultural landscape of the state as a whole.

## 6.3 Limitation and Challenges

The journey of conservation efforts, while marked by dedication and progress, has not been without challenges. Still, each serves as a learning opportunity and a stepping stone to resilience and adaptive preservation.

### 6.3.1 Community Engagement

While the involvement of the SUBMJJCFL was paramount in conservation efforts for the whole cultural landscape, ensuring genuine, consistent, and widespread community participation was challenging at times. There have been varying levels of challenges and understanding among them regarding the importance of preserving their natural and cultural heritage.

For instance, some villages with living root bridges have not yet formed cooperatives, for various reasons, including differing priorities and lack of awareness.

### 6.3.2 Remoteness of the Location

The areas where the Jingkieng Jri/Lyu Chrai and its associated attributes are found are often in remote regions with a rugged topography and steep slopes. Poor connectivity and the lack of proper roads make it difficult to access such locations.

### 6.3.3 Dispersed Locations

The JJLCCL sites are spread across various villages, that too, in two districts of the East Khasi Hills and West Jaintia Hills. While it is a testimony to diversity, this scatteredness also creates a complex of problems regarding community sensitisation activities, for each village has different traditions and governance structures. More so, levels of awareness vary because of differences in language and dialects.

### 6.3.4 Cultural Sensitivity

Striking the right balance between modern methodologies and indigenous knowledge can be difficult. This is particularly so when external expertise may clash with traditional ways of doing things.

### 6.3.5 Community Fear of Government Intervention

Many local communities are apprehensive about government-led documentation and intervention actions. Misconceptions about conservation efforts turning into land acquisition sometimes fuel distrust and even hostility to or rejection of conservation notions and practices.

### 6.3.6 Balancing Development and Conservation

Cultural influences worldwide have significantly led to an exchange of cultural ideas delving into modernity and developments contributing to improved livelihoods, lifestyles, health, and connectivity. While these changes bring significant benefits, they also present challenges in conservation and preserving the tangible heritage of a region.

For example, the adoption of reinforced concrete structures, metal wire fences and other modern infrastructure practices alters the JJLCCL and diminishes the authenticity and overall integrity of the cultural landscape.

Despite these challenges, the momentum of community-led conservation continues to grow, driven by a shared commitment to safeguarding both cultural and natural heritage. Addressing these challenges requires continuous dialogue, awareness building, and collaborative efforts that respect indigenous knowledge while incorporating sustainable solutions.

## 6.4 Conclusion

The Jingkieng Jri/Lyu Chrai Cultural Landscape (JJLCCL) stands as an outstanding example of how indigenous communities, through deep-rooted traditions and knowledge combined with contemporary science, can lead environmental actions while preserving their cultural identity.

Through a community-led landscape management project, communities came up with a sustainable approach to environmental stewardship, blending traditional wisdom with contemporary methods to safeguard the landscape. From the restoration and protection of living root bridges to the establishment of nurseries for native plant species, their efforts have not only contributed to ecological balance but have also fostered socio-economic growth by promoting sustainable tourism and livelihood opportunities.

The success of these initiatives lies in the active engagement of the local people, guided by their belief in the sacredness of nature. The formation of cooperatives and the creation of a state-level federation strengthened their collective action, laying the foundation for long-term sustainability. Collaborations with research institutions have brought scientific rigour to the conservation process. The inclusion of the JJLCCL in the tentative list for UNESCO World Heritage status is a testament to the importance of preserving this cultural and ecological treasure.

Despite facing challenges such as balancing growing development, modernity and maintaining cultural integrity, the communities continue to adapt and evolve their strategies. Their efforts to implement guidelines for conservation through the Dorbar system and engage in policy discussions with the government ensure that the landscape's integrity is protected for future generations. The active involvement of government agencies and the strengthening of the Heritage Act have contributed to the protection of the JJLCCL.

For the community, protecting the landscape is not just about conservation; it is about preserving their identity and honouring the wisdom of their ancestors. By working together, passing down knowledge and encouraging environmental guardianship, indigenous communities prove that their traditions, resilience, and commitment to conservation continue to safeguard their sacred land, so much so that their deep-rooted connection with nature remains unbroken.

# 7

# SUPPORTING GRASSROOT INNOVATIONS

*Veveane Sayo, Delanmi Massar, and Gordon Steven Pde*

## 7.1 Introduction

### 7.1.1 Purpose, Scope and Significance

Innovation is essential to achieve sustainable development, especially in ecologically and culturally rich regions such as Meghalaya. The state has a unique land tenure system, where traditional institutions, clans and communities own and manage over 90% of the land. This makes community-led interventions crucial for long-term sustainability.

Unlike most parts of India, where land is state-controlled, Meghalaya's governance structure follows the Sixth Schedule of the Indian Constitution. It grants autonomous status to local tribal councils in managing land, resources and customary laws. This decentralised governance model empowers communities but also creates challenges in policy implementation, resource management and economic development.

The Meghalaya Community-Led Landscape Management Project (CLLMP) was set up with support from the World Bank. It prioritised grassroots innovations through its Innovations Fund, recognising their potential in addressing environmental challenges, preserving traditional knowledge and strengthening rural livelihoods.

Given Meghalaya's high vulnerability to climate change, deforestation, soil erosion and water scarcity, grassroots solutions are vital in ensuring sustainable land use and livelihood security. The state realised the need for a dedicated platform to identify, support and scale community-driven natural resource management (NRM) initiatives. It initiated the fund to bridge the gap between policy and on-the-ground solutions. This way, local

DOI: 10.4324/9781003735380-8

conservation efforts would definitely receive the visibility and resources necessary for long-term success.

The project aimed at enabling communities to take an active role in managing their natural resources while enhancing natural resource management (NRM) by integrating traditional ecological knowledge with modern sustainability practices. It further sought to strengthen rural economies by supporting low-cost, high-result innovations. It would also **provide financial and technical support** to scalable, community-led initiatives that tackle local environmental and socio-economic issues.

Recognising the untapped potential of community-driven solutions, the CLLMP *introduced the Innovations Fund to*:

- Encourage community-led conservation efforts by providing financial and technical support to promising initiatives.
- Preserve and revitalise traditional ecological knowledge by integrating indigenous practices into modern sustainability frameworks.
- Foster rural entrepreneurship and strengthen local income development through innovations in agriculture, eco-tourism and environmental management.
- Enhance climate resilience by promoting scalable, nature-based solutions that mitigate the impact of environmental degradation.

Meghalaya's hilly terrain, high annual rainfall and rich biodiversity present both opportunities and challenges for grassroots innovations. The dependence on community-based governance required an approach that respects tribal autonomy, simultaneously considering sustainable development. By supporting indigenous innovations, the project aimed to demonstrate that traditional knowledge, when combined with modern technology and financial assistance, can lead to sustainable, scalable solutions for Meghalaya's developmental and environmental challenges.

In this chapter, **various** grassroots innovations will come to the fore that have received funding under this Fund. They cover diverse sectors, including sustainable agriculture, biodiversity conservation, waste management, eco-tourism and traditional knowledge preservation. By documenting these cases, we can highlight how community-driven interventions can lead to systemic change. In the bargain, they are strengthening the economic and environmental resilience of Meghalaya's indigenous communities.

### 7.1.1.1 Why Does the CLLMP Support Grassroots Innovations?

The project works with the realisation that indigenous communities predominantly manage the state's forests and natural resources. But rapid deforestation, soil erosion and the growing consequences of climate change are now posing significant threats to their traditional ways of life. For centuries, they had relied on age-old environmental stewardship practices. Now, despite their effectiveness, these practices struggle to keep pace with modern environmental and socio-economic challenges. Recognising the need to support these community-driven solutions, CLLMP launched the Innovations Fund as a dedicated platform to provide financial and technical assistance to local innovators working on conservation, livelihood generation and climate resilience.

A core initiative in the context of grassroots innovations under the CLLMP is to **promote climate resilience** by enabling communities to adapt to environmental challenges through sustainable, low-cost solutions. Many of these innovations focus on **disaster risk reduction, soil conservation, sustainable farming and biodiversity conservation**. This has been helping them remain resilient in the face of climate-related adversities such as erratic rainfall, floods and soil degradation.

By supporting these initiatives, the **fund** has been promoting **adaptive capacities, strengthening food security and fostering long-term environmental sustainability** for Meghalaya's indigenous communities.

The fund serves multiple objectives. Firstly, it aimed to empower rural communities by strengthening their conservation efforts. While many villagers already practise sustainable resource management, they often lack the financial means or institutional support to scale its effect. Through targeted funding and mentorship, the project has been supporting initiatives in this context to expand and become self-sustaining models of environmental protection.

Secondly, the fund was designed to revive and modernise indigenous knowledge by integrating it into contemporary sustainable development frameworks. Meghalaya has a wealth of traditional wisdom related to forest management, water conservation and biodiversity protection, much of which is at risk of being lost due to urbanisation and changing socio-economic conditions. By identifying and supporting grassroots innovations that carry on this knowledge, the project ensures that time-tested ecological practices remain relevant and beneficial in today's context.

Another key motivation behind this initiative is the need to boost rural economies through innovative livelihood strategies. Many of the funded projects focus on sustainable agriculture, eco-tourism and environmentally friendly enterprises, enabling communities to generate income while maintaining ecological balance. By supporting such models, the fund contributes to economic self-reliance in rural Meghalaya, reducing dependency on external aid and unsustainable resource extractions.

Besides, climate resilience is a critical component of the initiative. Meghalaya's fragile ecosystem is highly vulnerable to extreme weather events, including landslides, floods and erratic rainfall. Many grassroots innovations supported under the CLLMP focus on reinforcing local adaptability to climate shocks, whether through alternative farming techniques, water conservation efforts or disaster risk management strategies. Strengthening these grassroots solutions helps communities develop long-term coping mechanisms against environmental uncertainties.

Finally, the Innovations Fund was designed to bridge the gap between traditional wisdom and contemporary policy. As a consequence, successful community-led models are not only supported but also replicated on a larger scale. By integrating these grassroots innovations into broader state and national frameworks, Meghalaya has set a precedent for inclusive, bottom-up development, where communities are not just beneficiaries but also active partners in shaping sustainable futures.

### 7.1.2 Organisation of the Chapter

The chapter is structured as follows:

- **Section 7.1** provided an overview of the purpose and scope of the grassroots innovation initiative.
- **Section 7.2** details the research methodology to document, evaluate and analyse selected innovations.
- **Section 7.3** presents key findings, highlighting the effect and success of various grassroots innovations.
- **Section 7.4** discusses the broader implications of these findings, including their replicability, policy relevance and alignment with existing literature.
- **Section 7.5** summarises key takeaways and outlines potential directions for future research and scaling of grassroots innovations.

### 7.1.3 Review of Literature

The importance of grassroots innovation in sustainable development has been widely recognised in academic and policy literature. Grassroots innovations, in which local communities often take the lead, address pressing environmental, social and economic issues by leveraging traditional knowledge and low-cost, sustainable solutions. Studies revealed that community-led approaches to natural resource management (NRM) often yield more sustainable and effective outcomes than any other intervention.

In Meghalaya, a unique community-based NRM system governs 90% of the state's forests, with traditional tribal institutions playing a central role. Research suggests that empowering these institutions with financial and technical support can significantly enhance environmental conservation efforts (Agrawal & Chhatre, 2006). Additionally, studies on innovation diffusion indicate that funding, documentation and policy integration are crucial for scaling grassroots solutions beyond their initial communities (Smith et al., 2014).

The support for grassroots innovations aligns with Meghalaya's broader efforts in fostering local development through initiatives like the Meghalaya District Innovation Fund (DIF), an initiative under the 13th Finance Commission. The DIF has provided crucial financial assistance to district-level projects focusing on socio-economic development, technological advancements and environmental conservation. The fund has played a pivotal role in supporting community-led solutions and ensuring that local governance structures have the resources to implement meaningful projects (Gupta 2019).

By linking the Innovations Fund with existing mechanisms like the DIF, Meghalaya has been creating a robust ecosystem for grassroots-led sustainable development, ensuring that innovative ideas are not just piloted but also scaled effectively across districts.

### 7.1.4 Identification of Gaps

A major challenge in scaling grassroots innovations is their limited scalability because there are no mechanisms that facilitate expansion and standardisation at the state or national level. Many innovations remain confined to local contexts, since communities often lack the necessary resources and institutional backing to implement them on a broader scale.

A key issue observed during field visits was insufficient financial and technical support on an ongoing basis. While the Innovations Fund provides crucial initial assistance, many communities struggle to sustain their projects due to the absence of long-term funding, mentorship and skill-building initiatives. Without sustained investment and structured training programmes, several promising innovations face the risk of stagnation or even fizzling out.

Another significant concern is the lack of documentation and knowledge transfer. Traditional practices are often passed down orally and lack formal records, making it difficult for these innovations to be replicated in different regions. Without systematic documentation, many valuable techniques and useful know-how risk being lost over time. This is particularly so as younger generations tend to migrate to urban areas and seek other livelihood opportunities.

A final and crucial gap is the lack of awareness about funding mechanisms such as the Meghalaya District Innovation Fund. Many communities, particularly those in remote areas, have proven unaware of such financial support systems. This is most likely a result of limited publicity and small outreach of such funds.

Besides, some villages appeared hesitant in venturing beyond traditional agriculture and government schemes, preferring familiar economic activities over experimental innovations.

A low literacy level and lack of exposure to alternative livelihood strategies further restricted participation in innovation-based projects, leading to underutilisation of available resources.

Addressing these gaps is essential to ensure that grassroots innovations move beyond pilot projects and become catalysts for widespread socio-economic and environmental change (Luca et.al 2013).

## 7.2 Methodology of the Case

### 7.2.1 Description of Methodology

The methodology for identifying and implementing grassroots innovations under the CLLMP Innovations Fund was structured to ensure community participation, transparency and the selection of viable innovations. The process began with a statewide awareness campaign aimed at encouraging participation from diverse community innovators. Awareness initiatives included community engagement programmes, newspaper advertisements and workshops at block and village levels to educate local stakeholders about the Innovation Fund's objectives and application process.

Once applications were received, a multi-tiered selection process started. Initially, a technical team reviewed applications to assess feasibility, innovation and potential impact. Then, shortlisted proposals were presented before a State-Level Innovation Committee. It included government officials, subject-matter experts and representatives from the CLLMP. The final two stages involved approval from the Additional Project Director and the Project Director.

This way, the most promising and viable innovations were sure to receive funding and implementation support.

After approval, selected innovations were subjected to a structured implementation and monitoring process. This included field visits, financial assessments and impact evaluations to ensure proper fund utilisation and scalability support. Progress tracking was conducted through structured reporting mechanisms, allowing for continuous assessment of challenges and best practices.

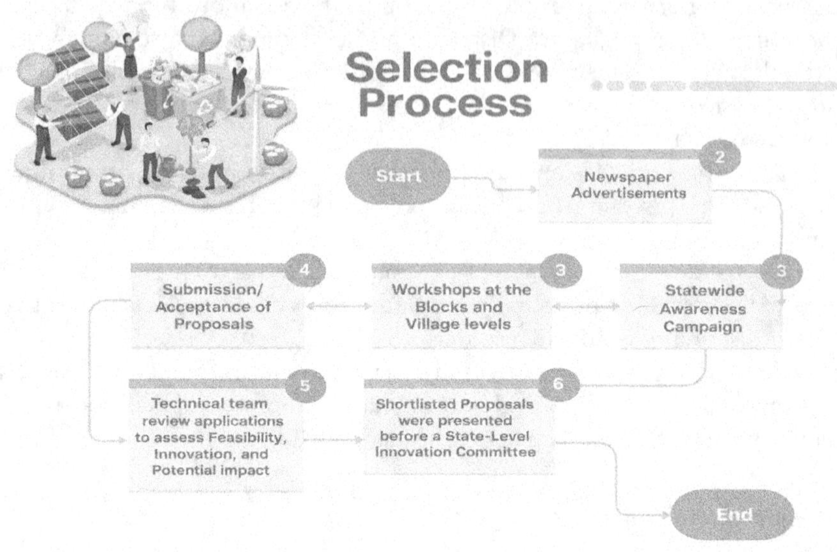


**FIGURE 7.1**   Key stages of selection process

*Source:* This image has been developed by the authors themselves, highlighting the various steps of the selection process.

The participatory nature of this methodology ensured that communities played an integral role in both the innovation process and decision-making. This brought about a sense of ownership and commitment to long-term sustainability on their part.

### 7.2.2 Justification of Methodology

The selection and implementation of grassroots innovations under the CLLMP Innovations Fund followed a structured and transparent methodology to ensure fairness and efficiency in the allocation of resources. A State-Level Innovation Committee oversaw the selection process to eliminate any potential bias and ensure that promising projects received adequate support. This committee comprised experts from diverse fields, effectuating a multidisciplinary approach to evaluating applications.

A key aspect of this methodology was that innovators were not restricted to CLLMP villages. Instead, the initiative sought to identify and support the best ideas, regardless of their origin. As a result, applicants from both CLLMP and non-CLLMP villages, who met the eligibility criteria, were considered. This way, the initiative remained inclusive and merit-based, allowing for broader participation and encouraging individuals and communities from various backgrounds to contribute innovative solutions.

Once selected, a contract signing process formalised the engagement between the CLLMP and the innovators, clearly defining expectations, deliverables and the financial structure. Funding came in two tranches, with the first instalment allowing the innovators to initiate their projects.

The second tranche was only released after thorough monitoring and verification by the State Project Management Unit (SPMU) and the District Project Management Unit (DPMU). Such evaluations made certain that the project was progressing as planned, achieving the expected result, while using funds effectively.

This stringent monitoring framework emphasised accountability and responsible fund utilisation, reinforcing the credibility of the initiative.

A key component of financial accountability was the rigorous verification of Utilisation Certificates (UCs). Every fund disbursement required proper documentation and justification of expenditures, ensuring that the allocated funds were being used efficiently and for their intended purpose. The SPMU and DPMU teams meticulously reviewed each UC before proceeding with subsequent funding. This way, they secured transparency and prevented financial mismanagement.

Importantly, stakeholder representation in the selection process played a crucial role in maintaining fairness and objectivity. The committee included professionals from natural resource management, entrepreneurship, environmental conservation and social impact assessment. This made for a well-rounded evaluation process. The diverse composition of the selection panel ensured that projects were assessed from multiple perspectives, strengthening the legitimacy of the selection process.

This approach guaranteed that grassroots innovations were selected on merit, nurtured through structured financial support and monitored to achieve long-term success. By embedding strong evaluation mechanisms, unbiased selection criteria as well as continuous engagement with innovators, this approach not only facilitated the identification of high-impact innovations but also created a scalable model for future initiatives in Meghalaya and outside the state.

### 7.2.3 *Limitations and Challenges*

Despite this comprehensive approach, there were several challenges:

1. **Limited access to data** – Many grassroots projects lacked proper financial records, requiring reliance on self-reported information.
2. **Geographical constraints** – Several innovations were located in remote areas, making field visits logistically difficult and time-consuming.
3. **Short timeframe** – Some initiatives were in the early development stages, making it challenging to assess long-term impact.

4. **Lack of standardisation** – Innovations varied widely in scope, making direct comparisons difficult.
5. Initially, when these innovations were brought up, most of them were **still at the budding stage**. The project had to work from scratch to make sure their ideas and innovative inventions would prove successful.

Despite these challenges, the methodology successfully provided a detailed, authentic and community-centred understanding of grassroots innovations under the CLLMP.

## 7.3 Results

### 7.3.1 Main Case Findings 1: Economic and Social Consequences of Innovations and Initiatives

#### 7.3.1.1 Enhance Local Growth and Livelihood Generation

The CCLMP-supported innovations have had a profound effect on local economies. They helped create income-generating opportunities, reduce wastage and promote financial stability for families across Meghalaya. Many innovations introduced cost-effective and scalable solutions accessible to rural entrepreneurs, self-help groups and small-scale farmers.

#### 7.3.1.2 Plastic-to-Fuel Conversion (Just Synrem): Turning Waste into Revenue

Just Synrem's plastic-to-fuel innovation **has converted** over 2,500 kg of plastic waste into fuel since this project was undertaken with the assistance of CLLMP , tackling the severe issue of plastic waste accumulation in Shillong. What makes this project remarkable is not just its environmental benefit, but also its economic potential. By producing fuel locally, Synrem established a self-sustaining revenue model, earning Rs. 400,000 annually. Moreover, the project created 10 full-time jobs for unemployed youth, many of whom had been casual labourers or in financial distress.

#### 7.3.1.3 Mud Beehives: Sustainable Beekeeping for Profit

Lamuni Sangma's mud beehives have revolutionised beekeeping in Meghalaya: They offer a cost-effective alternative to wooden beehives while boosting honey production by 40%. The ability of these beehives to regulate internal temperature has led to healthier bee colonies, increasing the annual yield per beekeeper by Rs.30,000. Farmers like Lamuni, who once struggled to make ends meet, now have a reliable source of income. It also allows them to invest in better farming tools and education for their children.

### 7.3.1.4 Stephan Shadap's Beekeeping Innovation

Stephen Shadap's beekeeping has had significant social, economic and environmental consequences. His eco-friendly straw hive technique promotes sustainability while he is preserving traditional knowledge through his Traditional Bee Museum cum Training Centre, fostering community engagement and providing new livelihood opportunities. Training and outreach programmes have empowered local beekeepers, bringing them skill development and knowledge.

Economically, the initiative improved honey production and market access and supported rural incomes, with income generated from honey sales steadily increasing. With the production of over 100 eco-friendly bee hives, the project has enabled sustainable beekeeping.

Environmentally, this eco-friendly beekeeping innovation reduces harm to ecosystems and promotes biodiversity. It has received positive feedback for its sustainability and practicality. Beekeepers noted that the basic components of the design extended a hive's lifespan while maintaining consistent honey yields. According to a local practitioner: 'This beehive, made of straw and mud, is sustainable because it eliminates the need for wood, reducing deforestation. I have been using these materials for the past two years and continue to harvest honey without disturbing the bees.'

Such testimonials bring out this hive innovation's reusability. It is an effective model for long-term, eco-conscious apiculture in Meghalaya and beyond.

### 7.3.1.5 Zero-Energy Cold Storage Initiative: Reducing Losses and Increasing Profits

For farmers in Meghalaya, one of the biggest challenges has been the spoilage of fresh produce due to the absence of refrigeration. The introduction of a zero-energy cold storage system has provided a game-changing solution.

Over 50 farmers adopted this low-cost storage method, cutting their post-harvest losses by 30% and increasing their annual earnings by Rs.15,000 per farmer. The system does not require electricity. This makes it ideal for remote villages with an unreliable power supply. Moreover, it is suitable for a wide range of vegetables like spinach, tomatoes, aubergine, carrots, cabbage and leafy greens. Onions and potatoes require low humidity, so it would not work.

## 7.3.2 Main Case Findings 2: Environmental Benefits and Sustainability – MBMA's Initiatives

### 7.3.2.1 Turning Waste into Usable Resources

Many grassroots initiatives have transformed waste material into functional, eco-friendly products, directly addressing issues of waste management and environmental degradation in Meghalaya.

### 7.3.2.2 Bottle-Brick Initiative: A Community-Driven Approach to Recycling

A bottle-brick initiative repurposed over 3,000 kg of plastic waste: It turned discarded bottles into construction bricks for eco-friendly infrastructure. This initiative has been instrumental in reducing non-biodegradable waste, particularly in areas where proper waste disposal is lacking.

The success of this project has encouraged local schools and communities to participate in waste collection efforts, contributing to a culture of environmental responsibility.

### 7.3.2.3 Organic Pest Control: Reducing Chemical Dependency in Agriculture

Jorsing Syngkli had developed an organic bio-pesticide using indigenous plant extracts. This proved a sustainable alternative to chemical pesticides. Over 200 farmers adopted this product, and they experienced a 20% increase in crop yield and significantly improved soil health.

This initiative has reduced farmers' dependency on costly chemical pesticides, helping them save money while protecting the environment.

### 7.3.2.4 Reclamation of Mining-Affected Land

Smt. Kyrsiew Ryngkhlem has been practising organic farming for nearly 16 years, driven purely by her passion for sustainable agriculture. In 2014, she purchased a plot of land that had been severely degraded due to coal mining activities. Determined to restore its fertility, she began experimenting with various reclamation methods, including the use of lime and planting different types of vegetation.

Through continuous trials, she discovered that the most effective method was the use of organic manure, which she produced herself. This manure was made from a mixture of grass, weeds, areca nut husks, animal waste and other biodegradable materials she collected. Composting takes over a month, but results have been transformative. By using this organic manure, she has successfully improved soil fertility, demonstrating the power of sustainable farming in land restoration.

### 7.3.2.5 Raid Buam Environmental Protection

Mr Donbok Buam, Secretary of the Raid Buam Environmental Protection Association, driven by passion and curiosity, constantly worked to localise and implement ideas that would benefit his village. His dedication not only paved the way for success but has also given inspiration to many others who share his vision for environmental conservation.

One of his key initiatives focused on restoring Amrawan, a forest that serves as the primary source of drinking water for the villages of Thangbuli and Umlatkur. This forest had been protected for generations, but certain areas became barren over time. To address this, Mr Buam and his team undertook large-scale tree planting. They successfully planted over 3,000 trees to restore degraded areas.

In 2018, Amrawan received official recognition for its ecological significance from the Governor of Meghalaya. It was especially important, since the primary goal of this conservation was and is to preserve the environment as well as drinking water sources. Because of this, the villages continue to enjoy clean water and fresh air.

In addition, the forest is also the source of the *Krang Suri* Waterfall, a popular tourist attraction.

### 7.3.3  Main Case Findings 3: Preserving Traditional Knowledge through Innovation

Beyond economic and environmental consequences, several grassroots innovations have strengthened traditional governance and preserved indigenous knowledge.

Some are showing significant potential in addressing key environmental and socio-economic challenges in the state. For instance, they focused on medicinal plant conservation and eco-tourism. On their part, various traditional governance systems have contributed to sustainability and economic growth.

Bitherson Sangma's initiative successfully revived over 50 endangered medicinal plant species through community-managed herbal nurseries, stimulating a collaboration between local healers and modern scientific researchers. The De·chraowe Sam A·chik Association established greenhouses and nurseries to cultivate and preserve traditional medicinal plants, ensuring their availability for future generations. It also undertook processing these plants into liquid decoctions and publishing a manual on traditional medicines and their uses. This was a significant contribution to knowledge retention and accessibility.

Further, eco-tourism has emerged as a viable strategy for balancing conservation with economic growth. For instance, a turtle sanctuary successfully merged wildlife conservation with tourism, attracting over 500 visitors annually and generating Rs.200,000 in revenue. This revenue is reinvested into habitat restoration and public awareness programmes, substantially helping achieve long-term sustainability of conservation.

### 7.3.3.1 Integration of National Innovations: ALC and Sadhana Forest and Immersion Centres

- **Agro-ecology learning circles** have trained over 2,040 farmers, significantly improving pest management and promoting organic agriculture.
- **Sadhana Forest has restored degraded land in 25 villages,** improving groundwater recharge and soil quality.
- **Immersion centres have trained almost 3,700 individuals in sustainable NRM,** equipping communities with knowledge and skills about long-term environmental conservation.

Many of these initiatives have involved traditional governance structures, such as the *Dorbar Shnong* (village councils). They integrate sustainability projects into community-led decision-making practices. This has strengthened local governance by formalising conservation strategies within traditional institutional frameworks. As a result, environmental and economic initiatives align with community priorities and receive widespread local support.

The active participation of such governance structures has reinforced the long-term sustainability of grassroots innovations, demonstrating that community ownership is a key driver of successful environmental and economic intervention.

## 7.4 Discussion

### 7.4.1 Interpretation of Findings for Replicability

The findings from the grassroots innovations indicate a strong potential for replication in similar ecological and socio-economic contexts. Many of the projects studied have achieved measurable economic, environmental and social benefits. It suggests that with the right support mechanisms, these models can be expanded beyond their pilot regions.

For instance, the plastic-to-fuel conversion project by Just Synrem can be implemented in other urban and semi-urban areas where plastic waste management remains a challenge. Given that the technology has low operational costs and high economic returns, other local entrepreneurs could be attracted. Then they can be trained and financially supported to set up similar units.

In the bargain, both employment opportunities and environmental benefits are expanding.

Similarly, Lamuni Sangma's mud beehives and Bah Stephan's eco-friendly beehives, both proving to increase honey yields while reducing colony mortality, could be introduced in other high-altitude beekeeping regions.

The scalability of this innovation lies in its affordability, ease of construction and immediate benefit to rural farmers.

The bottle-brick initiative – which uses waste plastic for construction – can be expanded to urban housing projects, eco-tourism infrastructure and public school construction. Since the project has already demonstrated cost efficiency and community engagement, governments and NGOs can integrate it into sustainable building programmes.

The agroecology learning circles (ALCs) have successfully trained over 2,000 farmers in sustainable farming. This could be replicated in other regions with soil degradation and climate-related agricultural problems. Given that ALCs encourage participatory learning and farmer-to-farmer knowledge exchange, similar programmes can be institutionalised in different agro-climatic zones.

The Sadhana forest project for reforestation and soil conservation, with its success in 25 villages, encourages expansion to other regions experiencing land degradation. This is particularly significant, where groundwater recharge and climate resilience are key concerns. The low-cost and high-impact nature of the intervention makes it a viable solution for large-scale environmental restoration.

Immersion centres have given effective training in sustainable NRM, climate adaptation and eco-friendly farming. Thousands have benefitted from it already. Establishing similar centres in other states can provide long-term capacity-building support to communities looking to integrate sustainability into their local economies.

A word of caution: Replication is not without its challenges. Some initiatives, such as medicinal plant conservation, require region-specific knowledge that cannot be easily transferred. Likewise, eco-tourism models like the Turtle Sanctuary project depend heavily on local biodiversity and community interest. So, while the core principles of innovations can be scaled, adaptations will be necessary to align with local contexts to achieve the ultimate goals fully.

Overall, the success of replication efforts will depend on strong community participation, financial investment and institutional backing. So, it is important that these innovations do not remain isolated success stories but become part of larger, statewide and national strategies for sustainable development.

### 7.4.2 Implications for Practice and Policy

The findings from these grassroots innovations have profound implications in two key areas: Practical applications and policy formulation.

### 7.4.2.1 Implications for Practice

On a practical level, these findings emphasise the importance of integrating traditional knowledge with modern technologies. For example, the medicinal plant conservation project reveals that traditional healing practices can be preserved and enhanced through scientific collaboration, paving the way for more holistic healthcare solutions.

Moreover, the study highlights the need for long-term financial sustainability. While the Innovations Fund provided critical seed funding, many innovators expressed concerns about access to continued funding for scaling their projects. This underscores the need for microfinance options, cooperative business models and integration into mainstream markets.

### 7.4.2.2 Implications for Policy

At the policy level, the study makes a compelling case for institutionalising grassroots innovations within government programmes. The plastic-to-fuel project and bottle-brick initiative, for example, fit in quite well with national policies on waste management and sustainable construction. Policymakers should consider creating subsidies, incubation programmes and training modules to assist in the expansion of such projects. Eco-tourism models, such as the Turtle Sanctuary, demonstrate how community-led conservation efforts can become viable income sources. By including such projects in state tourism policies, local governments can promote biodiversity conservation while boosting rural economies.

Overall, the study brought home clearly that grassroots innovations should be integrated into broader development frameworks, ensuring continued funding, technical support and legal recognition of community-driven solutions.

### 7.4.3 Comparison with Existing Literature

The findings from this study align with global research on grassroots innovation and sustainable development, while also presenting unique insights specific to Meghalaya's ecological and cultural landscape.

In comparison to similar community-led innovation models documented in India and other developing countries, the study reaffirms several key principles:

- **Community ownership fosters sustainability:** Much like the Self-Employed Women's Association in Gujarat, which empowered women entrepreneurs, Meghalaya's community-led handicraft initiatives through these innovators have demonstrated how local ownership has a sustained effect.

- **Traditional knowledge can complement modern technology**: Studies in Africa and Latin America have shown that integrating indigenous knowledge with scientific advancements enhances agricultural productivity and resource conservation. This is evident in the organic pest control project, which mirrors bio-pesticide models in Kenya and Brazil (Muthee, D. W., & Goudian, K. (2019).
- **Eco-tourism as a conservation tool**: Similar to Costa Rica's community-led eco-tourism projects, Meghalaya's Turtle Sanctuary initiative has shown how conservation can become financially self-sustaining through responsible tourism.

Nevertheless, the study also reveals critical gaps in existing literature:

- Most research looks at national-level policies but falls short on insights into state-led interventions. Meghalaya's model of grassroots funding through the Innovations Fund is a rare example of a government agency directly fostering grassroots enterprise.
- The role of traditional governance in sustainability remains underexplored. Unlike many regions where state authorities oversee environmental efforts, Meghalaya's innovations are deeply tied to local governance systems such as the *Dorbar Shnong*, offering a unique model of decentralised environmental management.

These insights highlight the need for more localised, context-driven research that recognises the diversity of grassroots innovations across different cultural and ecological landscapes.

## 7.5 Summary and Conclusion

### 7.5.1 *Key Findings Summary*

The study has shed light on the transformative impact of grassroots innovations in Meghalaya. It has demonstrated its ability to contribute to local growth, environmental conservation and cultural preservation.

Key findings include significant income generation, such as the Rs.400,000 annual revenue from plastic-to-fuel conversion and increased farmer earnings from zero-energy cold storage. Environmental benefits include reforestation, waste management and sustainable agriculture, reducing chemical dependency and enhancing biodiversity.

Additionally, traditional knowledge preservation through medicinal plant conservation and community-led eco-tourism showcases the power of indigenous solutions in modern contexts. These innovations not only empower local communities but also present scalable models for sustainable rural development.

### 7.5.2 Significance Restatement

Grassroots innovations are more than just local success stories: They represent **a new paradigm of sustainable development** where community knowledge meets modern science. The initiatives supported by the CLLMP illustrate how small-scale interventions can lead to large-scale effects when given the right resources.

The projects are critical in bridging policy gaps, fostering entrepreneurship and enhancing resilience against socio-economic and environmental challenges. By integrating traditional wisdom with innovative solutions, Meghalaya is setting an example for bottom-up, participatory development. The success of these innovations underscores the importance of continued investment, documentation and policy support to sustain and scale their ultimate effect.

### 7.4.2 Future Research and Practical Applications

To maximise the impact of grassroots innovations, further research is needed in areas such as long-term financial sustainability, cross-sector integration and scaling mechanisms. Understanding how these models can be replicated across different geographies will be key to expanding their reach and contributing to increasingly large-scale environmental and economic improvements.

Additionally, research on market linkages can help bridge the gap between local innovators and larger economic systems, ensuring financial viability beyond initial funding.

On a practical level, collaborations with academic institutions, policy-makers and private enterprises can enhance knowledge sharing and innovation scaling. The establishment of innovation hubs and incubators would provide technical mentorship and investment opportunities for emerging grassroots entrepreneurs.

By prioritising research, policy alignment and structured support systems, Meghalaya's grassroots innovations can serve as blueprints for other regions, creating a more inclusive and sustainable development framework.

#### 7.4.2.1 Incorporation of Innovation vs Initiative Differentiation

The Meghalaya Community-Led Landscape Management Project (CLLMP) has supported numerous grassroots interventions that address environmental, economic and social challenges. Based on peer review feedback, it is important to distinguish between **true innovations** – which introduce novel, scalable and replicable solutions – and **initiatives** that focus on sustaining and strengthening existing traditional practices or conservation efforts.

While this study initially classified all interventions as **grassroots innovations**, a closer analysis suggests that **many of these interventions are community-driven initiatives rather than new technological or process-based innovations**. Innovations such as the mud beehive, zero-energy cold storage and organic pest control demonstrate novel approaches that can be adapted and scaled in different contexts.

Conversely, initiatives such as the living root bridges restoration, reforestation and community-managed conservation projects contribute significantly to sustainability but do not necessarily introduce new methodologies or technologies.

By making this distinction, the study acknowledges the importance of both categories while formulating **clear policy recommendations and funding justifications**. The revised framework allows for a more structured approach to supporting and scaling effective projects under the CLLMP.

### 7.4.2.2 Objective: Promoting Climate Resilience

A core objective of grassroots innovations under the CLLMP is to **promote climate resilience** by enabling communities to adapt to environmental challenges through sustainable, low-cost solutions. Many of these innovations focus on **disaster risk reduction, soil conservation, sustainable farming and biodiversity conservation**. They would ensure that communities remain resilient in the face of climate-related adversities such as erratic rainfall, floods and soil degradation. By supporting these initiatives, the project promotes **adaptive capacities, strengthens food security and fosters long-term environmental sustainability** for Meghalaya's indigenous communities.

### 7.4.2.3 Sustainability of Grassroots Innovation Practice in the Present Days

The sustainability of grassroots innovations depends on multiple factors, including **financial viability, institutional support, community ownership and scalability**. While several initiatives have demonstrated long-term sustainability through income generation and self-sufficiency, others require continued funding and policy integration to remain effective.

Key elements contributing to the sustainability of these innovations include the following:

- **Community participation:** Ensuring that local stakeholders take an active role in managing and maintaining these innovations.

- **Market linkages:** Providing access to formal markets for products and services emerging from grassroots initiatives, such as organic farming and eco-tourism.

## References


Agrawal, A., & Chhatre, A. (2006). Explaining success on the commons: Community forest governance in the Indian Himalaya. *World Development, 34*(1), 149–166.

De Luca, L., Sahy, H., Joshi, S., & Cortés, M. (2013). Learning from catalysts of rural transformation, 135–167 (Chapter 7: Self Employed Women's Association (SEWA), India, - chapter under this book that was used as reference)

Gupta, S. (2019). Understanding the feasibility and value of grassroots innovation. *Journal of the Academy of Marketing Science, 48*, 941–965.

Muthee, D. W., & Goudian, K. (2019). *The role of indigenous knowledge systems in enhancing agricultural productivity in Kenya.* Article information.

Smith, A., Fressoli, M., & Thomas, H. (2014). Grassroots innovation movements: challenges and contributions. *Journal of Cleaner Production, 63*, 114–124.

# PART II

# Innovative Solutions: Harnessing Technology for Climate Resilience

# 8

# SLOPING AGRICULTURE LAND TECHNOLOGY

*John Kearney Wanniang*

## 8.1 Introduction

### 8.1.1 Purpose, Scope and Significance of the Case

The primary purpose of this case study is to assess the effectiveness of the Sloping Agricultural Land Technology (SALT) as a sustainable solution to soil erosion, declining soil fertility and environmental degradation in Meghalaya's hilly terrain. Given the adverse effects of Jhum cultivation and monoculture plantations, this study examines how SALT can serve as an alternative farming system that enhances soil conservation and ensures long-term agricultural sustainability. It also explores the socio-economic benefits of SALT, particularly in improving livelihoods and food security for indigenous farming communities.

The scope of this study encompasses agricultural practices, soil and water conservation techniques and policy frameworks that support the adoption of SALT. It evaluates SALT's applicability across various farming systems, from subsistence agriculture to commercial farming, while considering its environmental and economic consequences. Additionally, the study highlights the role of community participation, farmer training and indigenous knowledge systems in the successful implementation of SALT.

The significance of this study lies in its potential to guide policy interventions, sustainable land management strategies and climate-resilient farming practices. By demonstrating SALT's ecological and economic benefits, this case contributes to broader discussions on agroforestry, soil conservation and rural development. This makes it highly relevant for policymakers, researchers and farming communities.

DOI: 10.4324/9781003735380-10

### 8.1.2 Organisation of the Chapter

The structure of this chapter aims to provide a comprehensive understanding of the case. After the introduction, a review of literature looks at existing research on sustainable agriculture, soil conservation and agroforestry. An identification of gaps highlights areas needing further investigation. A theoretical framework presents project objectives and implementation strategies.

The methodology section details the approach, its justification and challenges encountered, while the results section presents key case findings.

The discussion interprets results in relation to replicability, theoretical implications and policy relevance. Finally, the summary and conclusion synthesise key findings, highlight significance and propose directions for future research and practical applications.

### 8.1.3 Review of Literature

Several studies bring out the importance of sustainable agriculture in hilly terrains, particularly in regions like Northeast India. Lal (1998) discusses soil erosion and its effect on agricultural productivity, emphasising the role of contour farming in reducing topsoil loss. Sanchez (1999) explores agroforestry as a strategy to improve soil fertility, noting its effectiveness in sloping landscapes.

Ramakrishnan (1992) examines shifting cultivation (Jhum) and its ecological consequences in North-East India. He finds that while Jhum supports biodiversity in the short term, long-term sustainability is compromised due to excessive soil erosion and nutrient depletion. Kumar et al. (2010) analyse alternative farming techniques. They conclude that contour bunding and SALT-based agroforestry significantly improve soil retention and moisture conservation.

Meghalaya-specific research by Tiwari (2017) outlines socio-economic challenges in adopting new agricultural technologies, particularly because of customary land ownership and scepticism towards government interventions. Nevertheless, successful community-led initiatives incorporating SALT principles demonstrate the potential for large-scale adoption.

The effectiveness of SALT in soil conservation has also been documented in Southeast Asian contexts, where similar geo-climatic conditions prevail. Mercado (2016) studied its implementation in the Philippines, showing improved crop yields and improved soil stability. These findings provide valuable insights for adapting SALT to Meghalaya's landscape.

### 8.1.4 Identification of Gaps

Despite the promising potential of the Sloping Agricultural Land Technology (SALT) in Meghalaya, several gaps remain in research and implementation. Firstly, there is a lack of long-term empirical studies on soil fertility improvements and yield sustainability. While short-term benefits have been observed, longitudinal research is necessary to assess the effectiveness of this technology over the decades.

Secondly, farmer adoption remains limited due to financial constraints, insufficient training and the unavailability or high cost of ideal nitrogen-fixing plant seeds and alternative species suited to the region. This points to a need for stronger support systems and incentive mechanisms.

Thirdly, there has been minimal research on integrating SALT with indigenous agricultural knowledge systems or on its socio-economic impact on marginalised communities.

Addressing these gaps will be crucial to strengthening the scalability and sustainability of SALT in Meghalaya.

### 8.1.5 Theoretical Framework

Soil erosion and declining soil fertility are major challenges in the hilly regions of Meghalaya, affecting agricultural productivity and farmers' livelihoods. The Sloping Agricultural Land Technology (SALT) project was initiated with the following objectives:

#### 8.1.5.1 Objectives

- To reduce soil erosion and enhance soil fertility using SALT interventions.
- To improve agricultural productivity through contour hedgerows and integrated crop farming.
- To train farmers in sustainable land management practices and organic farming techniques.
- To promote agroforestry and crop diversification for long-term environmental and economic benefits.

#### 8.1.5.2 Project Implementation

Site selection and farmer engagement: After farmer identification, awareness programmes informed them of the benefits of SALT. One thousand eight hundred farmers received training on land development, cultivation and harvesting techniques.

Land preparation and hedgerow establishment: A-frame tools marked contour lines. Hedgerows composed of nitrogen-fixing plants such as *Tephrosia candida* and *Indigofera arrecta* were planted along these lines. These plants are selected as they are perennial and can withstand pruning. Hedgerow maintenance involved regular pruning, mulching and weeding to promote soil stabilisation.

Crop cultivation and agroforestry integration: Various crops were planted between the hedgerows, viz.:

- Seasonal crops (2–6 months): Cereals, legumes, vegetables, pulses.
- Short-term crops (12–18 months): Yam, tapioca, ginger, turmeric.
- Medium-term crops (2–5 years): Pineapple, papaya, banana, perennial pigeon pea.
- Long-term fruits: Citrus species, Moringa (Drumstick Tree), Litchi, Mango, Jackfruit, *Parkia* (Tree Bean).

Farmer training and capacity building: Farmers received initial seeds, tools and financial support for hedgerow maintenance (Rs.19,980 per ha). Fifty farmers and staff participated in exposure visits to successful SALT farms in project villages as well as outside (Aben, Manipur) to encourage adoption.

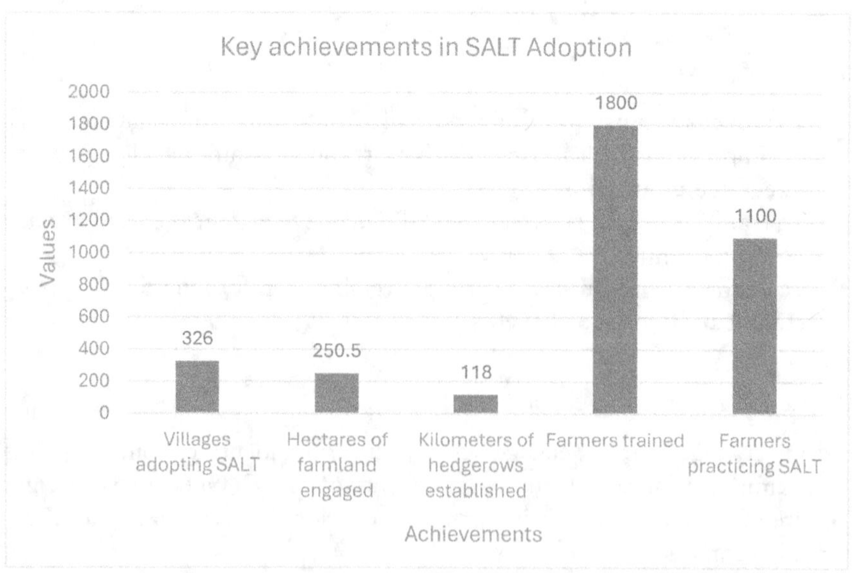


**FIGURE 8.1**   Key achievements in SALT adoption

*Source:* Extracted from the MIS

Monitoring and evaluation: Regular farm inspections were conducted to assess soil quality, crop performance and hedgerow maintenance. Monitoring focused on reducing soil erosion, improving fertility and evaluating farmer adoption rates.

### 8.1.5.3 Key Achievements

Despite receiving training, approximately 700 farmers did not adopt SALT. Key barriers included financial constraints, as many could not afford the initial investment even with partial support; labour shortages, with the intensive setup clashing with existing agricultural workloads; delayed returns, as SALT yields benefits over the medium to long term, discouraging those seeking quicker gains; and limited availability and high cost of suitable planting materials, which further hindered adoption.

## 8.2 Methodology of the Case

### 8.2.1 Description of Methodology

Implementation of the Sloping Agricultural Land Technology (SALT) followed a structured approach. It involved community engagement, site selection, farm preparation, hedgerow establishment, crop cultivation and regular monitoring. In some areas, particularly those above 1,100 m, alternative species like citronella were used.

### 8.2.2 Justification of Methodology

The methodology was designed to address the dual challenges of soil erosion and declining soil fertility while promoting sustainable agriculture. Nitrogen-fixing hedgerows improve soil structure, enrich nutrients and minimise run-off. Contour farming reduces topsoil loss in hilly terrain, ensuring long-term agricultural viability. The inclusion of multiple crop types – seasonal, short-term and perennial – brought farm diversification and food security.

Providing initial seeds, tools and financial incentives for hedgerow maintenance removed entry barriers. The structured monitoring framework ensured compliance and effectiveness.

The methodology's integrated approach, combining technical interventions with farmer capacity building, resulted in increased adoption and sustainable land management practices, benefiting over 1,100 practising farmers across Meghalaya.

### 8.2.3 Limitations and Challenges

The SALT initiative faced several challenges. The initial setup was labour-intensive, requiring significant effort to clear land, establish contour lines and plant hedgerows. Many farmers lacked awareness or technical knowledge, which necessitated extensive training before adoption.

Farmer adoption remains limited on account of the unavailability or high cost of ideal nitrogen-fixing plants (NFP) and alternative species suited to the region. Notably, the cost norm of Rs.19,980 per ha for hedgerow establishment and maintenance required external funding and subsidies, which were not always accessible to all farmers. Some areas, particularly those above 1,100 m, required alternative species like citronella, a more expensive species.

Continuous monitoring and maintenance were essential, but logistical challenges and resource constraints sometimes limited field inspection. While 1,100 farmers have successfully adopted SALT, wider adoption requires further financial and policy support.

Addressing these challenges through sustained training, farmer incentives and improved extension services and funding could ensure a long-term effect of SALT.

## 8.3 Results

### 8.3.1 Main Case Findings 1

#### 8.3.1.1 Introduction

This case is an examination of soil health improvements observed on Yoomeki Tangliang's farm following the adoption of SALT, in Mooriap in East Jaintia Hills.

It involves a comparative analysis using historical data from Swer and Singh (2004), which revealed the adverse effects of coal mining on soil properties in East Jaintia Hills. The comparison aims to assess the extent of soil recovery and the consequences of sustainable land management intervention.

The key soil parameters assessed include pH, nitrogen, phosphorus, potassium and organic carbon content.

#### 8.3.1.2 Site Characteristics

Yoomeki Tangliang's farm is situated in the hilly terrain of Mooriap, East Jaintia Hills. It is an area characterised by heavy rainfall, which poses risks of soil erosion and nutrient loss. To counteract this problem, the farmer placed seven contour hedgerows spanning 118 running metres. These hedgerows were composed of nitrogen-fixing plants, primarily *Indigofera arrecta* and *Tephrosia candida.*

Given the limited availability of data, this study took a comparative approach using historical research on mining-affected soils in the region. A 2004 study by Swer and Singh documented severe soil degradation, increased acidity and depletion of essential nutrients due to coal mining.

**TABLE 8.1** Analysis of soil properties, farm in East Jaintia Hills

| Soil parameter | Research findings (Swer & Singh, 2004) – mining areas | Recent soil testing (2024) – Yoomeki Tangliang's farm | Comparison |
| --- | --- | --- | --- |
| pH Value | 4.5–5.5 (Highly acidic) | 5.6 (slightly acidic) | Improvement in pH, indicating a less acidic soil environment |
| Nitrogen (ppm) | Low nitrogen levels due to nutrient depletion | 426 (high fertility) | Significant increase in nitrogen, promoting plant growth |
| Phosphorus (ppm) | Deficient phosphorus levels | 16 (moderate availability) | Improved phosphorus content, supporting root development |
| Potassium (ppm) | Deficient potassium | 1.81 (adequate for plant growth) | Enhanced potassium availability, boosting crop resistance |
| Organic carbon (%) | Low due to degradation and mining impact | 1.81 (good for soil structure and fertility) | Higher organic carbon, improving soil microbial activity |

*Source:* Extracted from MIS.

### 8.3.1.3 Recommendations for Future Interventions

- Capacity building programmes: Conduct training and workshops to educate farmers on the benefits of SALT and best practices for implementation.
- Promotion of integrated farming: Encourage farmers to adopt integrated farming practices, combining SALT with agroforestry, organic farming and mixed cropping to maximise soil health and productivity.
- Financial and technical support: Develop incentive programmes, subsidies and access to affordable NFP seeds to encourage wider adoption of sustainable agricultural practices.

## 8.3.2 Main Case Findings 2

### 8.3.2.1 Introduction

East Khasi Hills in Meghalaya is a region marked by steep slopes and high annual rainfall, making it vulnerable to soil erosion, acidification and nutrient leaching. The district's agricultural soils are generally classified as slightly to moderately acidic, with uneven nutrient distribution due to topography, cultivation practices and past land-use systems.

In 2022, Das et al. conducted a district-wide soil fertility mapping using Soil Health Card (SHC) data. This study highlighted trends in key soil parameters – such as pH, phosphorus, potassium and organic carbon – providing a baseline for understanding fertility status across the East Khasi Hills.

As a sustainable intervention, Sloping Agricultural Land Technology (SALT) has gained prominence for its ability to control erosion and enhance soil health in sloped terrains. One such successful implementation is found on Aihunlin Nongspung's farm, located in Tynring, Mawryngkneng Block of East Khasi Hills. Here, five contour hedgerows stretching across 100 running metres have been established, integrating SALT principles with local cropping systems.

In 2024, a comprehensive soil test was conducted on Aihunlin Nongspung's farm to assess the impact of SALT-based land management. The findings were compared against district-level data reported by Das et al. (2022), as shown in Table 8.2.

The case highlights the effectiveness of agroecological practices like SALT in restoring degraded soils, making them more resilient and productive over time. It also affirms the need for localised, data-driven soil management strategies in hilly regions like East Khasi Hills.

### 8.3.2.2 Lessons Learned and Future Considerations

Challenges such as the continuous maintenance of hedgerows remain. Increased farmer training and the integration of organic farm management practices can overcome these and further enhance the effectiveness of SALT.

### 8.3.3 Main Case Findings 3

Jhum has long been practised in the Garo Hills of Meghalaya, but its environmental impact has raised concerns about long-term soil health and sustainability. The study by Yadav et al. (2012) provides critical insights into how Jhum cultivation affects soil parameters, reporting significant degradation in soil fertility due to repeated burning and cropping cycles. Forest soils in the region were found to be moderately acidic (pH 4.8–5.3) with relatively good levels of nitrogen, phosphorus, potassium and organic carbon (up to 2.5%). However, under Jhum cultivation, these values declined sharply, indicating nutrient loss, erosion and reduced soil structure.

The case of Seroni Marak, a farmer from Kalak Dorek village in Samanda block, East Garo Hills, illustrates a shift towards sustainable land use

**TABLE 8.2**  Analysis of soil properties, farm in East Khasi Hills

| Soil parameter | District-wide average (Das et al., 2022) | Recent soil testing (2024) – Aihunlin Nongspung's farm | Comparison |
| --- | --- | --- | --- |
| pH Value | 68.66% of soils slightly acidic (approx. pH 5.0–6.5) | 5.97 (slightly acidic but within a suitable range for most crops) | Consistent with district trend; near-neutral, favourable for crops |
| Nitrogen (ppm) | Not reported in absolute terms; generally low to medium | 276 (moderate fertility) | Higher than average; supports vegetative growth |
| Phosphorus (ppm) | High availability in 35% of district soils | 22 (good availability for root development) | Reflects improved soil fertility and nutrient management |
| Potassium (ppm) | High levels in only 7.54% of district soils | 2.34 (adequate for plant health and disease resistance) | Better than district average; supports plant vigour |
| Organic carbon (%) | 98.9% of soils with high OC (typically >0.75%) | 2.34 (good for soil structure and microbial activity) | Significantly high; enhances soil moisture retention and biological life |

*Source:* Extracted from MIS

through Sloping Agricultural Land Technology (SALT). Her farm incorporates four contour hedgerows extending across 200 running metres, supporting the cultivation of mango, arecanut and broom grass – a diverse mix of perennial and commercial crops. This layout helps reduce soil erosion, improve moisture retention and gradually restore soil fertility.

Recent soil testing (2024) from Seroni Marak's farm reveals the following values (Table 8.3):

While the 2024 soil data still reflects significant nutrient depletion – particularly in phosphorus and organic carbon – compared to undisturbed forest soils, the implementation of SALT on Seroni Marak's farm marks a step towards more sustainable land management. The improved pH level suggests reduced acidity, likely due to better ground cover and erosion control from contour hedgerows. Though nutrient levels remain low, the long-term application of SALT practices holds promise for gradual soil restoration and improved resilience in upland farming systems. This case demonstrates the adaptability of SALT for perennial and commercial crops in the hilly terrains of Meghalaya.

**TABLE 8.3** Analysis of soil properties, East Garo Hills

| Soil parameter | Research findings Yadav PK, Kapoor M, Sarma K (2012) | Recent soil testing (2024) – Seroni Marak) | Comparison |
| --- | --- | --- | --- |
| pH Value | 4.8–5.3 (strongly to moderately acidic under Jhum cultivation) | 5.9 (moderately acidic) | Slight improvement in acidity levels |
| Nitrogen (ppm) | 200–400 (moderate in forest, low in Jhum fallows) | 142 (low) | Lower than forest soils; comparable to degraded Jhum soils |
| Phosphorus (ppm) | 3–10 (moderate in forest, reduced under cultivation) | 1 (extremely low) | Significantly depleted |
| Potassium (ppm) | 0.3–0.6 (higher in forest soils) | 0.25 (very low) | Very low; below forest and fallow levels |
| Organic carbon (%) | 0.8–2.5 (high in forest, declines with cultivation) | 0.25 (severely low) | Severely depleted organic matter |

*Source:* Extracted from MIS.

## 8.4 Discussion

### 8.4.1 Interpretation of Findings in Relation to Their Replicability

The findings from the implementation of the Sloping Agricultural Land Technology (SALT) in Meghalaya demonstrate a sustainable and effective approach to addressing soil erosion, improving soil fertility and enhancing farmers' livelihoods. The successful adoption of SALT by farmers like Yoomeki Tangliang from Mooriap, Aihunlin Nongspung from Tynring and Seroni Marak from Kalak Dorek underscores its potential for replicability across similar agro-climatic regions.

Key factors that contribute to its replicability include the use of locally available resources and the ability to generate benefits such as improving soil fertility and reducing soil runoff. Farmers have successfully implemented key techniques under SALT, including contour farming, hedgerow planting, intercropping and agroforestry. This has led to measurable improvements in farm productivity and soil quality. The integration of nitrogen-fixing plants in hedgerows, in particular, has proved to be an effective method to enrich soil fertility without dependence on chemical fertilisers.

The findings suggest that with the right institutional support and capacity-building initiatives, SALT can be successfully replicated in other hilly, high-rainfall regions where soil degradation is a concern.

### 8.4.2 Implications for Theory, Practice and Policy

The findings of this study contribute significantly to the understanding of sustainable land management in sloping terrains. The evidence from this project emphasises the role of nitrogen-fixing plants in promoting long-term soil sustainability and land restoration.

From a practical standpoint, the study provides a roadmap for smallholder farmers to transition from environmentally harmful practices such as Jhum cultivation to sustainable agroforestry systems. The successful adoption of SALT demonstrates that conservation-based farming is not only feasible but also economically viable for small-scale farmers.

The model has brought out the importance of combining indigenous knowledge with scientific practices to develop region-specific solutions for soil conservation and agricultural productivity improvement.

Integrating SALT into state-level agricultural programmes can provide the necessary political framework for long-term sustainability. Incentives for sustainable practices, such as subsidies for nitrogen-fixing plant seeds and hedgerow maintenance, could further encourage adoption. Moreover, research and extension services should be expanded to assess the long-term effects of SALT on soil health and farm incomes, also with a view to ensuring continued policy relevance and refinement.

### 8.4.3 Comparison with Existing Literature

The findings from the implementation of SALT align with existing literature on sustainable agriculture, agroforestry and soil conservation. Studies on agroforestry systems have long highlighted their effectiveness in reducing soil erosion and enhancing soil fertility. For instance, a study by Garrity (1996) on contour hedgerow intercropping in Southeast Asia found similar benefits in soil retention, water conservation and yield improvement, reinforcing the outcomes observed in the SALT implementation in Meghalaya.

Research on conservation agriculture practices by Pretty et al. (2006) also supports the notion that integrated land management practices lead to long-term productivity gains while reducing environmental degradation. The use of nitrogen-fixing hedgerows in SALT aligns with findings from Lal (2015). They emphasise the importance of biological nitrogen fixation in restoring degraded soils, particularly in tropical and subtropical regions.

Further, literature on climate adaptation in agriculture suggests that sustainable land-use practices play a crucial role in strengthening resilience to climate variability (Altieri & Nicholls, 2017).

The SALT model in Meghalaya contributes to this body of work by demonstrating how simple interventions can significantly improve soil health and farm productivity in regions prone to high rainfall and erosion.

While previous studies have highlighted the effectiveness of agroforestry and conservation agriculture, this project adds valuable insights specific to Meghalaya's socio-ecological conditions. Unlike some global studies that recommend large-scale commercial adoption, this study underscores the importance of smallholder-focused approaches, community engagement and culturally sensitive interventions for sustainable agricultural transformations in indigenous farming communities.

## 8.5 Summary and Conclusion

### 8.5.1 Key Findings Summary

The implementation of Sloping Agricultural Land Technology (SALT) in Meghalaya has demonstrated significant benefits in addressing soil degradation, erosion and declining fertility in hilly terrains. The introduction of nitrogen-fixing hedgerows, intercropping and agroforestry has improved soil conservation, farm productivity and livelihoods.

Case studies, such as that of Yoomeki Tangliang from the East Jaintia Hills, Aihunlin Nongspung from East Khasi and Seroni Marak from the East Garo Hills, highlight the practical success of SALT in real-world farming conditions. Key measurable outcomes include reduced soil erosion, increased organic matter and improved crop yields.

In short, SALT presents a viable, replicable and sustainable alternative to traditional shifting cultivation and monoculture.

### 8.5.2 Significance Restatement

SALT is a transformative approach to sustainable agriculture in Meghalaya, where traditional farming methods have led to soil degradation. Its eco-friendly techniques reduce erosion, increase soil fertility and ensure long-term agricultural productivity. By empowering farmers with knowledge and resources, SALT strengthens community resilience and food security. The approach also supports initiatives aimed at sustainable land management, demonstrating its broader relevance beyond Meghalaya to other regions facing similar environmental and agricultural challenges.

### 8.5.3 *Future Research and Practical Applications*

Future research should focus on scaling up SALT adoption through region-specific customisation, farmer training and economic incentives. Long-term studies on soil health improvements and crop yield sustainability are necessary to assess their full effect. Additionally, integrating digital tools like GIS mapping and remote sensing can contribute to even more effective monitoring and evaluation.

From a policy perspective, promoting SALT through government subsidies, financial incentives and cooperative farming models can drive wider adoption. Practical applications also include incorporating SALT into agroforestry models and integrating it into rural development projects.

Future initiatives should explore market linkages for diversified crops produced under SALT, ensuring economic sustainability for farming communities. Strengthening community engagement, knowledge-sharing networks and partnerships between farmers, research institutions and policymakers will be critical in maximising SALT's long-term benefits and scaling its impact across hilly terrains globally, contributing to a long-term sustainability of conservation efforts.

## References


Altieri, M. A., & Nicholls, C. I. (2017). The adaptation and mitigation potential of traditional agriculture in a changing climate. *Climatic Change, 140,* 33–45.

Das, P. T., Saikia, B., Lakiang, T., Majaw, R. M., & Saiborne, I. B. (2022). District wise soil fertility mapping from Soil Health Card data using GIS in Meghalaya. International Journal of Scientific Research and Engineering Development, 5(4), 1184–1195.

Garrity, D. P. (1996). *Tree-soil-crops interaction on slopes. Tree-crop interactions.* CABI.

Kumar, A., Whitbread, A., & Rao, K. P. C. (2010). International crops research institute for the semi-arid tropics. *Hyderabad.*

Lal, R. (1998). Soil erosion impact on agronomic productivity and environment quality. *Critical Reviews in Plant Sciences, 17*(4), 319–464.

Lal, R. (2015). Restoring soil quality to mitigate soil degradation. *Sustainability,* 7(5), 5875–5895.

Mercado, A. R. (2016). Application of Sloping Agricultural Land Technology (SALT) in the Philippines: Implications for sustainable upland farming. *International Journal of Agricultural Sustainability, 14*(3), 256–272.

Pretty, J. N., Noble, A. D., Bossio, D., Dixon, J., Hine, R. E., Penning de Vries, F. W., & Morison, J. I. (2006). *Resource-conserving agriculture increases yields in developing countries.* Environmental Science & Technology.

Ramakrishnan, P. S. (1992). Shifting agriculture and sustainable development: An interdisciplinary study from north-eastern India.

Sanchez, P. A. (1999). Delivering on the promise of agroforestry. *Environment, Development and Sustainability, 1,* 275–284.

Singh, O. P., & Swer, S. (2004). Diminishing life-sustaining role of water in Jaintia Hills of Meghalaya due to coal mining. *Development and Environment*, 258–272.

Tiwari, B. K. (2017). Challenges and opportunities of sustainable agriculture in Meghalaya: Socio-economic and environmental perspectives. *Indian Journal of Traditional Knowledge, 16*(4), 512–520.

Yadav, P. K., Kapoor, M., & Sarma, K. (2012). Impact of Slash-And-Burn Agriculture on Forest Ecosystem in Garo Hills Landscape of Meghalaya, North-East India. *Journal of Biodiversity Management & Forestry Research, 1,* 1.

# 9
# NEUTRALISATION OF ACID MINE DRAINAGE CONTAMINATED WATER

*Effie Gwyneth L. Kharjana, Wansah Pyrbot,*
*Justin Pathaw, Raynold Marwein, Bandarihun Kharlor,*
*Fidellia Sun and Cheryl Bernice Pohrmen*

## 9.1 Introduction

### 9.1.1 Purpose, Scope and Significance of the Case

Coal mining activity is widely practised and considered by many to be the primary source of income before its prohibition by the NGT bench in 2014. Although coal mining has brought employment opportunities, tribal sustenance and economic development, it has also led to extensive environmental degradation and the erosion of traditional values in society.

Environmental problems associated with mining have been felt severely because of the region's fragile ecosystems and richness of biological and cultural diversity. A major drawback that affects livelihoods is the contamination of water bodies due to acid mine drainage (AMD). Acid mine drainage is the outflow of acidic water from metal or coal mines (Caruccio, F. T. 1975).

The state receives an average annual rainfall ranging from 10,000–11,000 mm, and acidic runoff pollutes rivers, streams and groundwater. It renders them unsuitable for drinking, farming or supporting aquatic life, with pH levels dropping to as low as 2–3. In Meghalaya's fragile ecosystems, AMD disrupts biodiversity by killing living organisms unable to tolerate the low pH and toxic metal concentrations like iron, aluminium and manganese.

The contamination also leads to visible degradation, such as orange-red discolouration from iron precipitation, and affects livelihoods dependent on clean water, increasing environmental and social challenges in the region (Tiwary, 2001).

Various approaches have been taken to treat these contaminated resources, but the Open Limestone Channel (OLC) method has gained favour as a

DOI: 10.4324/9781003735380-11

passive treatment option, using locally sourced materials that communities can easily implement and duplicate. The OLC was found to be cost-effective and technically feasible in rural areas, so it prompted the construction of more such projects for the improvement of water quality and eco-restoration of degraded streams (Pyrbot et al., 2019).

### 9.1.2 Organisation of the Chapter

Coal mining has historically driven economic growth but severely disrupted ecosystems, particularly in regions situated in the southern belt of the state. The successful implementation of the Open Limestone Channels (OLCs) in 31 sites highlights the importance of community contribution in the successful implementation of the OLCs and improving water quality for domestic and agricultural use. This chapter is structured into three main sections:

- **Section 9.3** showcases the treatment's successful implementation, improving water quality and promoting water security in the community. The community's active participation significantly enhances the treatment's efficiency and sustainability.
- **Section 9.4** discusses on findings, their replicability and comparison with existing literature.
- **Section 9.5** provides the key findings and recommendations for improving and upscaling the technology.

### 9.1.3 Review of Literature

In Meghalaya, the effects are most severe in the Jaintia Hills, where many streams have become highly acidic, such as the Leshka, Kopili and Lukha rivers. In Garo Hills, only a few rivers have turned acidic due to mining activities.

The Karst topography in Meghalaya means few perennial water bodies exist in coal and limestone mining areas, worsening water scarcity during dry periods. Mining has disturbed the land and caused many streams to become seasonal. The water in coal mining areas has been found highly acidic. The pH of streams and rivers varies between 2.31 and 4.01. This indicates the serious condition of the water bodies of the area that can hardly support any aquatic life such as fish, amphibians and insects (Singh, 2012).

Chakraborty et al. (2002) found that opencast mining disrupts surface topography and watercourses, while underground mining lowers the groundwater table. Acid mine drainage is also common in underground mining. Swer and Singh (2003), Sahoo et al. (2012), Blahwar et al. (2012) and Chabukdhara and Singh (2016) reported low pH (3–5), high electrical conductivity and elevated sulfate and metal concentrations in mining-affected areas. The neutralisation of acidic water using Open Limestone Channel (OLC) has proven effective for treating AMD-contaminated water in mining areas. The present study found that OLC successfully neutralised acidic water, restoring the stream's pH to normal, as reported in restoration studies by Wansah et al. (2019) on the Moolawar Stream in East Jaintia Hills, Meghalaya. The water quality has significantly improved and been utilised by the community for domestic and agricultural purposes, creating a sense of water security.

### 9.1.4 Identification of Gaps

Water contamination in Meghalaya poses significant challenges for residents who rely on springs and spring-fed streams for their daily needs. Limited access to potable water forces them to travel long distances or purchase water, which causes them financial and physical burdens.

People have experienced problems of skin irritation and ineffective soap lathering; discolouration of clothes and chlorosis, also known as the yellowing of plants, occurs when they are unable to absorb nutrients. More critically, the consumption of polluted water has led to severe health risks.

The lack of a sustainable water management system points to an urgent need for community-driven interventions to provide safe and accessible drinking water, ensuring public health and environmental preservation.

### 9.1.5 Theoretical Framework

A comprehensive framework for addressing acid mine drainage (AMD) involves community-driven initiatives. Awareness programmes educate local populations on AMD causes, effects and sustainable practices, fostering collective responsibility. Community involvement is crucial: focal group discussions enable local people.

Open Limestone Channel, also called Oxic Limestone drains, are simple constructions for the treatment of acid mine drainage (Dadhwal, K. S.1999). The acidic mine water flows through a channel filled with coarse limestone, where the calcium carbonate reacts with the acidic water and raises the pH of the water. However, many factors play an important role, such as velocity, channel slope, acidic nature of water and the contact time between the acidic water and the limestone. During the neutralisation process, a metal hydroxide precipitate is formed on the surface of the limestone, a process known as limestone armouring. This phenomenon reduces the reactivity of the limestone surface with the acidic water. A laboratory container study showed that armoured limestone was 90% as effective in neutralising AMD as unarmoured limestone (Ziemkiewicz et al., 1997).

Open limestone channels (OLCs) aid AMD treatment by increasing pH and precipitating heavy metals through oxidation. The *vetiver* system excels in phyto-remediation, tolerating and accumulating heavy metals and organic pollutants (Truong & Luu, 2015). These strategies promote sustainable, long-term AMD solutions through community participation and nature-based interventions. It is cost-effective and easy to maintain, making it convenient for treatment in rural areas.

## 9.2 Methodology of the Case

### 9.2.1 Description of Methodology

The construction and implementation of Open Limestone Channels (OLCs) follows a structured and site-specific approach. The local community assisted in the identification of sites, in which training was provided to create mass awareness. They also carried out the implementation of the OLC with technical expertise from the engineers. A key step was establishing a village water committee, which appointed a village water facilitator to oversee the treatment process and ensure everything operated smoothly.

By sourcing all materials locally, the setup was designed for ease of maintenance, strengthening the community's ability to sustain the system over time.

The OLCs were constructed in two ways: Water was diverted for treatment from their source, and in some sites, on-site treatment was carried out based on selection criteria. The water flowed into chambers filled with limestone to provide sufficient reaction time for neutralisation. The treatment covered a span of 25–30 m. Vetiver grass was planted in the chambers for its ability to absorb heavy metals and other pollutants. Filters were installed at certain sites to capture suspended particles contributing to turbidity.

Altogether, 31 Open Limestone Channels were set up covering all three regions of the state (Khasi, Jaintia and Garo Hills) and funded by an external agency and the Government of Meghalaya.

Figure 9.1 illustrates the method.

### 9.2.2 Justification of Methodology

The OLC method is based on scientific principles, environmental considerations and community involvement, ensuring both effectiveness and sustainability. This approach offers a solution for treating AMD-affected water to meet the needs of local residents.

The abundance of local limestone in the state offers a readily available, natural solution to deal with AMD. Communities can efficiently use this resource in treatment technologies without concerns about scarcity or cost. This contributes to sustainable and cost-effective mitigation efforts in general.

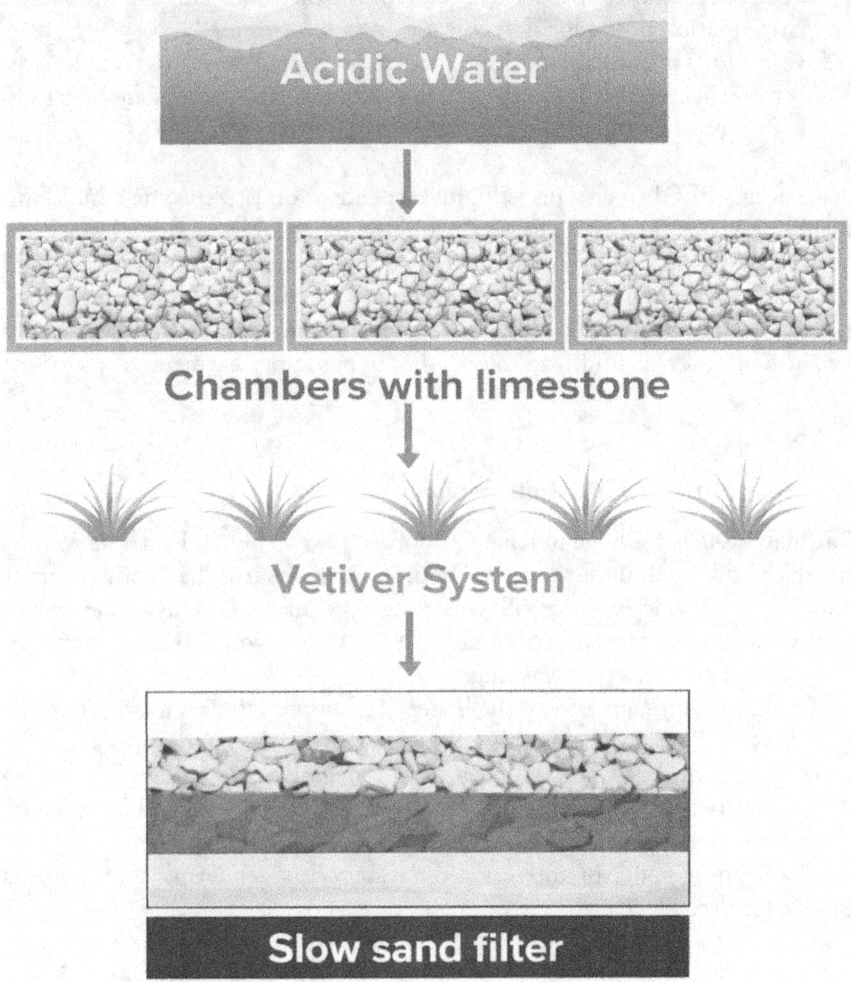


**FIGURE 9.1**  Working principle of the OLC

*Source:* Institute of Natural Resources, Meghalaya

### 9.2.3 Limitations and Challenges

Here are some limitations and challenges met on implementation of the OLCs:

Unavailability of community land: This presented a challenge, as it would compromise the OLC's design and restrict accessibility.

Armouring of limestones: Armouring of limestone can reduce the dissolution and acid neutralisation in an OLC. Periodic maintenance is required depending on the environmental factors of each site. This requires active participation of the local people.

Active mining sites: Active mining sites pose a challenge as treated acidic water still contains high levels of pollutants and metal precipitates, leading to more frequent coating of the limestone.

The success of OLC systems is highly dependent on site-specific conditions, including water chemistry, flow rate and the topography of the mining area. In areas with highly acidic or complex contamination, this system must not be as effective. Then it requires additional or alternative treatments. Besides, the topography in some regions can complicate the design of OLC systems because of underground water flows that bypass the treatment.

## 9.3 Results

### 9.3.1 Main Case Findings: Laitmawsiang

Laitmawsiang is a village in Khatarshnong Laitkroh block, East Khasi Hills. It is blessed with abundant water resources and a charm that draws tourists from far and wide, yet the village struggles to meet its daily water needs. Many water resources are contaminated by AMD, which makes them unfit even for simple tasks like washing.

Though mining has ended, the lingering runoff continues to seep into the water, exacerbating the contamination and leaving the community in a very difficult situation. As a result, they rely on neighbouring villages, purchase water weekly and travel long distances for basic requirements like washing clothes.

The ongoing pollution intensifies the water crisis, rendering many sources unsafe for use.

Without proper remediation, the legacy of past mining activities continues to threaten the water resources, particularly springs which serve as a primary source of drinking water.

In response, the community sought to reclaim a viable water supply polluted by AMD. They found an unused spring nestled in a forested area, with a pH of 2.3. Under the leadership of the Village Natural Resource Management chairman, the villagers selected this source for remediation and connected it via a pipeline with the existing Open Limestone Channel (OLC).

Made as a serpentine structure – 0.3 m wide, 0.3 m deep and spanning 10 m – the channel maximised contact time between the acidic water and limestone, facilitating neutralisation through carbonate dissolution. This reaction elevated the pH by releasing alkalinity, counteracting the acidity as the water was seeping through the system. To increase purification, vetiver grass

(Chrysopogon zizanioides) was strategically planted with the channels. Its robust root system excels at phyto-remediation: It can absorb heavy metals such as iron and manganese, which are prevalent in AMD-affected waters (Chadwick, M. J. 1973).

Once deemed unusable, this revitalised spring now delivers water to various parts of the village, easing the burden financially and physically. The transformation has created a sense of water security within the village.

Through dedicated community involvement, coupled with innovative local solutions, Laitmawsiang has not only secured a reliable supply but also strengthened its resilience.

The graph indicates that the pH rose by a value of 3 to 4 units from the source post treatment. Despite the pH increase, the final reading consistently remains within the limits defined by the BIS. This ensures that neutralisation of the acidic water does not exceed the standard limit due to the composition of the limestone locally available.

Plantation of the vetiver grass in these channels has proven to improve the quality of the treated water by absorption of total dissolved solids present. This design appears to be more efficient by enhancing the reaction time between the water and limestone, while also minimising maintenance issues associated with armour formation.

### 9.3.2 Main Case Findings 2: Mooliat Bri Sutnga

Mooliat Bri Sutnga is a village in East Jaintia. Active coal mining was extensively carried out here in the last decade. According to narrations from elderly people, before coal extraction began, the village had thrived with abundant flora and fauna. Residents had benefited from these natural resources, relying on them for food, shelter, water and other daily needs. But

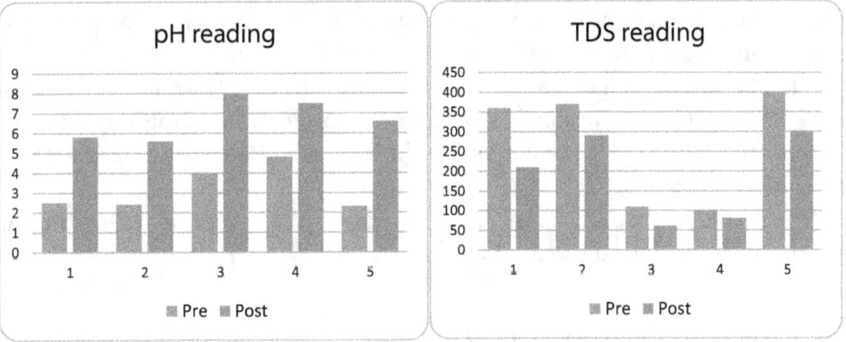


**FIGURE 9.2**  pH and total dissolved solids (TDS) before and after treatment

*Source:* Institute of Natural Resources, Meghalaya

due to unscientific and unbridled coal mining, irreversible damage had come to the entire ecosystem in their area.

Coal extraction had been carried out as rat-hole and opencast mining within and surrounding the village, which had severely depleted the natural resources there. Also, a number of springs and streams had dried up, and most of the agricultural land had been left barren due to coal mining seepage.

As a result, for the past decade, villagers had been facing a profound scarcity of water, especially during the lean season. At that time, each and every household would have to spend an amount of Rs. 1,500 a month to buy water from neighbouring villages for daily water needs. With most villagers living below the poverty line, the financial burden on the family had significantly increased.

Despite the village's abundance of water resources, it remained unusable due to continuing high acidity. Given the prevailing situation, restoring and rejuvenating the highly degraded ecosystem presented a significant challenge, requiring well-planned strategies for effective rehabilitation. The primary objective was, obviously, to neutralise the acidic spring, making it suitable for use for the community's water needs.

For the implementation of the OLC method, villagers had selected a spring called Thanglooh for treatment. The spring originates from the extensive rat-hole mining in the area, which gave it a distinct reddish colour with mine-spoil sediment accumulating at the bottom. It is a depression spring situated below the catchment area, surrounded by sedimentary rocks and a few unconsolidated formations. During the lean season, it used to maintain a recorded discharge of 1.65 L per minute.

The OLC itself had been thoughtfully designed into three chambers. This was a plan to which the community had given input to ensure practicality and effectiveness.

The first two chambers had been packed with locally sourced limestone, crushed into small, medium-sized gravel – each holding about 2 m$^3$ – while the third chamber contained a lighter mix with less limestone. This, too, was a decision villagers had made to fine-tune the treatment process.

This hands-on collaboration had not only harnessed local resources but also deepened the community's stake in the project, blending their knowledge of the land with technical know-how to breathe new life into a once-discarded spring.

The result has shown an increasing value of pH of around 3.14 from the source of the spring to the final chamber.

### 9.3.3 Main Case Findings 3: Amguri

Amguri, a village in the Selsella block of West Garo Hills, faces a unique challenge: It has no history of mining activities; its water sources are naturally

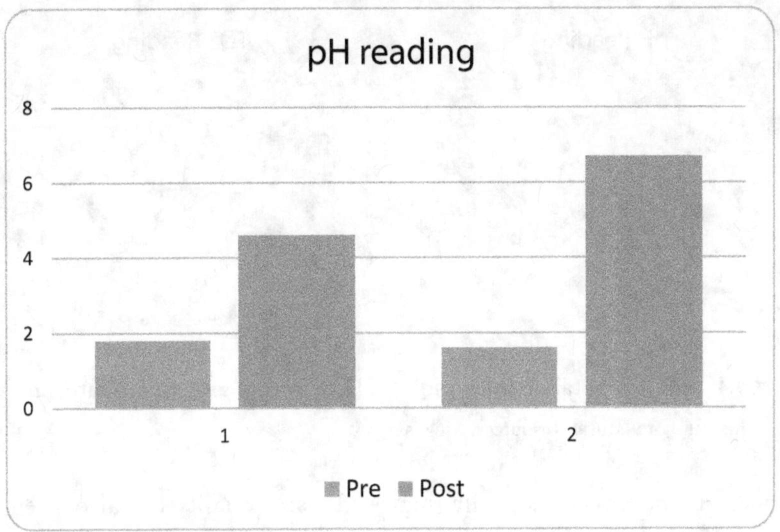



**FIGURE 9.3**   pH reading pre and post treatment

*Source:* Community-led landscape project

acidic due to the geological composition of the area. Situated within the southern belt of the state, which is rich in coal deposits, the groundwater is influenced by these natural conditions. With 120 households relying heavily on local springs for domestic use, the acidic water has forced villagers to travel long distances for drinking water and laundry, creating significant hardships.

Determined to find a solution, local people came together to address the issue collectively. They identified the Chikongbak spring as the ideal site for an Open Limestone Channel (OLC) initiative. With active participation from other villagers, they began by testing the water with a portable kit. This test revealed high iron levels, low pH and significant turbidity. Under the guidance of technical experts, people worked collaboratively to redirect the spring's flow into specially designed channels. These channels were constructed to adjust the water's pH naturally by passing it through five limestone-filled chambers, using gravity to facilitate the process.

To enhance the water quality, the villagers also planted vetiver grass around the site. To address turbidity, they built a slow sand filter – a simple yet effective solution   to capture suspended particles and improve water clarity.

The entire project was a testament to the villagers' unity and resourcefulness, as they contributed their labour, local knowledge and materials to achieve this success.

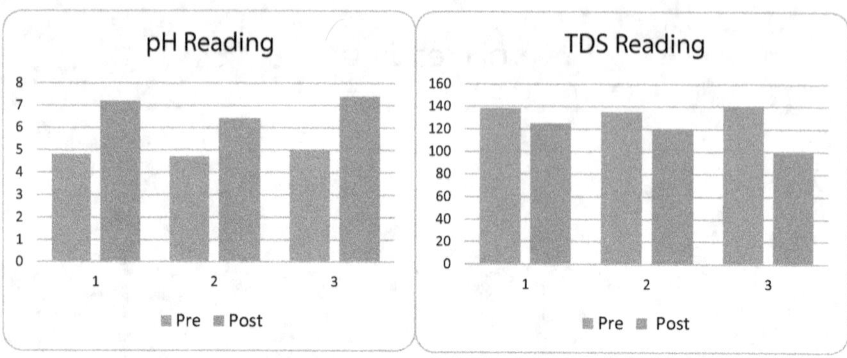

**FIGURE 9.4**   pH and total dissolved solids (TDS) before and after treatment

*Source:* Institute of Natural Resources, Meghalaya

In short, the project not only improved water quality but also strengthened community bonds, as villagers worked side by side to create a sustainable solution for their shared challenge. Their collective efforts have set an example of how community-driven initiatives, supported by technical expertise, can effectively address environmental and resource-related issues.

### 9.3.4 Results

Going by the data provided, a noticeable pH increase was observed from the water source to the post-treatment stage. Readings indicate significant improvements in water quality after it traverses through the chambers, resulting in a clearer appearance as it undergoes filtration within the treatment process. The pH value recorded falls within the standard defined by the Bureau of Indian Standards.

## 9.4 Discussion

### 9.4.1 Interpretation of Findings in Relation to Their Replicability

Data collected from these sites indicate that the OLCs with chambers are more effective, with a 90% success rate in neutralising acidic water to the desired standard. In contrast, on-site treatments using drainage ditches have a 60% success rate, because sedimentation over time buries the limestone, reducing its efficiency and complicating maintenance. While both methods are successful, the effectiveness of each depends on multiple factors, including ongoing mining activities in the surrounding area (Centre for Mining Environment. 2001–2002)..

In sites with active coal mining, the reaction between limestone and highly polluted water results in a greater accumulation of heavy metal precipitates.

This leads to frequent armouring of the limestone, necessitating more regular maintenance to sustain its effectiveness. Conversely, in areas without active mining, the presence of metal precipitates is significantly lower, since AMD originates from abandoned mines only. Heavy metals dissolved by acidic water are more prevalent in older sites than in newly established ones. While these sites still require maintenance, the intervals are longer and the treatment process is comparatively more efficient due to lower pollutant levels in the water.

In sum, the effectiveness of the OLCs is highly dependent on mining activity in the region. Chambered OLCs demonstrate a pH increase of up to 3.5, whereas drainage ditch designs exhibit a 2.6 unit increase. In a chambered OLC, discharge can be effectively controlled, ensuring the attainment of optimal standards at the treatment's conclusion without sedimentation issues. In contrast, a drainage ditch lacks these capabilities and proves difficult to maintain, particularly during the monsoon season.

### 9.4.2 Implications for Theory, Practice and Policy

Our study contributes to the growing body of literature on passive AMD treatment by highlighting the effectiveness of the OLCs in neutralising acidic water and improving overall water quality. It reinforces the theory that natural, locally available materials such as limestone can serve as cost-effective and sustainable remediation methods. Additionally, the findings emphasise the role of phyto-remediation, particularly the use of vetiver grass, in heavy metal adsorption.

These results validate environmental remediation models that advocate for community-driven, low-cost and scalable solutions to water contamination issues.

The implementation of the OLCs in various villages demonstrates their feasibility and adaptability in different environmental and socio-economic contexts. Our study underscores the importance of site-specific designs, where factors such as iron content, water flow rates and limestone armouring influence the efficiency of the system. The success of chambered OLCs in maintaining higher pH levels and reducing sedimentation highlights the need for controlled discharge mechanisms in future applications.

Further, the integration of vetiver grass and slow sand filtration in treatment chambers offers additional improvements in water quality. The findings advocate for community participation in operation and maintenance, with local water facilitators playing a crucial role in sustaining these systems. Training and capacity-building initiatives should be prioritised to ensure long-term success.

Our study emphasises the necessity of integrating passive AMD treatment strategies into broader water resource management policies. Policymakers should recognise OLCs as viable alternatives to conventional water treatment methods, particularly in resource-constrained regions. Incentives for community-driven water management projects, such as subsidies for limestone procurement and technical training programmes, should be introduced.

### 9.4.3 Comparison with Existing Literature

Based on a pilot study conducted in Mukhaialong, an upgraded version of the technology has been implemented across most of the OLC sites. This upgrade aimed to increase understanding of the factors that affect the technology's performance, while also improving the operation and maintenance processes within the community. The implementation of this advanced technology provides valuable insights into its sustainability, ensuring more efficient management and ensuring benefits for the local population.

## 9.5 Summary and Conclusion

### 9.5.1 Key Findings Summary

Based on the foregoing analysis, we can infer that chambered OLCs prove more effective in treating streams affected by acid mine drainage (AMD) when compared to drainage ditch designs. They demonstrate greater efficiency and offer economic advantages with reduced and simplified maintenance requirements. The vetiver system can be integrated into separate chambers to increase adsorption and absorption time for heavy metal precipitates. This way, it improves the water quality further by reducing total dissolved solids, biochemical oxygen demand and water hardness.

This approach addresses common issues associated with AMD-affected streams and springs, such as elevated metal concentrations, increased turbidity and poor water quality, which negatively affect aquatic life and human health.

### 9.5.2 Future Research and Practical Applications

1. Explore alternative neutralising agents, such as alkaline by-products or microbial inoculants that promote more effective neutralisation.
2. Integrated treatment methods: Combining the OLC method with technologies like constructed wetlands could offer a comprehensive solution to address both acidity and toxic metal levels.
3. Reclamation and ecosystem restoration: Future applications of AMD neutralisation systems could involve not only water treatment but also ecosystem restoration. Research could focus on integrating OLC systems with efforts to rehabilitate affected terrestrial ecosystems, such as

reforestation or soil remediation, to restore biodiversity and improve overall ecosystem health.

4. Long-term sustainability and monitoring: Research could focus on the long-term sustainability of OLC systems, considering factors such as the degradation of limestone over time, the consequences of seasonal changes and the effects of climate change (e.g., increased rainfall or drought). Developing more robust monitoring techniques using remote sensing, sensor networks and data analytics could enable real-time monitoring of water quality, allowing for proactive management and adjustments to the treatment process.

## References


Blahwar, B., Srivastav, S. K., & Smeth, J.B. (2012). Use of high-resolution satellite imagery for investigating acid mine drainage from artisanal coal mining in North-Eastern India. *Geocarto Intl, 27,* 231–247.

Caruccio, F. T. (1975). Estimating the acid potential of coal mine refuse. In M. J. Chadwick & G. T. Goodman (Eds.), *The ecology of resource degradation and renewal* (pp. 197–205). Blackwell Scientific Publications.

Chabukdhara, M., & Singh, O. P. (2016). Coal mining in northeast India: An overview of environmental issues and treatment approaches. *International Journal of Coal Science & Technology, 3*(2), 87–96.

Chadwick, M. J. (1973). Methods of assessment of acid colliery spoils as a medium for plant growth. In R. J. Hutnik & G. Devis (Eds.), *Ecology and reclamation of devastated land* (pp. 81–91). Gordon and Breach.

Chakrabotry, M. K., Singh, R. S., Chaulya, S. K., Ahmad, M., & Dhar, B. B. (2002). *Water quality simulation of Damudar River.* Proceedings on Environment Management Capacity Building.

Dadhwal, K. S. (1999). Rehabilitation of limestone mine spoils with reference to agroforestry. *Indian Journal of Agroforestry, I*(2), 141–148.

Pyrbot, W., Shabong, L., & Singh, O. P. (2019). Neutralisation of acid mine drainage contaminated water and eco restoration of a stream in a coal mining area of East Jaintia Hills, Meghalaya. *IMWA, 38,* 551–555.

Sahoo, P. K., Tripathy, S., Equeenuddin, S. M., & Panigrahi. M. K. (2012). Geochemical characteristics of coal mine discharge vis-a-vis behaviour of rare earth elements at Jaintia Hills coalfield, North-Eastern India. *Journal of Geochemical Exploration, 112,* 235–243.

Singh, O. P. (2012). *Impact of coal mining on water resources and environment in Meghalaya.* North Eastern Hill University. www.researchgate.net/publication/303487611

Swer, S., & Singh, O. P. (2003). Coal mining impacting water quality and aquatic biodiversity in Jaintia Hills District of Meghalaya. *ENVIS Bulletin Himalayan Ecology, 11,* 26–33.

Centre for Mining Environment. (2001–2002). *Technical assistance project: Mining sector sub-continent.* Indian School of Mines.

Tiwary, R. (2001). Environmental impact of coal mining on water regime and its management. *Water Air and Soil Pollution, 132,* 185–199.

Truong, P., & Luu, T. T. (2015). Vetiver system for wastewater treatment. *Pacific Rim Vetiver Network Technical Bulletin,* 2015/2.

Wansah, J. F., Nongkynrih, J. M., & Kharpran-Daly, B. K. (2019). Restoration of the Moolawar Stream in East Jaintia Hills, Meghalaya

Ziemkiewicz, P. F., Skousen, J. G., Brant, D. L., Sterner, P. L., & Lovett, R. J. (1997). Acid mine drainage treatment with armored limestone in open limestone channels. *Journal of Environmental Quality, 26*(4), 1017–1024.

# 10

# COMMUNITY LED HYDROLOGICAL SPRING MAPPING AND SPRINGSHED MANAGEMENT IN MEGHALAYA

*Gaurav Singh and Jillianda Kharjana*

## 10.1 Introduction

### 10.1.1 Purpose, Scope and Significance

In the hilly terrain of Meghalaya, where the landscape is marked by lush green forests, cascading waterfalls and steep slopes, water sources are both abundant and diverse. Springs, in particular, play a crucial role in supplying water to rural communities, especially in remote and inaccessible areas of the state. These natural sources are often the lifeblood for local populations, providing water for drinking, irrigation, sanitation and other domestic needs.

In fact, springs are the primary water source in the state for about 80% of people living in rural areas, who rely on these natural springs for their daily needs. During dry seasons, people face a critical water issue, despite having regions with extreme rainfall such as Cherrapunji and Mawsynram. Meghalaya receives an average annual rainfall of 2,818 mm, while these two areas experience over 11,000 mm of rainfall each year.

According to a Niti Ayog report, almost all springs have witnessed a decline in discharge due to land-use changes, including increased diversion, as well as pumping, groundwater exploitation, pollution of surface and groundwaters, degradation of natural recharge areas and, possibly, climate change too.

An increasing demand for water next to a decreasing supply (as attributed to these factors) underscores the necessity to focus on the accessibility of water resources, specifically with respect to springs, since they have limited

DOI: 10.4324/9781003735380-12

resources. The higher water demand has led to a reduction in spring discharge and, consequently, to their drying up in the dry season. This, in turn, has led to significant difficulties for women in the villages, because they are forced to travel often long distances to fetch water.

To address this and ease the burden on women (and children), the Ministry of *Jal Shakti* launched the *Jal Jeevan* Mission. It aimed to provide tap water connections to every household. But locating sustainable water sources for this initiative appeared problematic due to a lack of accurate data on locations and the availability of springs.

In this mission, a water quality test was conducted across six districts to assess the safety of various water sources. It examined total coliform, specific conductance, *E. coli*, the presence of *Salmonella* and yeast or mould in the samples. The results revealed that most water sources were contaminated with harmful bacteria. This made them unsafe for consumption.

So, the idea of spring mapping and springshed management was introduced to establish baseline data for future reference and identify critical springs that might or would require intervention.

To ensure sustainable springshed management, community-led initiatives in Meghalaya paid attention to mapping, monitoring and preserving water resources. Spring mapping was the first step, where local communities, NGOs and researchers collaborated to document spring locations, flow patterns and water quality using participatory methods and Geographic Information System (GIS) technology.

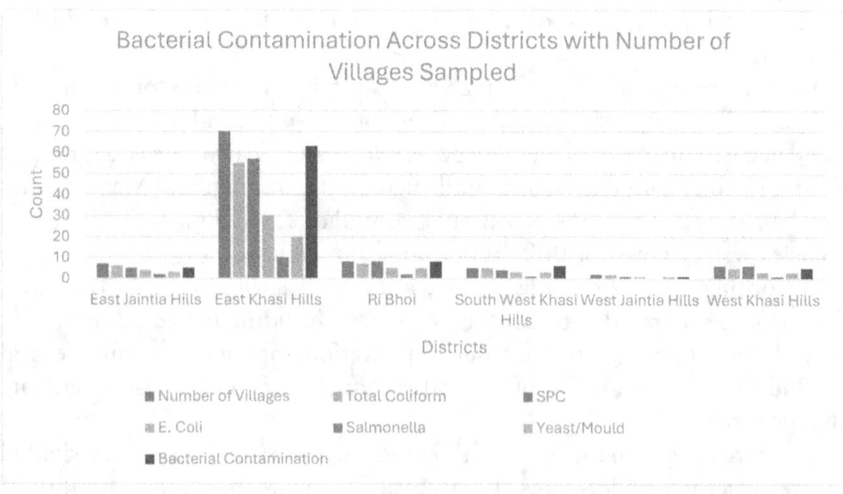


**FIGURE 10.1**   Contamination report on springs across six districts of Meghalaya

*Source:* Status Report of Bacterial Contamination of Spring Water Samples Tested at the State Food Testing Laboratory, Meghalaya

Community knowledge and participation were essential, since their traditional insights into seasonal changes and water usage proved to augment scientific assessments. Data collection and monitoring enabled villagers to track spring health, detect issues like declining flow or contamination and inform resource management decisions. This participatory approach turned out to foster ownership, which would lead to long-term conservation and the resilience of water sources.

### 10.1.2  Organisation of the Chapter

In this chapter, we explore the significance of springs in rural water supply in Meghalaya. It sheds light on their ecological significance and contribution to climate resilience, the challenges they face and the community-led efforts to map springs and develop sustainable springsheds for effective water resource management.

The idea is to provide a comprehensive understanding of hydrological spring mapping and springshed development. The latter comes to the fore as a key strategy for water security and community empowerment.

First comes an outline of the process of hydrological spring mapping, including data collection, recharge zone identification and discharge assessment. It is followed by a discussion on community-led interventions such as afforestation, soil conservation and sustainable land-use practices to improve groundwater recharge.

Then it is time to bring out challenges like community participation, funding constraints and technical limitations, at the same time proposing potential solutions. Next, the role of policies and institutions in supporting springshed initiatives receives attention, preceding case studies of successful projects.

We conclude with recommendations for scaling up sustainable springshed management, strengthening community engagement and addressing policy gaps to bring about long-term water security.

### 10.1.3  Review of Literature

Springs play a crucial role as water sources, especially in rural and mountainous areas, yet their importance in water resource management is often overlooked. Rural communities rely heavily on springs for drinking water, agriculture and livestock (Shah, 2017; Negi et al., 2019).

Erratic rainfall, changing geology, warming climate and increased water demand, compounded by deforestation, grazing, soil erosion and urbanisation, contribute to declining spring discharge across the Himalayas (Valdiya

& Bartarya, 1991, 1989; Negi & Joshi, 2002; Mahamuni & Kulkarni, 2012). Deforestation and reduced precipitation are identified as primary causes of spring decline (Singh & Pande, 1989; Negi & Joshi, 2002, 2004; Agarwal et al., 2012; Kumar & Ansari, 2012; Mahamuni & Kulkarni, 2012; Vashisht & Bam, 2012).

Systematic monitoring is crucial for effective spring management in the Himalayas (Bruijnzeel & Bremmer, 1989; Alford, 1992; Negi, 2002).

Unsustainable land-use practices, deforestation and climate change are exacerbating the decrease in spring discharge, posing water security risks in many regions (Tambe et al., 2012). This has spurred interest in hydrological spring mapping and springshed development as sustainable management approaches.

The ICIMOD manual (by Shrestha et al., 2018) outlines protocols for revitalising springs in the Hindu Kush Himalayas, emphasising hydrological methods and community involvement. Kharsyntiew (2017), in turn, highlights the effectiveness of spring mapping in Meghalaya to deal with rural water scarcity and to enhance livelihoods. Research by Asoka et al. (2018) underscores the link between precipitation and groundwater recharge, stressing the vulnerability of springs to climate variations.

Comprehensive spring inventories and hydrological mapping are advocated to identify recharge zones and assess anthropogenic impacts (Singh et al., 2015). Springshed development, supported by resource books like those from the Department of Water Resources (2023) and the Swiss Agency for Development and Cooperation (2023), focuses on afforestation, soil conservation and participatory strategies for sustainable management.

Successful interventions in springshed management, as documented by Tiwari et al. (2020) and Rawat et al. (2021), demonstrate the value of local knowledge and community engagement in improving spring flow and groundwater recharge.

Besides, collaboration among policymakers, scientists and local communities is crucial to take on water challenges comprehensively, especially amid current environmental uncertainties (Ministry of Jal Shakti, 2023).

### 10.1.4 The Gaps in Spring Mapping and Springshed Development

Gaps in spring mapping and springshed development included a lack of data concerning spring locations, recharge zones and flow dynamics. Seasonal variability presented challenges, since there was a limited understanding of how seasonal variations affected spring discharge. Community awareness was lacking, as was engagement in the scientific aspects of recharge zone management. Sustainable practices such as soil conservation, afforestation and sustainable agriculture have not been widely adopted yet. Monitoring and evaluation were not available properly, hindering ongoing assessments of spring health and recharge efforts. Scaling successful interventions remained challenging across larger regions.

### 10.1.5 Theoretical Framework

The theoretical framework for hydrological spring mapping and springshed development is based on watershed management, hydrogeology and community-based natural resource management. Springs function as discharge points for groundwater systems, which are recharged through precipitation and infiltration in specific zones.

Meghalaya's springshed initiative integrated hydrogeological, ecological and socio-economic approaches to ensure sustainable groundwater conservation. The watershed-based approach focused on aquifer recharge, land-use effects and climate resilience. Ecosystem services refer to the role of forests and soil conservation in maintaining spring discharge. The common-pool resource concept looks at community participation and collective ownership, fostering long-term stewardship.

An adaptive co-management model allows local knowledge to guide sustainable interventions, while integrated water resource management promotes collaboration among policymakers, scientists and communities.

These principles aligned with global frameworks such as the sustainable development goals (SDGs), emphasising water security, climate action and ecosystem sustainability. Meghalaya's approach can serve as a scalable model for similar water-stressed regions.

## 10.2 Methodology

### 10.2.1 Description of Methodology

There was a comprehensive approach to map, manage and monitor springs. It included the development of a spring manual and training for village community facilitators, green volunteers, engineers and GIS experts, along with additional master trainers. Advanced technologies such as the Geographic Information System (GIS), Global Navigation Satellite System and other mobile applications were also being used.

Over 55,000 springs were mapped under the External Aided Projects scheme. A central dashboard was developed to monitor spring-related activities in real time. Spring mapping and springshed management were regarded as the best approach for several reasons.

Despite significant investments in natural resource management, the lack of baseline data was a major problem. Spring mapping addressed this gap by providing crucial information on springs. This enabled the development of sustainable management plans.

Various parameters were captured, including discharge measurements, water quality indicators such as pH, total dissolved solids (TDS), electrical conductivity, salinity and temperature, along with socio-economic factors. Monthly monitoring of critical springs supported springshed development activities aimed at improving spring discharge through recharge interventions.

Community engagement played a key role in this approach with participatory rural appraisal (PRA) exercises followed by a hydrogeological field survey and interventions. This involvement has empowered local people in their understanding of conservation.

The approach also proved cost-effective, ensuring efficient resource utilisation while achieving significant benefits in water management. Further, it provided government departments with a clearer understanding of on-the-ground realities, simultaneously improving water management strategies and interventions. Geospatial technology was incorporated to create a visual representation of spring locations via the dashboards. These enhanced accessibility to data on spring locations, water quality and flow rates, which aided decision-making, in turn.

### 10.2.2 Justification of Methodology

The methodology integrates a participatory approach, actively engaging local communities to incorporate traditional knowledge and ensure ownership. Scientific rigour is maintained through detailed recharge area demarcation based on geological assessments, contributing to intervention effectiveness. Initiatives are customised to local contexts, focusing on region-specific challenges for greater relevance.

Sustainability is promoted through capacity-building and scientific interventions, empowering communities for long-term resource management. Evidence-based monitoring ensures data-driven decision-making and adaptive management. The approach is scalable and replicable, making it applicable to other water-stressed regions. Moreover, it aligns with global and national policies, including the SDGs and the National Rural Drinking Water Programme 2013.

### 10.2.3 Limitations and Challenges

Despite extensive efforts, comprehensive spring mapping remains incomplete, because land in Meghalaya is community-owned under the 6th Schedule. Some village leaders resisted participation because of a lack of awareness and apprehension. This slowed down the progress of the intervention.

Remote villages with poor road access hindered mapping and monitoring, particularly during the monsoon when many areas became inaccessible.

Also, land acquisition for recharge interventions appeared to be a problem, since most of the land is privately owned and used for livelihoods.

Besides, deforestation and shifting rainfall disrupted groundwater recharge, exacerbating seasonal water scarcity, even in high-rainfall areas like Cherrapunji and Mawsynram. Meghalaya has seen a 27% increase in surface water (Ashutosh. 2024) and a loss of 270 km$^2$ of forest (Forest Survey of India, 2011–2021). This indicates potential long-term effects on groundwater availability.

With over 80% of the population relying on agriculture and forests, dealing with these challenges effectively is critical to ensuring sustainable water security and adequate springshed management in the region.

## 10.3 Results

### 10.3.1 Springshed Approach under MBDA

The Meghalaya Institute of Natural Resources (MINR) had initially conceptualised spring mapping and springshed management under the Meghalaya Basin Development Authority (MBDA). This concept was subsequently expanded through various externally funded projects, including the JICA-funded MegLIFE, the World Bank-funded Meghalaya Community-Led Landscapes Management Project (MCLLMP) and the IFAD-funded Meghalaya Livelihoods and Access to Markets Project (MLAMP).

Now, the initiative is advanced through MegLIFE, under which a springshed management and development initiative was adopted. It was also taken up further in the Kreditanstalt für Wiederaufbau-funded MegARISE project. This one is particularly for the rejuvenation of two key catchments in the state: The Umiew catchment in the East Khasi Hills and the Ganol catchment in the West Garo Hills.

### 10.3.2 Case findings

#### 10.3.2.1 Meghalaya Springs Protection Initiative by MINR

In 2015, the MINR launched the "Meghalaya Springs Protection Initiative" with the goal of developing a climate-resilient springshed management plan for vulnerable springs. Under this initiative, Mawphanlur village in Mawthadraishan Block, West Khasi Hills, was identified for this project. It led to the mapping of 23 springs and regular monitoring of 10 priority springs to assess their condition before, during and after project implementation.

As part of efforts to improve spring recharge and water availability, several conservation measures were carried out. These included the excavation of 1,260 contour trenches, peripheral bunding and afforestation of 10,827 trees in the catchment area. Additionally, two check dams were built for water conservation, a water filter tank was installed and eco-water filters were distributed to 30 households to ensure access to safe drinking water. Check dam constructions play a crucial role in improving springs by conserving surface runoff and facilitating groundwater recharge. When water is stored in the check dam reservoir, a portion gradually percolates into the subsurface, thereby enhancing soil moisture and replenishing aquifers. If the structure is located within a spring recharge zone, this percolation process directly contributes to sustaining or augmenting spring discharge. The stored water is processed through eco-water filters and filter tanks to ensure its safety and potability. This dual approach not only supports spring rejuvenation by improving recharge dynamics but also provides communities with access to safe drinking water.

There is a misconception that constructing recharge structures (e.g., check dams, trenches, ponds) will automatically and effectively rejuvenate any spring. Scientific evidence clarifies that the success of these structures is highly dependent on the hydrogeology of the specific location. Many springs, particularly those in hard-rock mountainous regions like the Himalayas, are fed by deep, fractured aquifers. Surface recharge structures, which primarily increase water infiltration in the topsoil, may have a minimal or no impact on these deep systems if the water cannot penetrate to the required depth. For recharge to be effective, it must be strategically implemented in the spring's specific springshed or recharge area, which can be far from the spring's outlet and does not always align with surface topography. This "springshed management" approach, which combines scientific hydrogeological mapping with local knowledge, has shown positive results in several case studies by targeting the actual recharge zones.

The project resulted in notable improvements in spring discharge, soil fertility and overall water security for the community. With enhanced water availability, households experienced better access to water for domestic and agricultural use, reducing the risk of crop failure and economic loss caused by moisture stress. Further, the intervention helped curb land degradation, contributing to aquifer recharge and long-term water sustainability in the region.

Thanks to community-driven efforts and nature-based solutions, Mawphanlur has taken a critical step toward climate resilience, ensuring sustainable water management and improved livelihoods for its residents.

### 10.3.2.2 Spring Mapping across Meghalaya under CLLMP

The Meghalaya Basin Management Agency, with funding from the World Bank for the "Community-Led Landscape Management Projec" (CLLMP), initiated a statewide initiative to map and document the state's springs for the first time. The objective was to identify and record all springs that

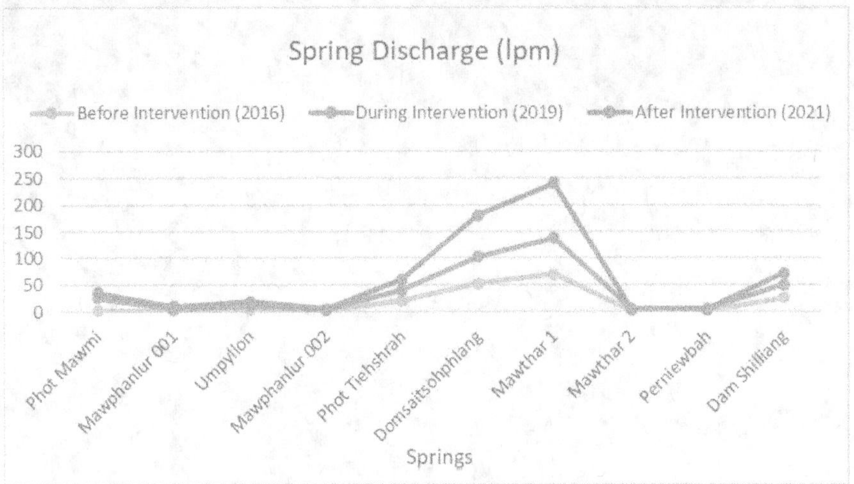


**FIGURE 10.2**    Springs discharge trend pre- and post-intervention

*Source:* Field data collected during the project period

communities depend on, while systematically measuring and monitoring key parameters such as discharge, pH, coliform levels, total dissolved solids (TDS) and electrical conductivity, among others.

The MINR facilitated and executed this initiative at the field level with the active involvement of village community facilitators (VCFs). Over 800 VCFs were trained to support sustainable water management practices and strengthen community participation. More than 55,000 springs were mapped across project and non-project villages.

A centralised public dashboard was created for visual representation of the spring locations, integrating critical data on water quality, flow rates and surrounding ecosystems. Among these, 3,000 springs located within CLLMP project villages were monitored monthly to assess discharge and water quality. This regular monitoring enabled planning and implementation of appropriate interventions under the project.

These are key findings:

1. Aminda Simsanggre, a village in Gambegre Community & Rural Development Block, is home to 95 households; the majority rely on agriculture-based livelihoods. A village survey conducted prior to the implementation of the CLLMP showed that only 14 households had access to water connections. The rest, comprising Aminda Kongsang Gittim, Aminda Ading and Bazaar Gittim, depended on distant water sources, often requiring up to an hour-long trip to fetch water.

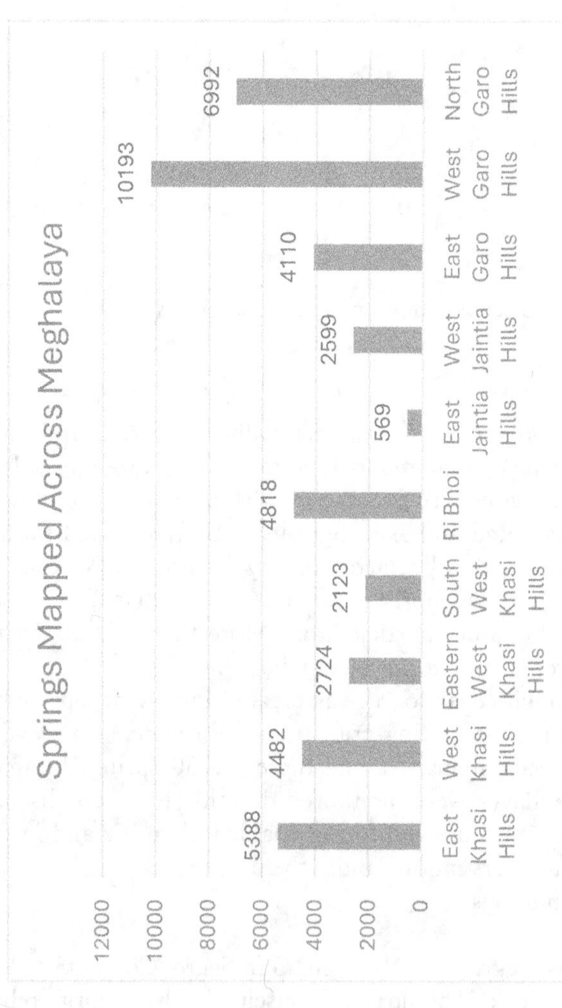


**FIGURE 10.3**   Springs mapped across Meghalaya

*Source:* Meghalaya State Geohub – Spring Dashboard

Recognising the urgent need for sustainable water management and easily accessible water sources, people came together to formulate a community natural resource management plan (CNRMP) in a participatory village meeting. It was attended by men and women and officials, including the village chief, village natural resource management committee members and village facilitators.

The proposed interventions are aimed at improving water availability and restoring degraded ecosystems. They included the development of springs, the construction of check dams, a community nursery, afforestation activities and contour trenches.

Under this initiative, the community undertook the construction of cemented storage tanks and dams at four separate locations to conserve water. Contour trenches were dug to facilitate groundwater recharge and ensure a perennial flow in the springshed area.

A 12-bedded community nursery was set up at Aminda Chiring, from where transplantation of local tree species to forest reserve areas commenced. To improve agricultural productivity, the Berkeley method of hot composting was introduced at Aminda Chiring, enhancing both the quality and yield of crops.

Since the project's inception in 2018, the village has witnessed remarkable progress in just over three years (i.e., the time of this study). Springshed management interventions have directly benefited 12 households already, while soil and water conservation efforts have supported another 13. The community nursery has provided saplings to all households, with 40% of the saplings successfully sold, providing income for local growers and creating market opportunities within the community, which in turn helps strengthen the local economy and supports long-term economic sustainability. In addition, the village collectively declared its forested area a "community forest reserve," establishing rules and regulations to protect water bodies and ensure long-term conservation.

Through these strategic interventions, Aminda Simsanggre has successfully strengthened its natural resource management programme, securing water availability, restoring degraded landscapes and fostering environmental resilience for future generations.

2.  Chambagre village in Selsella C&RD Block is one of the 1,350 villages that participate in the IFAD-supported Meghalaya Livelihoods and Access to Markets Project. Earlier, the village struggled with severe water shortages for both domestic use and irrigation, significantly affecting paddy cultivation. The absence of an irrigation system had led to declining agricultural productivity, while women and children spent up to two hours daily fetching water from distant sources.

Recognising the need for sustainable water management, the community actively carried out a participatory rural appraisal to identify key challenges and develop solutions through natural resource management interventions.

This resulted in the construction of two water storage tanks in the upper and lower regions of Chambagre, with capacities of 14,000 L and 17,000 L, respectively. An irrigation canal covering one hectare of farmland was constructed, which significantly improved agricultural water access.

With respect to afforestation, 5,000 trees were planted across 10 ha of community land to restore the catchment area, enhance groundwater recharge and ensure long-term water availability.

Here too, a community nursery was established to promote the growth of indigenous plant species known for their role in water retention and ecological balance. Analogous to grass cover, afforestation and tree plantation contribute to water conservation by reducing surface runoff and enhancing the residence time of rainwater, thereby promoting infiltration. The root systems of trees further facilitate this process by developing preferential flow paths and macropores in the soil, which enable water to percolate to deeper subsurface layers. When implemented in the recharge zones of springs and other groundwater-dependent ecosystems, such plantation practices can augment aquifer recharge and subsequently enhance spring discharge.

These interventions have had a profound impact on the village: 42 households now benefit from the water storage tanks, while 15 households have directly benefited from the irrigation canal, enabling more efficient farming. The time spent fetching water has been drastically reduced to just 30 minutes per day, freeing up valuable time for other productive activities.

Beyond water conservation, the project has instilled a deep awareness of spring rejuvenation and the importance of preserving native vegetation among the villagers. Besides, the afforestation drive has not only improved ecological resilience but also created livelihood opportunities for unemployed youth in the village.

Chambagre's experience demonstrates how community-driven natural resource management can transform water security, enhance agricultural productivity and create sustainable employment opportunities, strengthening rural resilience and livelihoods.

Thanks to their dedicated efforts in community-led natural resource management, Aminda Simsanggre and Chambagre secured 2nd and 3rd place, respectively, in the "Best Village Panchayat" category (North-East Zone) at the 3rd National Water Awards. This recognition was for their proactive approach to water conservation and sustainable resource management.

The success of these villages stemmed from the strong leadership of their village employment councils and their community-driven conservation efforts.

3.  Mawkyrdep, a small village in the Ri Bhoi District, had had to grapple with growing concerns over water scarcity, forest degradation and biodiversity loss for quite some time. Recognising the need for a sustainable solution, a Mawkyrdep village natural resource management committee initiated efforts to restore and conserve their natural resources with support from the World Bank-funded CLLMP. This led to the development of a community natural resource management plan. It was a strategic framework that reflected the collective needs and aspirations of the villagers.

To ensure that interventions addressed the most pressing concerns, a participatory rural appraisal was conducted first.

The findings highlighted several key challenges: Water shortages affecting both domestic use and irrigation, deteriorating forest cover and the need for conservation of endemic and medicinal plant species. Most critically, two vital springs that served as the village's primary water sources required urgent restoration and recharge improvements.

In response, the community implemented a series of targeted interventions. Two check dams were constructed at strategic locations to improve water availability, while contour trenches were dug along the slopes to facilitate groundwater recharge and create spring sustainability. To restore the catchment area, extensive afforestation efforts were undertaken, with native tree species planted to stabilise the land and improve water retention. Then, a community nursery was set up to raise saplings of endemic and medicinal plants, ensuring a continuous supply for reforestation efforts.

Through these initiatives, the village successfully addressed its environmental challenges and promoted long-term sustainability. Their outstanding efforts in natural resource management earned them national recognition: Mawkyrdep was awarded 3rd place in the "Best Village Panchayat" category at the 4th National Water Awards held in New Delhi.

By combining traditional knowledge with scientific approaches, the people of Mawkyrdep not only revitalised their natural resources but also demonstrated the power of community-driven conservation. Their efforts serve as a model for other rural communities seeking to implement sustainable resource management.

## 10.4 Discussion

### 10.4.1 Interpretation of Findings in Relation to Their Replicability

Meghalaya's springshed interventions underscore the importance of springshed-based recharge strategies, soil conservation, afforestation and climate-resilient land-use practices. All these can be effectively adapted to other Indian Himalayan and global mountainous regions.

Successful replication requires community participation through local facilitator training, scientific monitoring for hydrological assessments, multi-stakeholder collaboration involving policymakers, scientists and funding agencies, and climate adaptability to counter changing precipitation patterns. Meghalaya's model integrates scientific knowledge with traditional practices, ensuring sustainable groundwater management and serving as a replicable framework for addressing water security challenges in hilly and rural landscapes (Rathod, R., et al. 2021).

The springshed development and management initiative is scaled up through various externally funded projects, including the JICA-funded Meghalaya Livelihood and Finance Ecosystem (MegLIFE), the World Bank-supported Meghalaya Community-Led Landscape Management Project (MCLLMP) and the IFAD-funded Meghalaya Livelihoods and Access to Markets Project (MLAMP). The most recent expansion is through the KfW-funded MegARISE project, which focuses on rejuvenating two key catchments: The Umiew catchment in East Khasi Hills and the Ganol catchment in West Garo Hills.

Under the MegLIFE project, the Meghalaya Basin Development Authority is implementing comprehensive springshed management activities across 500 villages, aiming to enhance water security and sustainability through targeted watershed and recharge interventions. The project is built on extensive collaboration with expert agencies to achieve its objectives and ensure a comprehensive approach to springshed management.

Key partners are the Central Himalayan Rural Action Group, the Advanced Centre for Water Resources Development and Management, Rajarha PRASARI and People's Science Institute. They provide essential technical expertise, capacity-building and support for the implementation of springshed interventions.

Further, the project has organised training and capacity-building sessions for a broad range of stakeholders, including village community facilitators, project staff and personnel from government departments such as those of Water Resources and Soil & Water Conservation. These training initiatives enhanced the knowledge and skills of participants working across the water sector, fostering effective collaboration and a cohesive implementation of springshed management activities.

As part of the MegLIFE project, over 1,700 springs across 500 villages in the state have been identified and are being monitored regularly. In all these villages, one critical spring is selected and eventually treated comprehensively through springshed management.

This approach is currently followed for 168 critical springs across 168 villages. It aims to enhance water security and sustainability through targeted recharge interventions within a village. Collectively, these initiatives have laid a strong foundation for the long-term viability of Meghalaya's springs, underscoring their essential role in sustaining rural communities and working towards climate resilience in the region.

### 10.4.2 Implications for Theory, Practice and Policy

Meghalaya's springshed management approach advances understanding of groundwater dynamics in mountainous regions. It emphasises an integrated hydrogeological and socio-ecological framework. Traditional models often overlook springs as dynamic systems influenced by climatic variability and anthropogenic activities. This project brings out the effectiveness of combining scientific hydrology with indigenous knowledge, reinforcing the value of interdisciplinary approaches in sustainable groundwater management.

Also, it underscores the role of participatory methodologies in natural resource conservation, contributing to community-based water governance theories and experiences. The large-scale mapping of over 50,000 springs, the preparation of 168 detailed technical reports for critical springs and the training of over 800 village community facilitators offer a replicable model for springshed management.

The project's success demonstrates that integrating watershed-based recharge strategies, afforestation and soil conservation measures can restore and sustain spring flows. The implementation of these activities under various externally funded projects highlights the importance of multi-stakeholder collaboration. The creation of a public dashboard by the MBDA for real-time monitoring of water quality, flow rates and ecosystem health set a precedent for data-driven decision-making in water governance.

Above all, perhaps, Meghalaya's experience confirms that community participation is key to ensuring long-term sustainability of water conservation efforts.

The Meghalaya model is a compelling case for integrating springshed management into state and national water policies. Policymakers should recognise springs as critical water sources and promote systematic spring inventories, hydrological mapping and climate-resilient recharge interventions.

The initiative also demonstrates the wisdom and efficacy of decentralised water governance, where local communities actively participate in decision-making. Surely, long-term financial commitments, as seen in the externally funded projects, are essential for ensuring the sustainability of water conservation programmes.

Replicating this model in other water-stressed, hilly regions requires institutional support, scientific research and policy-driven incentives to encourage participatory and adaptive water resource management.

In conclusion, Meghalaya's springshed management initiative provides valuable insights that can inform water conservation strategies at local, national and global levels, ensuring long-term water security in ecologically sensitive regions.

### 10.4.3 Comparison with Existing Literature

Meghalaya's springshed management approach builds upon existing research on Himalayan spring hydrology, watershed management and community-based conservation. Studies by Valdiya and Bartarya (1989, 1991) and Negi and Joshi (2002) have documented the declining discharge of Himalayan springs due to deforestation, land-use changes and climate variability. The state's interventions reaffirm these findings by highlighting the consequences of erratic rainfall and anthropogenic pressures on groundwater recharge.

Additionally, Mahamuni and Kulkarni (2012) and Tambe et al. (2012) have emphasised the need for systematic monitoring and recharge mapping. Meghalaya has operationalised this through large-scale spring mapping and a public dashboard for real-time monitoring.

Unlike earlier studies that focused on theoretical assessments of spring degradation, Meghalaya's initiative provides a scalable, applied framework for springshed management, integrating watershed-based interventions and participatory conservation strategies. This aligns with Singh et al. (2015), who advocated for comprehensive spring inventories and hydrological mapping to identify recharge zones and assess human-induced effects.

The emphasis on community-led conservation resonates with Bruijnzeel and Bremmer (1989) and Alford (1992), who stressed that effective water scarcity solutions require integrating local knowledge with scientific hydrology.

Meghalaya's interventions also align with global best practices, such as those documented by the Swiss Agency for Development and Cooperation (2023) and the Department of Water Resources (2023). Both look at afforestation, soil conservation and participatory governance. The work of Asoka et al. (2018) linking precipitation intensity with groundwater recharge correlates with Meghalaya's emphasis on climate-adaptive water management strategies. Finally, the ICIMOD manual by Shrestha et al. (2018) provides structured methodologies for spring revival, mirroring interventions in Meghalaya.

While existing literature has extensively documented the causes of spring degradation, Meghalaya's initiative advances the field by offering large-scale, replicable solutions that integrate scientific hydrology, policy-driven support and community participation, serving as a model for water-stressed regions worldwide.

## 10.5 Summary and Conclusion

### 10.5.1 Key Findings Summary

Springs are crucial for rural communities in Meghalaya, providing essential water for drinking, agriculture and daily needs. But climate change, deforestation and over-extraction threaten their sustainability. Community-led springshed management has mapped over 55,000 springs, prepared 168 detailed technical reports and trained 800 village community facilitators. Scientific monitoring, climate-resilient strategies and recharge interventions ensure long-term conservation. A public dashboard increases data accessibility, aiding informed decision-making. Active community participation has fostered sustainable practices, aligning with global best practices in afforestation and soil conservation. This model offers a replicable framework for other water-stressed regions, ensuring long-term water security.

### 10.5.2 Significance Restatement

Meghalaya's springshed management initiative demonstrates the effectiveness of integrating scientific approaches with traditional ecological knowledge to ensure sustainable water security. By promoting community participation and decentralised governance, the initiative strengthens local ownership and resilience. The project emphasises the importance of systematic spring inventories, hydrological monitoring and climate-adaptive recharge strategies in addressing water scarcity. This model serves as a blueprint for other mountainous and water-stressed regions, highlighting the importance of collaborative conservation. By aligning policy, research and community engagement, Meghalaya's approach is effectively addressing the need for integrated water resource management to mitigate the effects of climate change and environmental degradation.

### 10.5.3 Future Research and Practical Applications

Future research should focus on long-term hydrological monitoring to assess the effectiveness of recharge interventions on spring discharge patterns. Expanding GIS-based modelling and remote sensing techniques can contribute to better identification of springs and improve the understanding of recharge zones, facilitating more precise and data-driven conservation efforts. In addition, socio-economic studies are needed to evaluate the effects of community-led conservation on rural livelihoods. Practical applications should include scaling up springshed management initiatives to cover the entire state of Meghalaya, integrating them into broader water governance policies. Strengthening collaborations between government agencies, research institutions and local communities will enhance replicability in other water-scarce regions. Investments in nature-based solutions, such as afforestation and soil conservation, should be prioritised to improve groundwater recharge. Continued capacity-building programmes will empower local stakeholders, ensuring the sustainability of conservation efforts. Advancing research and implementation of springshed management will significantly contribute to climate resilience and long-term water security in vulnerable regions in the state and elsewhere.

## References

Agarwal, A., Kumar, R., & Ansari, M. A. (2012). Water resources management in the Himalayas: Challenges and strategies. *Journal of Mountain Hydrology*, 4(2), 112–125.

Alford, D. (1992). Hydrological aspects of the Himalayan region. *International Journal of Water Resources Development*, 8(1), 15–25.

Ashutosh, S (2024). *Mapping of waterbodies in Meghalaya and analysing their spatial &temporal changes over a period of eight years from 2014 to 2022.* National symposium on remote sensing for sustainable future. Indian Society of Remote Sensing, Dr. A. P. J.Abdul Kalam Technical University, Lucknow, 11–13 Dec. 2024.

Asoka, A., et al. (2018). Precipitation intensity and groundwater recharge: The impact of monsoon dynamics on water resources in India. *Hydrology and Earth System Sciences*, 22(6), 3391–3401.

Bruijnzeel, L. A., & Bremmer, C. N. (1989). Highland-lowland interactions in the Ganges–Brahmaputra River basin: A hydrological perspective. *Journal of Hydrology*, 114(3–4), 217–240.

*Forest Survey of India.* (2011). Forest Survey of India, 2013; Forest Survey of India, 2015; Forest Survey of India, 2017; Forest Survey of India, 2019; Forest Survey of India, 2021.

India Meteorological Department, Shillong. (2021). Ministry of Earth Sciences, Government of India.

Kharsyntiew, T. (2017). Springs mapping in Meghalaya: Addressing rural water scarcity. *Journal of Himalayan Ecology and Development*, 15(3), 45–56.

Kumar, R., & Ansari, M. A. (2012). Climate change and its impact on Himalayan water resources. *Environmental Research and Development Journal*, 6(3), 145–158.

Mahamuni, A., & Kulkarni, A. (2012). Springs: A common source of water in the Himalayas –Current status and conservation strategies. *Journal of Environmental Management*, 37(2), 129–140.

Ministry of Jal Shakti. (2023). *Springshed management in the mountainous and Himalayan regions.* Government of India.

Negi, G. C. S. (2002). Water management in the Himalayas: Issues and challenges. *Mountain Research and Development*, 22(1), 32–39.

Negi, G. C. S., et al. (2019). Spring discharge and water security: A case for springshed management in the Himalayas. *Current Science*, 117(8), 1345–1351.

Negi, G. C. S., & Joshi, S. P. (2002). Rainfall variability and its impact on springs in the Central Himalayas. *Journal of Earth System Science*, 109(3), 141–152.

Negi, G. C. S., & Joshi, S. P. (2004). Long-term monitoring of spring discharge and its relation to climate change. *Hydrological Processes*, 18(5), 987–998.

Rawat, G., et al. (2021). Community-based springshed management for improved water security. *Water Resources Research*, 57(2), 1–10.

Rathod, R., et al. (2021). *Resource book on springshed management in the Indian Himalayan Region: Guidelines for policy makers and development practitioners.* International Water Management Institute (IWMI); NITI Aayog, Government of India; Swiss Agency for Development and Cooperation (SDC).

Shah, T. (2017). Springs: The overlooked water sources in rural India. *Water Policy Research Highlights*, 12(4), 123–130.

Shrestha, U., et al. (2018). *A manual on springshed management: Reviving springs in the Hindu Kush Himalayas.* ICIMOD.

Singh, S., et al. (2015). Comprehensive spring inventories and hydrological mapping: A necessity for understanding spring dynamics. *Environmental Monitoring and Assessment*, 187(5), 265.

Singh, T., & Pande, R. (1989). Impact of deforestation on spring water availability in the Himalayan region. *Indian Journal of Forestry*, 12(2), 78–86.

Tambe, S., et al. (2012). Reviving dying springs: Climate change impacts and sustainable spring management in the Indian Himalayan Region. *Mountain Research and Development*, 32(1), 62–72.

Tiwari, P., et al. (2020). Springshed development and community participation: A case study from Uttarakhand. *Journal of Water and Climate Change*, 11(4), 1003–1014.

Valdiya, K. S., & Bartarya, S. K. (1989). Hydrological studies in the Kumaon Himalaya withemphasis on springs. *Himalayan Geology*, 10(1), 157–175.

Valdiya, K. S., & Bartarya, S. K. (1991). Geological and geomorphological controls on springs in the Himalayan region. *Current Science*, 61(6), 381–385.

Vashisht, R., & Bam, W. (2012). Understanding the causes of declining Himalayan springs and strategies for their revival. *Journal of Himalayan Studies*, 8(4), 211–225.

# 11

# SEED BALL INITIATIVES FOR BIODIVERSITY CONSERVATION

*Subhash Ashutosh, Tremie M. Sangma, and Rikmenlang Khongsng*

## 11.1 Introduction
### 11.1.1 Purpose, Scope, and Significance of the Case

Natural degradation and loss of forest cover are a worldwide problem; in Meghalaya, the story is no different. As trees fall, we lose water sources and fertile soil, and our ecosystems are ravaged by landslides, floods, and climate change.

Afforestation is one of the key ways to regenerate ecosystems, and Meghalaya is doing it differently. The government has launched multiple initiatives aimed at promoting afforestation and reforestation. One of them is the seed ball initiative, which comes under the Community-Led Landscape Management Project with the Meghalaya Basin Management Agency.

The seed ball programme is strategically designed to address critical environmental challenges through a multifaceted approach aimed at reforestation, habitat restoration and biodiversity conservation.

By reintroducing native plant species to degraded and barren landscapes using seed balls, the initiative catalyses the recovery of ecosystems and promotes biodiversity. This approach not only supports a variety of wildlife by providing essential habitats and food sources but also plays a crucial role in stabilising soil and preventing erosion. Additionally, the growth of these plants significantly contributes to carbon sequestration, actively reducing the concentration of $CO_2$ in the atmosphere and mitigating the effects of climate change.

DOI: 10.4324/9781003735380-13

The seed ball initiative is one of the unique initiatives to bridge a long-pending gap in the state's Natural Resource Management to engage children and youth in environmental conservation. This particular initiative was designed for the active participation of children across Meghalaya, to sensitise them about this issue, so they would positively contribute to it.

In addition to its environmental objectives, the seed ball programme placed strong emphasis on community engagement and education. It involved local communities, particularly students, in the entire process from seed ball creation to dispersal. By doing so, it fostered a sense of responsibility and pride in environmental conservation.

This hands-on involvement also helped raise awareness about the importance of ecological stewardship and promoted sustainable practices among participants.

The programme's design as a low-cost, low-tech solution ensures that it is accessible and practical. It enables adaptation to various environments, including those that are arid or traditionally difficult for cultivation.

Through various efforts, the seed ball initiative aims to cultivate a long-lasting impact, empowering communities to continue these practices and contribute to a healthier planet.

### 11.1.2 Benefits of Seed Balls

Seed balls are balls of soil mixed with compost, biochar, and paddy straw in which seeds are embedded, then dried in the shade.

Benefits:

1. A seed ball is one of the easiest and most economical ways of spreading greenery.
2. It is an easy way of enriching biodiversity.
3. Dispersal of seed balls is easy. It can be done by throwing them manually or using catapults to disperse them across a distance.
4. The wrapping mixture of clay and compost keeps the seed safe until it can properly germinate.
5. The seed ball size ranges from ½ inch to 3 inches in diameter and weight (depending on the size and nature of the seeds).
6. The best months to prepare seed balls and disperse them in the state are from March to September.

## 11.2 Review of Literature

With climate change and global warming, afforestation through different methods needs to be implemented. In Meghalaya, due to shifting cultivation

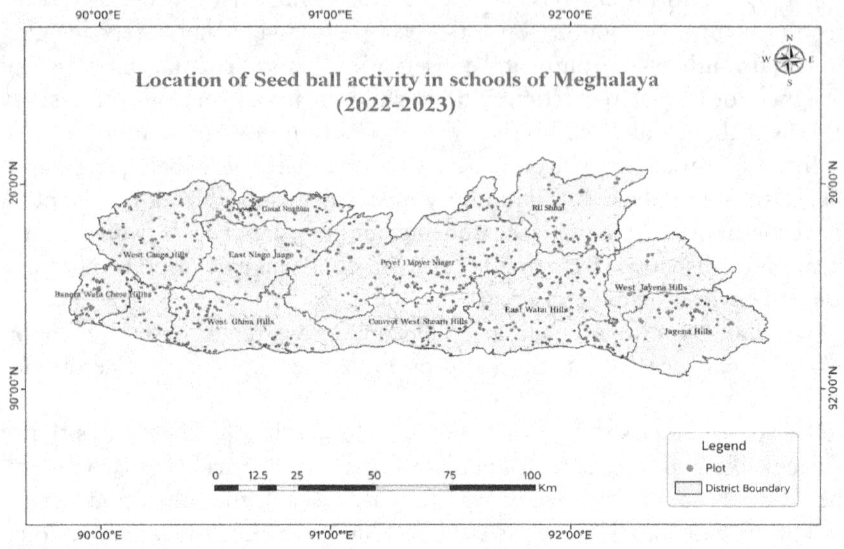



**FIGURE 11.1**   Location of seed ball activity in schools of Meghalaya

*Source:* GIS Lab, Centre of Excellence (CoE) MBMA

and coal mining, most of the forested areas have become fragmented and degraded. Seed ball use is an effective method of restoration and propagation. The soil protects the seeds from biotic and abiotic pressure and predation.

For sustainable land management, seed ball activities have been taken up in various parts of the country (Rawat et al., 2024; Kannan & Janani, 2021). It is a cost-effective activity, and various natural materials can be used. Kannan and Janani (2021) presented a future perspective of seed ball technology for the creation of a new ecosystem.

The seed ball technique to promote crops is also found in other parts of the world (Tamilarasan et al., 2021). Zubaidah et al. (2022) conducted a study for the reclamation of a mining area through a seed ball coating formulation to stimulate germination and the growth of fruit and forest seeds. The study revealed that it yielded successful results.

To secure livelihood and promote a thriving ecosystem, there is an urgent need to mitigate climate change problems through different afforestation methods (Shackelford et al., 2021).

## 11.3 Methodology

The implementation of the seed ball activity started with a training module. The training was a combination of presentations, group discussions, practical demonstrations, and hands-on activities. Trainers, including village community facilitators as Master Trainers and external experts, were involved in guiding the sessions. Practical demonstrations for teachers and students were to ensure effective knowledge transfer.

Activities were carried out across all districts, with both Master Trainers and staff from Meghalaya Basin Management Authority receiving training throughout the State. The Master Trainers in each district then took up activities for seed ball making at schools. They followed specific guidelines to prepare the seed balls.

Thirty per cent fruit trees and 70% forestry tree species were selected. Ingredients such as seeds, red soil, bamboo biochar, rice gruel, fresh cow dung, and wood ash, as well as organic materials, were used in the preparation of the balls. GPS coordinates of each dispersal site were determined.

To ensure the success of the initiative, seed balls were properly dispersed on the ground. In this regard, schools were requested to follow the guidelines provided. Videos, brochures, and pamphlets on seed ball preparation techniques in Garo and Khasi languages were circulated. Figure 11.2 (a) summarises the methodology.

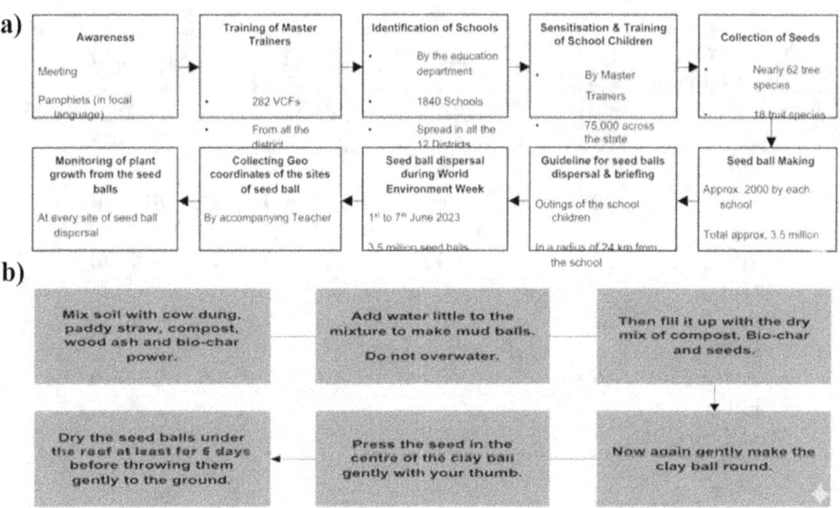


**FIGURE 11.2**    (A) Work flow for seed ball activity. (B) Steps to prepare seed balls.

*Source:* Centre of Excellence (CoE) MBMA

## 11.4 Results

From October 2022 to June 2023, a training programme on seed ball making was conducted in 1,840 schools across all 12 districts of Meghalaya. Approximately 3 million seed balls were prepared for World Environment Month 2023.

The activity was meant to be interactive, engaging, and age-appropriate for students of different levels, but mostly Upper Primary Schools were selected. It dealt with the importance of reforestation and biodiversity conservation, the role of seed balls in promoting sustainable planting, as well as seed ball preparation techniques and the actual preparation. Sixty-two tree species, including 18 fruit species, were used. The *Pongamia pinnata*, *Terminalia bellirica*, *Myrica esculenta*, and *Elaeocarpus floribundus* were among them.

**TABLE 11.1** Seed ball activity of schools in Meghalaya

| Sl. no. | District names | No. of schools | No. of students | No. of seed balls prepared | No. of seed balls dispersed |
|---|---|---|---|---|---|
| 1 | East Khasi Hills | 440 | 17,440 | 872,000 | 623,999 |
| 2 | West Khasi Hills & Eastern West Khasi Hills | 160 | 6,400 | 330,600 | 320,000 |
| 3 | South West Khasi Hills | 80 | 3,200 | 164,715 | 160,000 |
| 4 | Ri Bhoi | 160 | 6,400 | 362,800 | 320,000 |
| 5 | West Jaintia Hills | 120 | 4,800 | 240,000 | 232,378 |
| 6 | East Jaintia Hills | 80 | 3,240 | 162,000 | 161,901 |
| 7 | North Garo Hills | 120 | 4,800 | 240,000 | 240,000 |
| 8 | East Garo Hills | 120 | 4,800 | 241,300 | 240,000 |
| 9 | South Garo Hills | 160 | 6,400 | 320,000 | 320,000 |
| 10 | West Garo Hills | 280 | 11,840 | 592,000 | 504,700 |
| 11 | South West Garo Hills | 120 | 4,880 | 244,000 | 244,000 |
| | Total | 1,840 | 74,200 | 3,769,415 | 3,366,978 |

*Source:* Centre of Excellence (CoE) MBMA.

### 11.4.1  *Dispersal Activity*

The respective schools started dispersing seed balls in and around their localities, school compounds, and throughout their villages from May 2023 onward. In total, around 75,000 students participated in this activity across the selected schools in the State. The seed balls were scattered freely: The children dispersed and threw seed balls in different directions. While school children disbursed most seed balls by throwing, some also buried a few in shallow-surface soil. A small hole was dug double the size of a seed ball, then it was buried – through a 'throw two, bury one' – method.

The number of dispersal sites was 2,971.

### 11.4.2  *Survey for Survival of Seed Balls Dispersed*

After two months of dispersal of the seed balls, a survey to assess their germination and survival percentage was undertaken through a sample. All district units randomly selected 10% of the dispersal sites and conducted the survey over a plot of 100 m × 100 m.

The survey team assessed the dispersal sites for survival percentage with the coordinates recorded at the time of seed ball dispersal. A summary of the number of sample sites in each district and the average survival and germination percentage found is presented in Table 11.2. Using the data in the table, the mean of the germination and the survival percentage of the seed balls dispersed was calculated. It showed up as 55%.

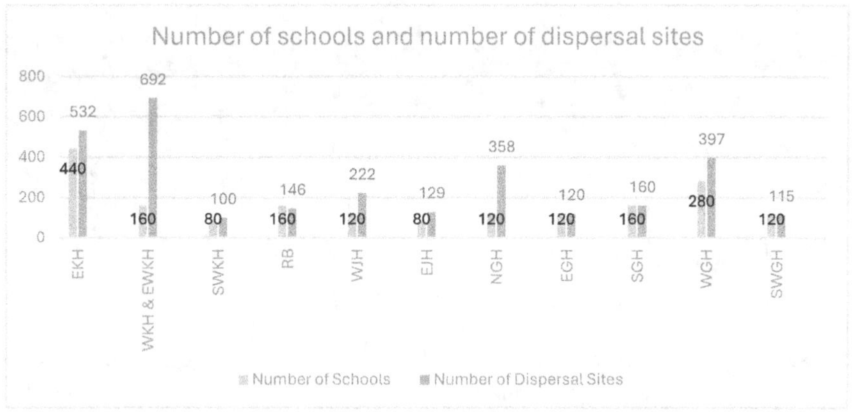


**FIGURE 11.3**   Number of schools and number of dispersal sites

*Source:* Centre of Excellence (CoE) MBMA

**TABLE 11.2** Dispersal sites and survival percentage of seed ball activity

| Sl. no. | Districts | No. of dispersal sites surveyed | Survival (%) |
|---|---|---|---|
| 1 | East Khasi Hills | 36 | 65 |
| 2 | West Khasi Hills & Eastern West Khasi Hills | 33 | 60 |
| 3 | South West Khasi Hills | 7 | 45 |
| 4 | Ri Bhoi | 17 | 60 |
| 5 | West Jaintia Hills | 12 | 35 |
| 6 | East Jaintia Hills | 16 | 45 |
| 7 | North Garo Hills | 12 | 70 |
| 8 | East Garo Hills | 12 | 78 |
| 9 | South Garo Hills | 18 | 50 |
| 10 | West Garo Hills | 106 | 49 |
| 11 | South West Garo Hills | 12 | 73 |
| | Total | 281 | 55 |

*Source:* Centre of Excellence (CoE) MBMA.

## Percentage of seed balls dispersal and percentage of seeds survived

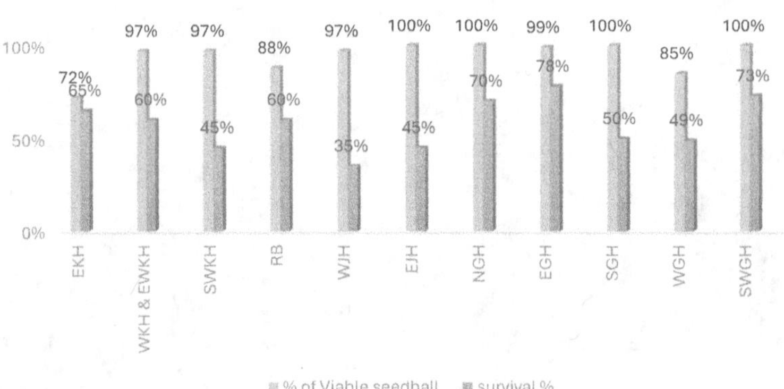


**FIGURE 11.4**  Percentage of seed balls for dispersal and percentage of seeds survived

*Source:* Centre of Excellence (CoE) MBMA

**TABLE 11.3**  Tree species used for seed ball activity in various districts

| Sl. no. | District | Species |
|---|---|---|
| 1 | East Khasi Hills | *Alder nepalensis, Elaeocarpus floribundus, Myrica esculenta, Citrus reticulata, Citrus latipes, Docynia indica, Exbucklandia populnea, Schima wallichii, Betula alnoides* |
| 2 | West Khasi Hills & Eastern West Khasi Hills | *Pinus kesiya, Schima wallichii, Myrica esculenta, Azadirachta indica, Cinnamomum verum, Toona ciliata, Prunus nepalensis, Terminalia bellirica, Emblica officinalis, Shorea robusta* |
| 3 | South West Khasi Hills | *Myrica esculenta, Prunus nepalensis, Artocarpus heterophyllus, Cinnamomum tamala, Citrus reticulata, Prunus domestica, Prunus persica, Castanopsis hystrix, Elaeocarpus robusta* |
| 4 | Ri Bhoi | *Toona ciliata, Michelia champaca, Azadirachta indica, Shorea robusta* |
| 5 | West Jaintia Hills | *Schima wallichii, Toona ciliata, Pinus kesiya, Prunus nepalensis, Shorea robusta, Azadirachta indica* |
| 6 | East Jaintia Hills | *Toona ciliata, Pinus kesiya, Terminalia bellirica, Shorea robusta, Azadirachta indica* |
| 7 | North Garo Hills | *Tectona grandis, Emblica officinalis, Azadirachta indica, Tamarindus indica, Syzygium cumini, Sterculia foetida* |
| 8 | East Garo Hills | *Tamarindus indica, Citrus maxima, Gmelina arborea, Emblica officinalis, Artocarpus heterophyllus, Cassia fistula* |
| 9 | South Garo Hills | *Pterocarpus santalinus, Tectona grandis, Delonix regia, Terminalia arjuna, Tamarindus indica, Pongamia pinnata, Bauhinia variegata* |
| 10 | West Garo Hills | *Citrus maxima, Pterocarpus santalinus, Azadirachta indica, Tamarindus indica, Acacia species, Bauhinia purpurea* |
| 11 | South West Garo Hills | *Pterocarpus santalinus, Tectona grandis, Pongamia pinnata, Tamarindus indica, Azadirachta indica, Emblica officinalis, Delonix regia, Terminalia bellirica, Citrus maxima, Saraca asoca* |

*Source:* Centre of Excellence (CoE) MBMA.

## 11.5 Discussion

In Meghalaya, the work on seed balls as a greening activity has yielded a successful outcome thanks to active community participation. From a total of around 3 million seed balls dispersed during this activity, the **germination** or survival rate is approximately 55%.

Seed ball is known to accelerate succession in degraded coal-mine land. The seed ball technique is effective for propagating plants from seeds because the seeds are protected from external stress and predation.

The present experiment investigated the applicability of this technique in afforestation programmes. This practice is also followed in other states (Hazarika et al., 2023). Deforestation and forest degradation are the biggest threats to our forests. Afforestation is one way to sustain life and combat global warming (Tamilarasan et al., 2021).

A key highlight of this initiative is its inclusive model of implementation. Unlike traditional afforestation strategies that often rely on top-down execution, the seed ball programme engages a wide range of stakeholders, especially children and youth.

Furthermore, the programme's low-tech, adaptable methodology makes it suitable for diverse terrains. Materials can be locally sourced, making the process both sustainable and replicable. The reported germination demonstrates a significant success rate, validating the efficacy of the approach in practical terms. However, while the results are promising, the long-term success of the initiative depends on continuous monitoring and post-dispersal care in some areas.

The seed ball initiative represents a sustainable, inclusive, and scalable model of land restoration. By blending traditional ecological knowledge with innovative, community-driven practices, Meghalaya is setting an example for environmentally conscious development that other regions may adapt and learn from.

## 11.6 Summary and Conclusion

Approximately 3 million seed balls were dispersed during World Environment Day in 2023. Seed ball activities are a practical, inclusive, and cost-effective method for ecological restoration. They help in increasing the green cover and sensitising school students about climate change and the ecological balance.

Seeds of 62 tree species, including 18 fruit species, were collected. About 75,000 children took part in the initiative. Teachers and students were actively involved in the training of seed ball making and dispersal. All students participated from pre-primary to upper primary levels, among them many female students.

Certain schools adopted making seed balls as part of their extracurricular activities in environmental management. Along with the school compounds, seed balls were dispersed in and around the village boundary, including private and community lands. The activity helped empower the village community facilitators to take on leadership roles.

Seed ball activity enhanced biodiversity, improved soil health, engaged communities, and school children contributing to climate change mitigation, making them a valuable tool in environmental conservation efforts. Significant progress has been made in promoting afforestation.

### 11.6.1 Outcome

1. Approximately 3 million seed balls were made successfully.
2. A total of 1,840 schools participated in all 12 districts of Meghalaya.
3. Approximately 75,000 students received training across the state.
4. A total of 282 VCFs were trained.

The training programme succeeded in enhancing students' understanding of reforestation and environmental conservation. Pre-training and post-training assessments revealed a significant improvement in students' knowledge about the importance of seed balls, the process of seed ball preparation, and suitable planting techniques. The training programme also fostered community engagement and participation. Students, along with their teachers, took the initiative to involve community members in the planting process.

This collaborative effort created a sense of ownership and responsibility among them, contributing to the long-term sustainability of the project. It is also a step towards developing resilience to climate change.

The State Council of Science, Technology & Environment, Meghalaya, took up the initiative in the schools of Meghalaya as a replicability and sustainability model in 2024. Notably, the seed ball activity received NITI Aayog recognition for Best Practices under Agriculture and Allied Services.

## References

Hazarika, P., Dutta, D., Hazarika, P., Giri, K., & Dutta, S. P. (2023). Seed balls accelerate succession of plant species in degraded coal mined land restoration: A case study on Tikak Colliery, Margherita, Assam, India. *Indian Forester, 149*(2), 197–206.

Kannan, R., & Janani, T. S. K. (2021). Future perspective of seed ball technology for creating new ecosystem. *International Journal of Plant and Environment, 7*(4), 293–296.

Rawat, D., Kohli, A., Khanduri, V. P., Singh, B., Riyal, M. K., & Sati, S. P. (2024). Seed ball technology: Facets and prospects for restoration of degraded lands. In G. Mishra, K. Giri, S. Singh, & KumarM. (Eds.), *Sustainable land management in India* (pp. 149–166). Springer Nature.

Shackelford, N., Paterno, G. B., Winkler, D. E., Erickson, T. E., Leger, E. A., Svejcar, L. N., & Suding, K. L. (2021). Drivers of seedling establishment success in dryland restoration efforts. *Nature Ecology & Evolution, 5*(9), 1283–1290.

Tamilarasan, C., Jerlin, R., & Raja, K. (2021). Seed ball technique for enhancing the establishment of subabul (*Leucaena leucocephala*) under varied habitats. *Journal of Tropical Forest Science*, *33*(3), 349–355.

Zubaidah, S., Mansur, I., Budi, S. W., & Yusmur, A. (2022). Seed ball coating material formulation to enhance germination and growth of fruit and forest seeds. In *IOP conference series: Earth and environmental science* (Vol. 959, No. 1, p. 012039). IOP Publishing.

# 12

# INTEGRATING TECHNOLOGY FOR NATURAL RESOURCE MANAGEMENT

*Raymond Wahlang and Fettleman Dohling*

## 12.1 Introduction

### 12.1.1 Purpose, Scope, and Significance of the Case

Meghalaya, a Sixth Schedule state, operates under decentralised, community-led governance. It faces significant challenges in natural resource management due to a lack of structured geospatial and non-spatial data. The absence of village and forest boundaries, cadastral maps, thematic maps, and centralised repositories has led to fragmented planning, duplication of efforts, and inefficient governance. Traditional data collection methods, such as manual land-use surveys, are slow, expensive, and imprecise, resulting in large historical data gaps, which are critical for planning and conservation.

Despite being part of the Himalayan biodiversity hotspot (Myers, 2003), Meghalaya struggles with deforestation, shifting cultivation, and unsustainable resource use on account of reactive rather than proactive conservation efforts. Reliance on third-party mapping services increases costs and limits local capacity building. To address this, geographic information systems (GIS), remote sensing (RS), and unmanned aerial vehicles (UAVs) have been introduced, but their success has depended on community ownership. By training Village Community Facilitators (VCFs) and Village Data Volunteers (VDVs), the initiative demystified geospatial technology, integrating it with Traditional Ecological Knowledge (Rai & Mishra, 2023). This participatory approach is empowering communities, ensuring technology is an enabler, not an external imposition. Moreover, it sets a scalable model for sustainable, community-driven conservation and governance.

DOI: 10.4324/9781003735380-14

## 12.2 Organisation of the Chapter

This chapter is structured to provide a detailed understanding of community-driven geospatial monitoring and its role in conservation. It begins with an introduction, outlining the context, significance, and objectives of the initiative. A review of literature explores existing research on participatory conservation, geospatial technologies, and Payment for Ecosystem Services monitoring. The methodology section details the approach used for data collection, community engagement, and Management Information Systems integration. The case findings are divided into boundary mapping, Land Use Land Cover mapping, and Measurement Reporting and Verification, highlighting the role of communities in each activity. The chapter concludes with policy implications, challenges, future research directions, and a summary of key takeaways.

## 12.3 Review of Literature

The integration of geospatial technologies, including geographic information systems (GIS), remote sensing (RS), and unmanned aerial vehicles (UAVs), has transformed community-led natural resource management. GIS is instrumental in mapping community forests, identifying vulnerable ecosystems, and optimising resource distribution in rural areas (Nemec & Raudsepp-Hearne, 2012). Coupled with decision-support systems, GIS contributes to participatory conservation strategies. This way, local communities actively engage in resource governance (Gonzalez-Redin et al., 2016). RS enhances community-driven environmental monitoring by offering high-resolution data on deforestation, land degradation, and water resources (Singh & Kumar, 2012). In biodiversity hotspots like Meghalaya, integrating RS with community participatory mapping has been highly effective, combining traditional land-use practices with modern conservation efforts (Machiwal et al., 2011). RS, combined with machine learning models, supports predictive ecosystem analysis, aiding proactive conservation planning (Shaik et al., 2024).

UAVs have emerged as a transformative tool for grassroots monitoring, enabling facilitators to assess forests, shifting cultivation, and water bodies in real time (Ventura et al., 2018). Their precision in mapping inaccessible terrains is valuable in mountainous regions, empowering stakeholders with actionable data (Acharya et al., 2021). In Meghalaya, UAVs have proven effective for afforestation monitoring and Payment for Ecosystem Services (PES) accountability (Mishra & Rai, 2020). Management Information System (MIS) improves data transparency by streamlining geospatial data collection and sharing, enabling decentralised governance (Hognogi et al., 2021). Even so, capacity-building initiatives are needed to address technical knowledge gaps among rural communities for the effective use of these tools (Quamar et al., 2023).

Despite advancements, there is limited research exploring how geospatial technologies integrate with traditional knowledge systems in the state. Moreover, studies on the role of community participation in promoting technology use are scarce. Addressing these gaps is vital for sustainable, inclusive natural resource management.

## 12.4  Identification of Gaps

Despite the growing use of geospatial technology in Meghalaya, communities still face barriers in fully leveraging these tools for natural resource management. The integration of Traditional Ecological Knowledge (TEK) with GIS, Remote Sensing (RS), and UAVs remains limited, preventing local stakeholders from actively shaping conservation strategies. Capacity gaps, limited data accessibility, and technical challenges further hinder community-driven decision-making. Strengthening training programmes, localised geospatial tools, and participatory mapping initiatives are essential to ensuring technology serves as an enabler rather than a barrier.

## 12.5  Methodology of the Case

### 12.5.1  Description of Methodology

The methodology focused on empowering local communities by equipping nominated young locals designated as Village Community Facilitators (VCFs) or Village Data Volunteers (VDVs) with geospatial technology skills for natural resource management. Through hands-on training, participants learned to use GIS for mapping village boundaries, forest resources, and land-use patterns. They were also trained in the use of Global Positioning System (GPS) devices and a customised mobile application to collect, validate, and upload geospatial data. This community-driven approach fostered ownership of conservation initiatives. The integration of Traditional Ecological Knowledge with advanced tools allowed communities to plan afforestation, monitor resources, and restore degraded landscapes. The participatory framework strengthened grassroots governance and promoted sustainable resource management.

### 12.5.2  Justification of Methodology

The methodology was designed to gather, decentralise, and democratise data, empowering communities and improving engagement. Rural participants, often lacking technical expertise, were introduced to foundational concepts of GIS and GPS, along with customised mobile applications, building their confidence regarding advanced techniques. Village Community Facilitators were key stakeholders in data collection and decision-making, bringing about ownership and ensuring solutions align with community priorities. Practical training emphasises hands-on use of GPS for mapping village boundaries and integrating data into GIS layers, equipping participants with real-world skills. UAV images, captured by trained officials, were used for generating precise maps of village boundaries and resources. The tiered training programme progressively deepens participants' expertise, culminating in advanced geospatial data management. By emphasising participatory data collection and local adaptability, this methodology balanced technical innovation with community empowerment, creating a sustainable and scalable model for natural resource management.

## 12.6  Results

### 12.6.1  Main Case Findings 1

#### Village Boundary Mapping

The village boundary mapping initiative in Meghalaya was a community-led effort undertaken by Village Community Facilitators to support natural resource management. With the state's unique system of decentralised governance under the Sixth Schedule of the Indian Constitution, this initiative ensured that communities played a central role in defining and managing their land and forest resources. A participatory approach integrated Traditional Ecological Knowledge with modern geospatial technologies, strengthening local ownership of conservation efforts. VCFs were trained in GIS and Global Positioning System (GPS) technologies, enabling them to map systematically village boundaries, land use, and forest cover. The use of geospatial tools improved boundary delineation precision, while local knowledge ensured the mapped territories reflected community-recognised land divisions rather than externally imposed administrative borders.

Essentially, this initiative aimed to address challenges related to resource conflicts, unclear territorial demarcations, and the absence of comprehensive geospatial data for decision-making. Many village boundaries in Meghalaya had been traditionally defined through informal agreements rather than official documentation, creating inconsistencies in planning afforestation, conservation, and land-use interventions. By involving trained VCFs in the mapping, the initiative ensured that the data collected were both scientifically accurate and contextually relevant. Handheld GPS devices were used to capture key boundary coordinates, which were later integrated into GIS platforms for digitisation. The resulting digital boundary maps were validated by community members to ensure alignment with their traditional territorial understanding.

Among the many VCFs who contributed to this initiative, two individuals stand out as success stories – Teiningstone Dkhar from Mawthungkper Village in the West Khasi Hills district and Asime Sangma from Kongtokpara Village in the West Garo Hills district. Their journeys highlight the transformative power of equipping local communities with geospatial knowledge and the significant effect this initiative has had on natural resource management. Teiningstone Dkhar had no prior experience with mapping tools before becoming a VCF. After receiving training, he quickly mastered GPS navigation and mapping techniques. His newfound skills enabled him to lead community discussions and ensure accurate documentation of village boundaries and resources. Through his involvement in village boundary mapping and land use and land cover (LULC) mapping, as well as bamboo resource assessments, he played a crucial role in sustainable land management. His expertise also extended to afforestation and reforestation efforts. He further participated in the Payment for Ecosystem Services (PES) and GREEN (Grassroot Level Response Towards Ecosystem Enhancement and Nurturing) Meghalaya initiatives. Today, he is recognised as a leader in his community, guiding others in the responsible use of technology for natural resource management. His dedication earned him the position of a Village Data Volunteer (VDV), where he continues to contribute to village development and conservation.

Similarly, Asime Sangma emerged as an inspiring figure in a traditionally male-dominated field. Initially facing challenges as one of the few female VCFs, she remained determined to make a difference. Through rigorous training, she became proficient in GIS tools, GPS technology, and field data collection, specialising in spring mapping and village resource mapping. Her efforts in tracking water resources have had a direct effect on daily life in her village, ensuring sustainable management of critical water sources. She played a crucial role in afforestation and reforestation projects, as well as in conservation initiatives under the PES and GREEN Meghalaya programmes. Her contributions strengthened environmental monitoring and inspired other women to take an active role in community-based natural resource management. Her journey from a community volunteer to an expert in GIS-based planning highlights the growing involvement of women in environmental conservation. She is now working with IORA Ecological Solutions, continuing her mission to support sustainable resource management.

The integration of community-based mapping into natural resource management produced several tangible benefits. It strengthened the ability of local governance structures, such as Village Natural Resource Management Committees (VNRMCs), as shown in Figure 12.1, to make informed

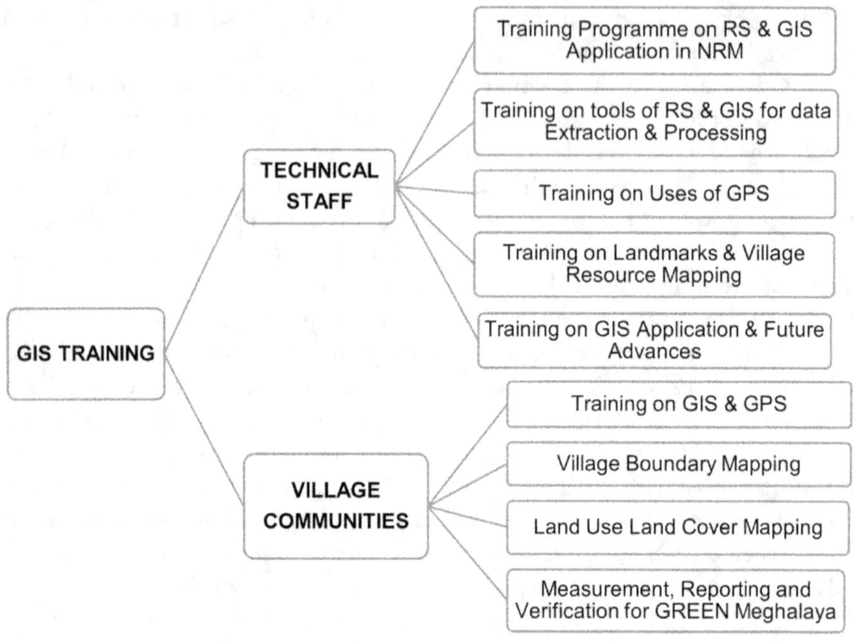


**FIGURE 12.1**  Methodology flowchart

*Source:* Authors

decisions regarding land use and conservation planning. Mapped village boundaries provided a reliable reference for identifying and designating forest conservation areas, ensuring afforestation and resource extraction activities were conducted sustainably. This was particularly relevant to PES schemes, where precise boundary demarcations were required to allocate benefits and incentives for conservation efforts. With accurate maps, communities were better equipped to engage in afforestation projects and forest management plans, ensuring equitable distribution of conservation benefits.

Another significant outcome of the initiative was its positive role in conflict resolution. Disputes over land and forest resources have been long-standing challenges in Meghalaya, particularly between neighbouring villages with unclear boundary demarcations. By documenting territorial divisions through a participatory process, the mapping exercise provided an objective basis for addressing disputes. Since community members collected the data rather than external agencies, trust in their validity was higher. Where conflicts arose, the maps served as evidence to facilitate negotiations and consensus-building, fostering cooperation between villages in managing shared resources.

The boundary mapping exercise also enforced local capacity for geospatial data collection and management. By equipping VCFs with technical skills, the initiative ensured communities had in-house expertise to update and refine maps as needed. This reduced dependence on external agencies for geospatial assessments and reinforced grassroots governance. The use of digital mapping tools also standardised data collection methodologies across villages, ensuring consistency in spatial data management. This enabled local authorities and policymakers to integrate the mapped boundaries into broader land-use planning, supporting sustainable development at district and state levels.

The participatory nature of the exercise was crucial to its success. Training provided to VCFs was designed to be accessible and practical, with a focus on hands-on learning. Community members actively participated in data validation, ensuring that the final maps reflected their lived experiences and historical knowledge. This bottom-up approach not only improved data accuracy but also fostered a sense of ownership and responsibility among participants. When communities see their knowledge reflected in conservation strategies, they are more likely to support and sustain such initiatives over the long term.

Beyond its technical outcomes, the mapping initiative had important socio-economic implications. In villages where land tenure systems are based on customary governance, having geospatially defined boundaries facilitated better access to conservation funding and development programmes. Many environmental and afforestation schemes require clear documentation of land ownership and resource management plans before disbursing financial support. With digitised maps, communities could more effectively advocate for their land rights and participate in programmes that promote ecological sustainability. This was particularly beneficial for villages engaged in PES initiatives, as it allowed them to demonstrate compliance with conservation guidelines and secure incentives for ecosystem preservation.

Despite its successes, the mapping initiative faced challenges. One of the key issues was varying levels of technological literacy among community members. While VCFs received extensive training, some initially struggled with operating GPS devices and mobile applications for data collection. This required additional capacity-building efforts to ensure consistency in data recording. Furthermore, in areas with rugged terrain or dense forests, GPS signal accuracy was sometimes compromised, leading to minor discrepancies in recorded boundary coordinates. These issues were dealt with through multiple validation rounds, where community members cross-checked mapped data against traditional boundary markers.

Another challenge was ensuring the long-term sustainability of the initiative. While initial training and data collection were successful, ongoing support and updates were necessary to keep the geospatial data relevant. The establishment of decentralised geospatial hubs within village clusters was proposed as a solution, where trained VCFs could continue refining maps and addressing emerging boundary-related issues. Strengthening collaboration between local governance institutions and technical experts was also seen as a way to maintain the initiative's momentum.

Overall, the village boundary mapping initiative demonstrated how a community-centric approach can effectively integrate traditional knowledge with modern technology for natural resource management. By placing local communities at the forefront of the mapping process, the initiative improved spatial data accuracy, strengthened local governance, and enhanced conflict resolution mechanisms. The benefits extended beyond conservation, supporting economic empowerment through improved access to environmental funding programmes. The participatory nature of the process ensured that the mapped boundaries reflected community realities, promoting trust and long-term commitment to sustainable resource management. The success stories of Teiningstone Dkhar and Asime Sangma illustrate how empowering community facilitators with technical skills can drive meaningful change. This makes this initiative a scalable model for other regions facing similar challenges in defining and managing communal lands.

### 12.6.2 Main Case Findings 2

The Land Use Land Cover (LULC) mapping initiative in Meghalaya was a community-driven and technology-assisted effort aimed at improving natural resource management, conservation planning, and sustainable land-use practices. Given Meghalaya's diverse ecological landscapes, including dense forests, agricultural lands, and shifting cultivation zones, accurate and community-validated land classification was necessary for informed

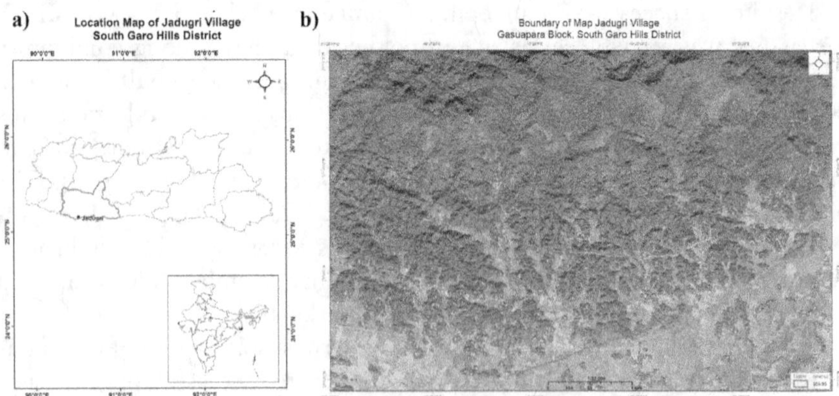


**FIGURE 12.2A** & b   Village boundary of Jadugri village

*Source:* Authors

**TABLE 12.1** Number of VNRMCs established, along with the number of VCFs trained

| Districts | VNRMCs | Number of VCFs trained |
| --- | --- | --- |
| East Garo Hills | 30 | 90 |
| East Jaintia Hills | 30 | 90 |
| East Khasi Hills | 93 | 279 |
| North Garo Hills | 20 | 60 |
| Ri Bhoi | 30 | 90 |
| South Garo Hills | 23 | 69 |
| South West Garo Hills | 49 | 147 |
| South West Khasi Hills | 40 | 110 |
| West Garo Hills | 40 | 91 |
| West Khasi Hills | 25 | 75 |
| West Jaintia Hills | 20 | 60 |
| Total | 400 | 1,161 |

*Source:* Authors.

decision-making at both local and policy levels. This initiative was designed to empower communities by integrating modern geospatial technology with indigenous knowledge, ensuring that land-use mapping remained relevant, participatory, and actionable.

A high-resolution LULC classification was achieved using unmanned aerial vehicles (UAVs) and satellite imagery, allowing for precise identification of forests, water bodies, agricultural lands, urban settlements, and degraded areas. Technical staff trained in UAV operations conducted flights, capturing high-resolution imagery that was later processed by trained GIS personnel. The integration of UAV data with satellite imagery provided a comprehensive and updated view of land-use patterns, allowing for detailed classification and analysis.

While trained UAV and GIS staff handled data collection and processing, Village Community Facilitators (VCFs) played a critical role in ground-truthing. They worked closely with local land users, farmers, and traditional land custodians to validate LULC classifications against actual field conditions. VCFs conducted field surveys using GPS devices and mobile GIS tools, confirming the accuracy of land classifications and traditional land-use boundaries. This participatory verification was particularly crucial in regions with complex land-use transitions, such as shifting cultivation, mixed agroforestry systems, and community-conserved forests.

The LULC maps generated through this initiative were not just technical outputs but valuable community resources for land-use planning, conservation efforts, and sustainable resource management. With detailed maps, communities could identify deforestation-prone areas, potential afforestation sites, and conservation zones, ensuring that land-use decisions aligned with both ecological sustainability and community needs. The maps also enabled better planning for watershed protection, because communities could visualise sedimentation-prone areas, water catchments, and potential wetland restoration sites. One of the key applications of LULC mapping was tracking land-use changes over time. By comparing historical satellite imagery with new UAV-generated maps, both technical teams and community members were able to assess deforestation trends, forest degradation, and encroachment on conservation areas. The data revealed notable land-use shifts in certain community forests, where canopy thinning and fragmentation indicated unsustainable resource extraction. These findings were crucial for communities engaged in community forestry initiatives, allowing them to strategise afforestation efforts and strengthen sustainable land governance. In addition to forest monitoring, LULC mapping provided insights into urban expansion and agricultural land-use patterns. High-resolution data helped track the spread of built-up areas, new settlements, and changes in cropland use, ensuring that land-use planning accounted for both development needs and environmental sustainability. This was particularly useful for managing land conversion near roads, market hubs, and rural townships, helping communities prevent unregulated expansion into ecologically sensitive zones.

The initiative also contributed to watershed and hydrological planning: LULC maps were used to identify areas prone to soil erosion, declining groundwater recharge, and potential wetland degradation. This information was critical for local water management groups and conservation committees, enabling them to prioritise reforestation efforts in erosion-prone areas and improve community-led watershed conservation initiatives. Another significant effect of this programme was its role in land tenure security and conflict resolution. Meghalaya's governance system is rooted in customary land ownership, where traditional governance structures manage land distribution. But unclear or undocumented territorial boundaries often led to land-use disputes between communities. The high-resolution LULC maps, combined with the ground-truthing data VCFs collected, provided verifiable spatial documentation of community lands. This helped resolve boundary conflicts, strengthened community land governance, and provided a reliable reference for documenting traditional land claims.

LULC mapping also had direct socio-economic benefits for rural communities. The initiative improved community access to conservation funding and Payment for Ecosystem Services (PES) programmes, where financial incentives were provided for forest conservation and sustainable land-use management. Many PES programmes require clear documentation of land ownership and conservation zones. The detailed UAV-generated maps enabled communities to apply for financial incentives under forest protection agreements. In addition, LULC mapping supported agroforestry planning, helping farmers and landholders optimise crop rotations, soil conservation techniques, and sustainable harvesting practices.

Despite its successes, the initiative faced several challenges, particularly in data processing, UAV accessibility, and capacity-building for GIS applications. Processing large volumes of UAV imagery required specialised GIS expertise, and while communities were involved in ground-truthing, data processing remained limited to trained GIS personnel. Additional efforts were made to train local governance institutions in GIS interpretation, ensuring that community leaders could understand and use geospatial data for decision-making.

While UAV operations and data processing were conducted by technical staff, the continued engagement of community facilitators in data interpretation and land-use planning was crucial for long-term sustainability. Capacity-building programmes focused on improving GIS literacy among community leaders, land-use committees, and forest management groups, enabling them to apply LULC data in conservation planning and land governance efforts.

This programme demonstrated that technology-driven conservation works best when it is community-led, participatory, and grounded in local knowledge. The integration of UAV-based mapping with traditional land-use governance created a scalable and replicable model that could be applied to other forest-dependent communities in Meghalaya and beyond. With continued investment in geospatial technology, participatory mapping, and local capacity-building, LULC mapping will remain a valuable tool for ecosystem resilience, sustainable land management, and community-driven conservation.

By prioritising community engagement, strengthening participatory governance, and ensuring that LULC mapping remains accessible to local decision-makers, the initiative has laid the groundwork for a more sustainable, inclusive, and data-driven approach to land management. This case study points out the transformative power of combining high-resolution geospatial technology with community leadership, ensuring that land-use decisions reflect the needs, values, and aspirations of the people who rely on these landscapes the most.

With ongoing collaboration between technical experts, local governance bodies, and VCFs, LULC mapping will continue to serve as a cornerstone for resource planning, forest conservation, and sustainable development. This will ensure that Meghalaya's rich ecological heritage remains protected for future generations. Figure 12.3 shows the LULC maps of Jadugri village in Gasuapara block of the South Garo Hills District.

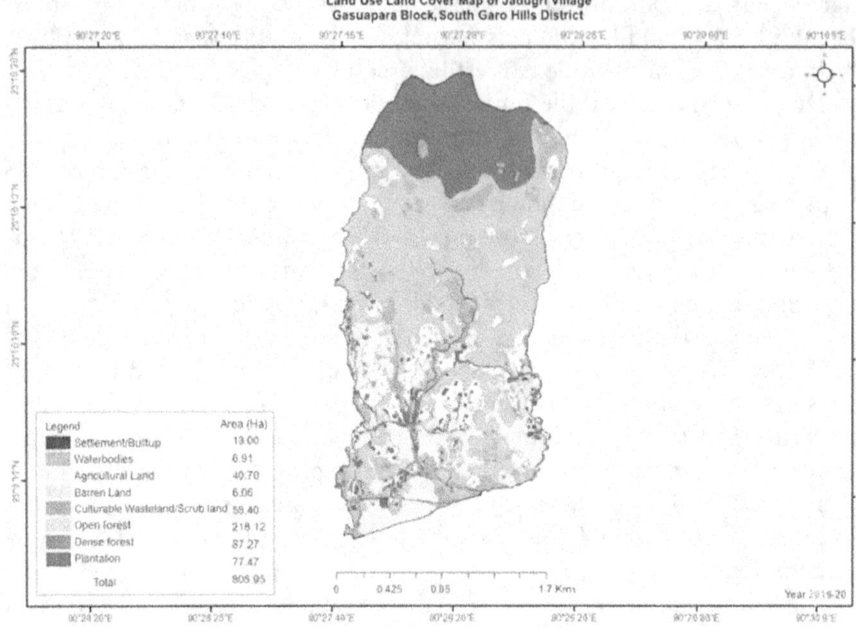


**FIGURE 12.3A**  Land use land cover classes of Jadugri village

*Source:* Authors

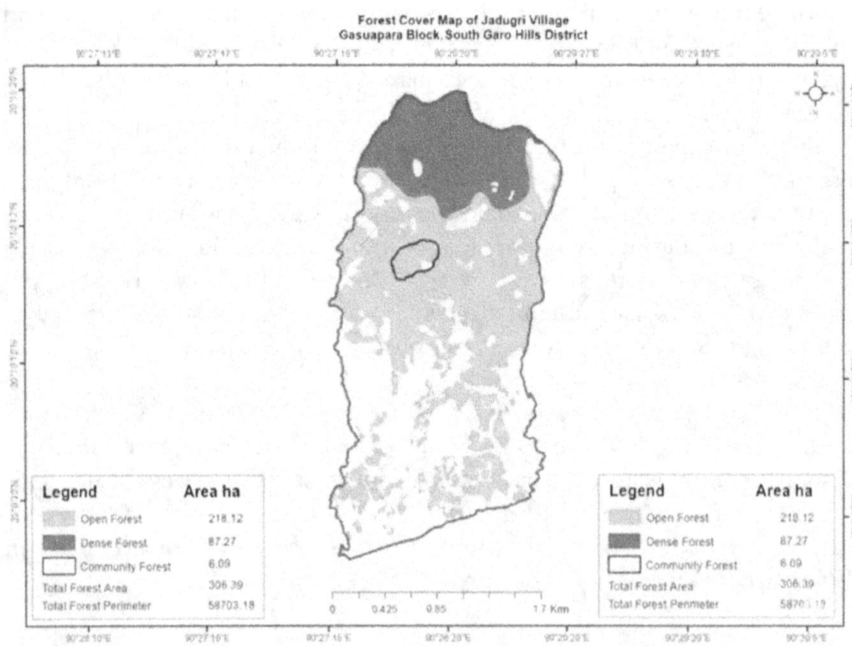


**FIGURE 12.3B**  Forest cover map of Jadugri village

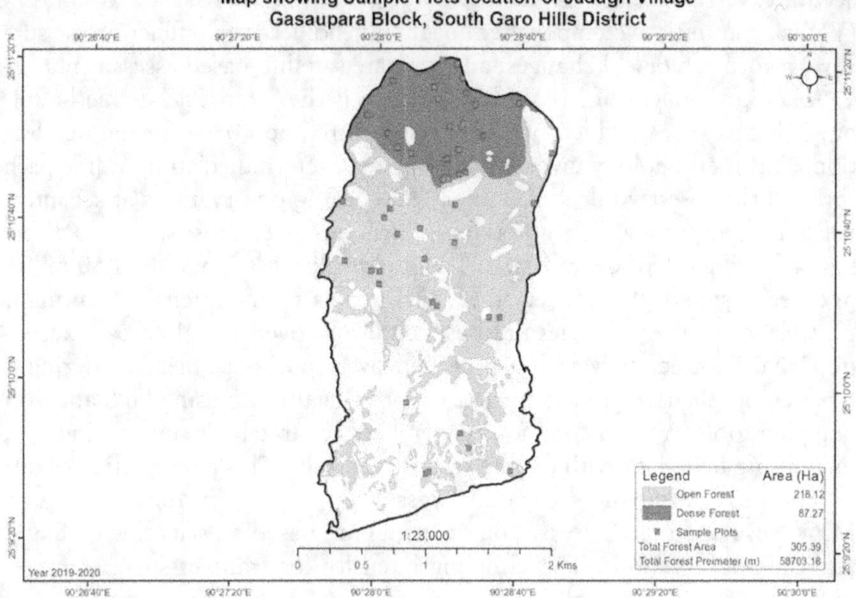


**FIGURE 12.3C**   Map showing sample plot locations of Jadugri village

### 12.6.3 *Main Case Findings 3*

The Measurement, Reporting, and Verification (MRV) protocol has become an essential tool in forest conservation efforts. This is essential for transparent tracking of environmental changes and fostering community participation. In Meghalaya, where forests are central to livelihoods and biodiversity, integrating geographic information systems (GIS) and remote sensing (RS) into MRV has transformed conservation practices. The GREEN Meghalaya Scheme, a pioneering initiative under this framework, utilises geospatial monitoring to assess forest cover, verify conservation activities, and facilitate Payment for Ecosystem Services (PES). This approach improves accountability in conservation efforts while strengthening community stewardship of natural resources.

Traditionally, MRV in forest conservation faced significant challenges due to limited technical expertise, a lack of standardised monitoring frameworks, and resource constraints. Field-based assessments were time-consuming, and many remote areas remained unmonitored due to difficult terrain. Additionally, land-use conflicts arose due to inconsistencies in forest boundary demarcation, often leading to disputes among communities. The introduction of geospatial technologies and community-based monitoring mechanisms helped bridge these gaps, ensuring that conservation data are accurate, verifiable, and locally relevant. By training Village Data Volunteers (VDVs), the initiative empowered local stakeholders to conduct forest surveys, record ecological changes, and validate satellite-based assessments.

A key component of MRV in Meghalaya is the "transect line methodology" (Figure 12.4), which provides a structured approach for data collection. This methodology involves walking a predetermined straight-line path through the forest while systematically recording observations at set intervals. Transect surveys help assess forest structure, tree density, species diversity, and evidence of degradation. The standardised strip width of 50 m (25 m on each side of the line) ensures consistent sampling intensity, making it possible to quantify changes in forest conditions over time. In areas exceeding 2 ha, transect surveys are preferred over spot assessments, offering a more comprehensive understanding of forest health. By using GPS and GIS mapping tools, trained community members accurately document findings, integrating field data with remote sensing analyses. This combination of on-ground verification and geospatial assessments strengthens the reliability of MRV, ensuring that conservation strategies are based on scientific data and real-time observations. The sampling intensity for transect surveys ranges from 5–15% of the total forest area, ensuring that representative data is collected without requiring full-coverage field assessments. This efficiency enables MRV to continuously track changes in forest cover, biodiversity, and carbon stock, providing insights into ecosystem health over multiple

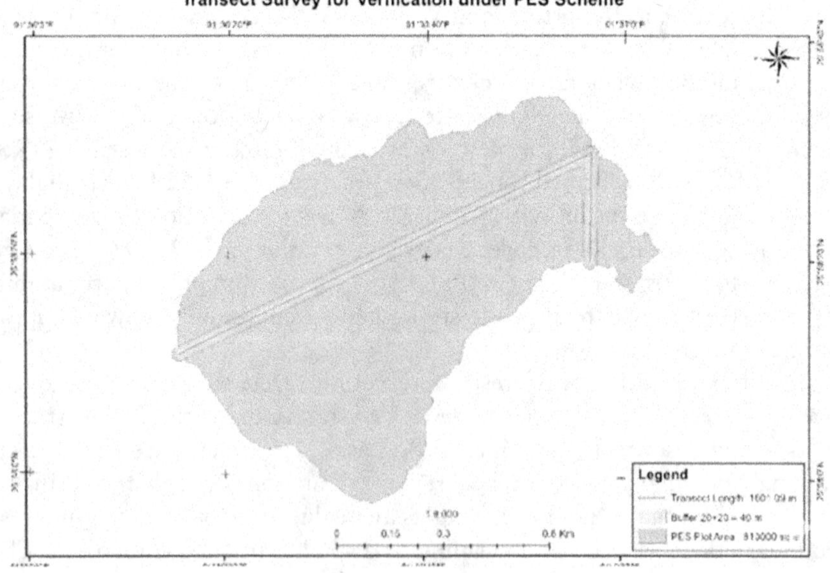


**FIGURE 12.4** Transect line methodology for MRV

*Source:* Authors

monitoring cycles. Additionally, a baseline reference period is established, allowing conservation progress to be measured against initial forest cover and biomass assessments.

This ensures that forest growth, afforestation, and carbon sequestration are quantifiable and verifiable, forming the basis for PES-based incentives. Carbon stock estimation follows scientific methodologies, using transect-based tree measurements, allometric equations, and satellite-derived biomass calculations to assess total carbon sequestration improvements over time.

The adoption of high-resolution satellite imagery further strengthens MRV by providing real-time insights into forest cover changes, land-use patterns, and deforestation trends. In 20 pilot villages of the Ganol Catchment area, geospatial assessments identified eco-sensitive zones, afforestation sites, and water recharge areas, allowing conservation efforts to be strategically planned and implemented. The ability to overlay historical land-use data with current remote sensing information also aids in detecting trends in forest degradation, enabling timely intervention and policy adjustments.

To improve the speed and accuracy of forest monitoring, MRV was integrated with Global Forest Watch (GFW) alerts, which provide near real-time updates on deforestation activities. GFW's Integrated Deforestation Alert Layer combines alerts from GLAD-L (Global Land Analysis and Discovery – Landsat), GLAD-S2 (Global Land Analysis and Discovery – Sentinel-2), and RADD (Radar for Detecting Deforestation), ensuring early detection of unauthorised forest loss. The State Forest Watch System receives GFW alerts, enabling communities to analyse deforestation trends and verify alerts through field-based ground-truthing, allowing rapid responses to illegal logging, forest degradation, or land encroachment. Alerts classified as "Highest Confidence" (confirmed by multiple detection systems) help local authorities prioritise high-risk zones, ensuring timely conservation action.

For transparency and efficiency in PES disbursement, a PES Forest Watcher Dashboard (figure 12.5) has been integrated into the MRV framework. This dashboard serves as a centralised monitoring tool that enables communities, forest managers, and policymakers to track conservation progress, measure ecosystem improvements, and verify PES-linked activities. The dashboard uses geospatial monitoring, remote sensing data, and community field reports to assess whether conservation efforts meet agreed-upon performance indicators. Since verification is conducted annually through spot or transect surveys, the dashboard provides an integrated platform where field data is cross-checked with satellite imagery and Global Forest Watch (GFW) alerts to confirm reported improvements in forest cover, biodiversity, and carbon sequestration.

Beyond its environmental benefits, community-driven MRV has had a positive socio-economic impact, particularly through the PES. The scheme

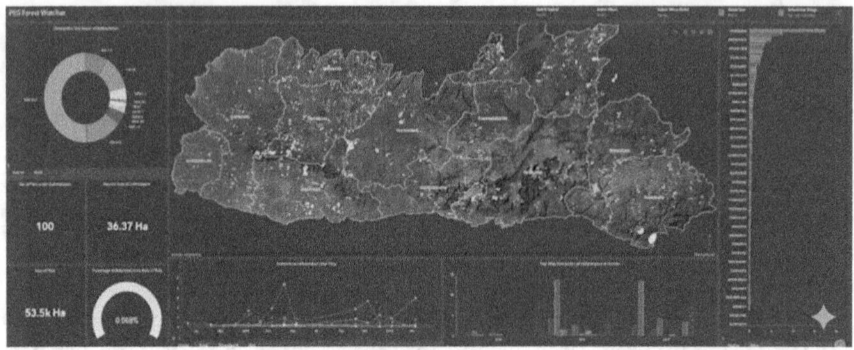

**FIGURE 12.5**    PES forest watcher dashboard

*Source:* Authors

incentivises local communities to engage in reforestation, watershed protection, and forest conservation by providing financial compensation based on verified conservation activities. By linking PES to geospatial monitoring and field assessments, the programme ensures that payments are based on measurable environmental outcomes rather than estimates. This has brought about greater accountability in conservation efforts while generating economic opportunities for forest-dependent communities. With 1,161 VCFs trained across 400 villages, the initiative has not only strengthened local governance structures but has also provided sustainable livelihood options through environmental stewardship.

Despite these successes, implementing MRV has not been without challenges. While training programmes have significantly improved technical capacity, some VDVs initially struggled with GIS tools and digital data collection techniques.

The policy implications of community-driven MRV are substantial, influencing state and national-level conservation strategies. The accurate mapping of sacred groves, reserve forests, and biodiversity hotspots has helped policymakers integrate community knowledge into broader land-use planning frameworks, ensuring that conservation strategies align with local priorities and governance structures.

Ultimately, the Meghalaya model of MRV demonstrates that technology and community knowledge, when integrated effectively, can drive transformative conservation outcomes. By creating an inclusive, transparent, and data-driven MRV framework, the initiative has not only improved forest governance but has also empowered local communities. As participatory geospatial monitoring gains recognition as a best practice in conservation, the lessons from Meghalaya can inform global forest management strategies, ensuring that conservation remains a people-driven, sustainable, and effective endeavour.

## 12.7 Discussion

### 12.7.1 Interpretation of Findings in Relation to Their Replicability

The findings from this study highlight the effectiveness of participatory geospatial monitoring in natural resource management. Meghalaya's MRV model demonstrates that a community-led approach, combined with technological tools like GIS, remote sensing (RS), and UAVs, can improve conservation accountability and empower local governance structures. This approach presents a scalable and adaptable framework that can be replicated in other socio-ecological landscapes, provided key enabling conditions are met.

A primary factor supporting replicability is the active involvement of community members in geospatial data collection and resource monitoring. Community participants were trained to conduct village boundary mapping, LULC mapping, and MRV activities, ensuring that technical knowledge was localised and that conservation strategies were built on community-driven data validation. By equipping local stakeholders with geospatial skills, the initiative strengthened community ownership over conservation processes, reducing dependency on external experts and increasing long-term sustainability. This participatory approach makes the model scalable in other regions where community engagement is central to resource governance.

The integration of GIS, RS, and UAVs has been instrumental in monitoring afforestation success, deforestation trends, and PES-linked forest conservation. The PES Forest Watcher Dashboard serves as an example of how digital tools can enhance transparency in conservation incentives, making this approach replicable in biodiversity-rich yet economically vulnerable regions. Nevertheless, regional adaptations are necessary, since variations in land tenure systems, policy frameworks, and socio-economic conditions may require tailored engagement strategies.

Despite its replicability, challenges persist. Literacy barriers, technical skill gaps, and infrastructure limitations affect community participation. Capacity-building programmes must be continuous, and geospatial tools need to be adapted for local accessibility and governance needs. Ensuring that communities have long-term access to data, decision-making autonomy, and institutional support is critical for successful implementation elsewhere.

Overall, this study provides a scalable, community-driven model that bridges Traditional Ecological Knowledge with modern geospatial monitoring, offering insights for other regions looking to enhance participatory conservation through technology.

## 12.8 Implications for Theory, Practice, and Policy

The findings of this study have significant implications for theory, practice, and policy, particularly in the context of community-led geospatial monitoring, participatory conservation, and decentralised governance models.

### 12.8.1 Implications for Theory

This study contributes to the theoretical understanding of community-driven geospatial monitoring by demonstrating that local participation improves the accuracy and acceptance of conservation initiatives. It challenges traditional top-down approaches by validating the role of Traditional Ecological Knowledge (TEK) in modern geospatial analyses. The integration of GIS, Remote Sensing (RS), and UAVs within community governance structures reinforces the theory that technological adoption in conservation is most effective when it aligns with existing sociocultural systems. Furthermore, the study expands the discourse on PES-linked conservation models, emphasising the need for localised verification mechanisms rather than reliance on centralised monitoring agencies.

### 12.8.2 Implications for Practice

The study provides a replicable framework for integrating geospatial tools into community-led conservation. By employing trained community members for village boundary mapping, LULC classification, and MRV activities, the initiative establishes a practical approach for decentralising resource management. The successful implementation of the PES Forest Watcher Dashboard brings out the importance of transparent, real-time conservation monitoring. Practitioners can use this model to enhance accountability in PES schemes, ensuring that incentives are directly linked to verifiable conservation actions. All the same, sustained capacity-building efforts and context-specific training programmes are essential to bridge technical skill gaps and ensure that local communities adopt geospatial tools on a long-term basis.

### 12.8.3. Implications for Policy

This study underscores the need for policies that formalise community-led conservation governance by institutionalising geospatial training within local governance frameworks. Policymakers must prioritise funding mechanisms that support grassroots-level technology adoption, ensuring that village councils and local forest management groups have continuous access to geospatial data and decision-support tools. Additionally, policy frameworks for PES verification should incorporate community-led MRV mechanisms, recognising local actors as legitimate contributors to environmental monitoring. This study also highlights the necessity of adaptive land-use policies that integrate TEK into geospatial conservation planning, ensuring that both scientific and indigenous knowledge systems guide sustainable resource management.

## 12.9 Summary and Conclusion

### 12.9.1 Key Findings Summary

This study aligns with existing literature on participatory geospatial monitoring, reinforcing the importance of community engagement in environmental governance. Prior work highlights that integrating GIS, RS, and UAVs improves data accuracy and efficiency (Ventura et al., 2018; Acharya et al., 2021). Building on this, our findings show that community members, rather than external agencies, can effectively conduct boundary mapping, LULC classification, and MRV, ensuring long-term sustainability.

Scholars have noted that Traditional Ecological Knowledge (TEK) enhances land-use planning when combined with scientific tools (Mishra & Rai, 2020). This study demonstrates that TEK is not just complementary but foundational, making mapping outcomes both scientifically sound and locally accepted.

In terms of Payment for Ecosystem Services (PES), previous research emphasised verification and transparency challenges (Rai & Mishra, 2023). The adoption of the PES Forest Watcher Dashboard in Meghalaya offers a scalable model for real-time monitoring, making PES disbursements more equitable and performance-based.

Importantly, this study fills a gap by showing that community-led monitoring can be institutionalised within decentralised governance, making geospatial technology accessible at the grassroots. While earlier studies focused on technical aspects, few examined how local governance structures can sustain these technologies.

Community participation has strengthened conservation by improving data accuracy, validation, and transparency. Their contributions have enriched the Management Information System (MIS), integrating real-time geospatial data for better decision-making and PES verification. At the block level, community-led monitoring enables timely interventions, and at the district level, it supports performance tracking and resource planning. This participatory framework makes MIS more effective, decentralised, and responsive to local conservation needs.

### 12.9.2 Significance Restatement

Our research demonstrates how integrating geospatial technology with community-driven conservation efforts strengthens environmental governance. By using GIS, remote sensing (RS), and participatory MRV, local stakeholders are contributing directly to data collection, decision-making, and forest monitoring. The PES-linked monitoring framework enhances accountability and transparency, making this model scalable and adaptable for conservation initiatives. Addressing technical capacity gaps will improve long-term adoption and results. The case shows a policy-relevant, replicable approach that bridges traditional governance with modern geospatial tools, reinforcing inclusive, transparent, and decentralised resource management across diverse socio-ecological landscapes.

### 12.9.3 Future Research and Practical Applications

Future research should focus on strengthening community ownership of geospatial monitoring, ensuring that local stakeholders can independently manage forest conservation efforts. Studies should explore how participatory MRV enhances long-term forest resilience, biodiversity protection, and carbon sequestration, while addressing technical skill gaps and accessibility challenges. Further research is needed to refine geospatial tools for easier use by communities, enabling them to collect, interpret, and apply data without relying on external experts.

Practically, this community-driven model can be expanded to other regions, ensuring that local knowledge and governance structures remain central to conservation planning. Strengthening capacity-building programmes will empower communities to use GIS, remote sensing (RS), and PES-linked monitoring systems effectively. Moreover, integrating Traditional Ecological Knowledge with modern geospatial tools will enhance locally led decision-making. Future research must prioritise co-designed approaches where communities lead conservation efforts with modern technology as a support system. By prioritising community participation, future applications of this model can support inclusive, sustainable, and self-sufficient conservation initiatives.

## 12.10 Limitations and Challenges

### 12.10.1 Technical Limitations

GPS and UAV signal interruptions in dense forests affected geospatial accuracy. Data validation errors, especially from untrained staff, compromise reliability. Remote sensing resolution may not have captured fine details, and mobile device compatibility issues hindered seamless data entry.

### 12.10.2 Human Resource Challenges

Despite training, Village Community Facilitators (VCFs) struggled with geospatial tools, reducing data quality. Community resistance due to land ownership concerns complicated inventory efforts. Technical support gaps led to inconsistent competence levels.

### 12.10.3 Methodological Constraints

Stratified sampling may have overlooked local forest variations. GPS-based village boundaries are functional, not legal, and this has caused disputes. Besides, Meghalaya's rugged terrain poses logistical difficulties.

### 12.10.4 Institutional and Governance Challenges

A lack of centralised data systems has limited integration across agencies. Long-term sustainability depends on uncertain funding.

### 12.10.5 Social and Cultural Barriers

Community engagement was difficult due to misalignment with traditional practices. The absence of cadastral maps had led to ownership conflicts.

## References

Acharya, B. S., Bhandari, M., Bandini, F., Pizarro, A., Perks, M., Joshi, D. R., Wang, S., Dogwiler, T., Ray, R. L., Kharel, G., & Sharma, S. (2021). Unmanned aerial vehicles in hydrology and water management: Applications, challenges, and perspectives. *Water Resources Research*, *57*(11). https://doi.org/10.1029/2021WR029925

Gonzalez-Redin, J., Luque, S., Poggio, L., Smith, R., & Gimona, A. (2016). Spatial Bayesian belief networks as a planning decision tool for mapping ecosystem services trade-offs on forested landscapes. *Environmental Research*, *144*(Pt B), 15–26. https://doi.org/10.1016/j.envres.2015.11.009

Hognogi, G.-G., Pop, A.-M., Marian-Potra, A.-C., & Someşfălean, T. (2021). The role of UAS–GIS in digital era governance. A systematic literature review. *Sustainability*, *13*(19), 11097. https://doi.org/10.3390/su131911097

Machiwal, D., Jha, M. K., & Mal, B. C. (2011). Assessment of groundwater potential in a semi-arid region of india using remote sensing, GIS and MCDM techniques. *Water Resources Management*, *25*, 1359–1386. https://doi.org/10.1007/S11269-010-9749-Y

Mishra, P. K., & Rai, A. (2020). Role of unmanned aerial systems for natural resource management. *Journal of the Indian Society of Remote Sensing*, *49*, 671–679. https://doi.org/10.1007/s12524-020-01230-4

Myers, N. (2003). Biodiversity hotspots revisited. *BioScience*, *53*(10), 916–917. https://doi.org/10.1641/0006-3568(2003)053[0916:BHR]2.0.CO;2

Nemec, K., & Raudsepp-Hearne, C. (2013). The use of geographic information systems to map and assess ecosystem services. *Biodiversity and Conservation, 22*, 1–15. https://doi.org/10.1007/s10531-012-0406-z

Quamar, M. M., Al-Ramadan, B., Khan, K., Shafiullah, M., & El Ferik, S. (2023). Advancements and applications of drone-integrated geographic information system technology—A review. *Remote Sensing, 15*(20), 5039. https://doi.org/10.3390/rs15205039

Rai, S. C., & Mishra, P. K. (2023). Traditional ecological knowledge and resource management: A conceptual framework. In S. C. Rai & P. K. Mishra (Eds.), *Traditional ecological knowledge of resource management in Asia* (pp. 1–11). Springer. https://doi.org/10.1007/978-3-031-16840-6_1

Shaik, A. S., Shaik, N., & Priya, C. K. (2024). Predictive modelling in remote sensing using machine learning algorithms. *International Journal of Current Science Research and Review, 7*(6), 4116–4123. https://doi.org/10.47191/ijcsrr/v7-i6-62

Singh, R. B., & Kumar, D. (2012). Remote sensing and GIS for land use/cover mapping and integrated land management: case from the middle Ganga plain. *Frontiers of Earth Science, 6*, 167–176. https://doi.org/10.1007/s11707-012-0319-x

Ventura, D., Bonifazi, A., Gravina, M., Belluscio, A., & Ardizzone, G. (2018). Mapping and Classification of ecologically sensitive marine habitats using Unmanned Aerial Vehicle (UAV) imagery and Object-Based Image Analysis (OBIA). *Remote. Sensing, 10*(9), 1331. https://doi.org/10.3390/rs10091331

# PART III

# Green Financing: Innovations in Funding Climate Action

# 13

# CONSERVATION OF FORESTS AND PAYMENT FOR ECOSYSTEM SERVICES

*James T. Kharkongor and Lavinia Mary Dkhar*

## 13.1 Background and Introduction

### 13.1.1 Purpose, Scope and Significance of the Initiative

The GREEN Meghalaya scheme is a Payment for Ecosystem Services (PES)-based initiative. It is built on the concept of compensating forest owners for the ecosystem services their forests provide. The forests should be natural. The services include regulating water flows, maintaining biodiversity, mitigating climate change through carbon sequestration and reducing soil erosion.

The initiative aligns economic incentives with environmental conservation, enabling communities to derive financial benefits from forest protection rather than exploitation. This approach is especially important in a state like Meghalaya, where 90% of the land is community-owned and managed, making community participation in conservation efforts crucial.

Those who have degraded their forest tend to become the beneficiaries of most initiatives and Natural Resources Management (NRM) projects. Those who have taken efforts to prevent degradation in the first place often miss out on benefitting from such investments. In terms of cost, rejuvenation of degraded land costs significantly more than payment for conserving an existing forest.

The net present value of forests as established by the Ministry of Environment, Forest and Climate Change has kept the rate between Rs.10–15 lakhs per hectare, depending on the class and quality of the forest. Moreover, afforestation cannot replicate the original biodiversity of lost natural forests.

DOI: 10.4324/9781003735380-16

Launched in 2022, **the initiative (GREEN Meghalaya)**, which is implemented by the Meghalaya Basin Management Agency (MBMA), has already made significant strides. A total of 3,300 beneficiaries, comprising villages, clans and individuals, have enrolled in the scheme, committing to the long-term conservation of more than 51,000 ha of natural forest. These beneficiaries have received financial incentives totalling Rs.48 crores to support their conservation efforts. The base incentive for participation in the initiative is Rs. 5,000 per hectare per year, with additional bonuses available for communities that register their forests as community reserves or for forests that are designated as sacred groves, eco-sensitive zones, living root bridges, or very dense forests.

Based on the success of this first phase, the government committed to scaling up the initiative under a new, expanded initiative called GREEN Meghalaya+ (GM+). This extension aimed to bring an additional 50,000 ha of natural forest under conservation, increasing the total area covered under the initiative to over 100,000 ha.

Under the new phase, the base incentive has been raised to Rs. 10,000 per hectare per year for five years. This reflects the government's commitment to enhancing the financial attractiveness of the initiative for participants. An incentive of Rs. 5,000 per hectare per year would be paid for 'very dense' and 'moderately dense' forests. For forests recognised as eco-sensitive zones, sacred groves, elephant corridors, or those with tourism potential such as waterfalls and caves, participants would receive a further payment of Rs. 5,000 per hectare.

The beneficiaries are required to conserve the forests for a period of at least 30 years under the initiative.

Meghalaya has emerged as the first state in India to implement a Payment for Ecosystem Services (PES) system at a significant scale. While some other states have introduced PES, they remain limited in scope and coverage. In contrast, the GREEN Meghalaya+ Scheme has already encompassed 2.30% of the state's total geographical area and 10.70% of its forest area. This marks a substantial and pioneering effort in ecosystem service-based conservation (Table 13.1).

The implementation of the initiative is guided by robust monitoring and evaluation. Also, a Measurement, Reporting and Verification (MRV) protocol has been developed to assess the effectiveness of conservation activities. Regular monitoring, using a Geographic Information System (GIS) and satellite imagery, ensures that forest conservation commitments are being met.

The Green Field Associates (GFAs), recruited to work at the block level, play a critical role in field-level monitoring, assisting communities with reporting and compliance. Data collected through MRV will be used to

**TABLE 13.1**  Difference between GREEN Meghalaya and GREEN Meghalaya+

| Sl. No. | Particular | GREEN Meghalaya | GREEN Meghalaya+ |
|---|---|---|---|
| 1 | Minimum eligible forest area for participation | 2 ha | 1 ha |
| 2 | Maximum reward per hectare per year | Up to Rs. 15,000 | Up to Rs. 20,000 |
| 3 | Base reward amount per hectare per year | Up to Rs. 8,000 | Up to Rs. 10,000 |
| 4 | Special conditions being considered | Dense forest or a traditionally recognised sacred grove or has a living root bridge or is in eco-sensitive zones around protected areas or wildlife corridors<br>**Up to Rs. 2,000** | a) Dense forest<br>**Up to Rs. 5,000**<br>b) Eco-sensitive zones around wildlife sanctuaries/national parks, sacred groves, living root bridges, elephant corridors and forests with tourism-potential sites such as waterfalls and caves.<br>**Up to Rs. 5,000** |
| 5 | Are there any terms and conditions on how the funds can be used? | Yes, funds must be used only for prescribed activities | No, funds can be used at the discretion of the beneficiaries |

*Source:* Guidelines of GREEN Meghalaya+, MBMA, 2024.

adjust future payments and to ensure that the initiative remains transparent and accountable.

A dynamic GIS-based dashboard has been developed to track and monitor the progress of the GM+ initiative systematically. This interactive platform uses technology to provide real-time spatial and temporal data on key project indicators, including forest cover changes, land area under conservation agreements, beneficiary participation and fund disbursement status.

The system contributes to transparency and accountability by offering up-to-date insights on project performance. It allows for data-driven decision-making, helping authorities identify areas requiring additional support, optimise resource allocation and assess the overall effect of the initiative.

The GM+ initiative is a pioneering effort, the first of its kind in India, and possibly across the entire subcontinent, to be implemented at such a large scale. This uniqueness presents both strengths and challenges in its execution.

One of its key strengths is its innovative approach to conservation, combining community-driven forest protection with financial incentives under the structured PES model. This not only has been fostering local ownership and environmental stewardship but has also set a precedent for large-scale forest conservation through direct beneficiary engagement.

Nevertheless, the novelty of the initiative also presented challenges. Since there are few, if any, existing cases from which the state can draw reference or guidance, the initiative had to develop its own methodologies for identifying, onboarding and managing beneficiaries. This requires continuous learning, adaptive management and real-time adjustments based on field experiences.

### **13.1.1**.1 Rationale

The GM+ initiative holds significant importance and relevance for the state of Meghalaya on account of several critical factors:

- Under the Sixth Schedule of the Indian Constitution, land ownership in Meghalaya primarily rests with local communities, clans and individuals, rather than the government. As a result, the government directly controls less than 5% of the state's forest area, making community-driven conservation efforts essential for sustainable forest management.
- Meghalaya has experienced a high rate of deforestation, leading to the loss of both flora and fauna over recent decades. The decline in forest cover has severely affected biodiversity, ecosystem services and climate resilience. Addressing deforestation is crucial to maintaining environmental stability and sustainable livelihoods for local communities.
- Despite conservation efforts, forest loss continues to be a challenge. According to a Forest Survey Report of 2023, Meghalaya's forest cover has decreased to over 75%, with a recorded loss of nearly 84 km$^2$ since 2021.
- This highlights the need for continuous monitoring, stronger conservation policies and additional support for local communities to ensure long-term sustainability.
- Many communities in Meghalaya have been protecting and preserving their forests for centuries, yet their contributions have largely gone unrecognised. GM+ aims to acknowledge and reward their conservation efforts formally, reinforcing traditional practices that have contributed to the state's rich biodiversity.

- Proactively conserving existing natural forests is significantly more cost-effective than large-scale afforestation and reforestation. By supporting and incentivising communities to maintain their forests, the initiative reduces the financial and ecological costs associated with restoring degraded landscapes.
- Once biodiversity is lost, restoring ecosystems takes decades, if not centuries. Protecting existing forests through GM+ helps preserve Meghalaya's unique ecological balance, prevents irreversible species loss and ensures long-term sustainability for both the environment and local communities.

### 13.1.2 *Review of Literature*

PES programmes are increasingly being adopted globally to enhance sustainability outcomes (Le Tuyet-Anh et al., 2024). PES were initiated in the early 1990s at different spatial scales, with the world's first national PES programme launched in Costa Rica in 1997 (Murillo et al., 2014; Blundo-Canto et al., 2018). More importantly, PES programmes have been considered a powerful economic instrument for conserving ecosystems in the face of threats from local and global **change** (Yang et al., 2018; Friess et al., 2015).

PES were instrumental in providing positive incentives for conservation (Yang et al., 2018; Bladon et al., 2016; Bullock et al., 2011). At the same time, they were facilitating socio-economic development and seeking to address sustainability requirements such as poverty reduction, efficiency and equity, along with ecological outcomes (Pascual et al., 2014; Martin, et al., 2014).

PES programmes have been increasingly adopted worldwide to address diverse ecosystem challenges. Globally, they have played a central role in soil conservation and erosion control, with China's *Grain for Green Program* significantly reducing sediment transport into rivers and improving water quality (Zheng Guo et al., 2024).

More recently, PES models in Himachal Pradesh have involved hydropower developers contributing to upstream catchment conservation.

PES has also been used to promote biodiversity conservation, such as Mexico's payments for maintaining wildlife corridors (Bladon et al., 2016) and India's support for protecting sacred groves, which are culturally significant and ecologically rich (Upadhyay et al., 2018). In agricultural landscapes, PES has incentivised pollination services and agro-biodiversity – for example, in Nepal and Vietnam, where PES is integrated into sustainable farming systems to support smallholders while conserving ecosystem functions (Le Tuyet-Anh et al., 2024).

Taken together, these experiences demonstrate that PES is a flexible economic instrument adaptable across ecological and cultural settings. While developed countries such as the United States and members of the European Union have integrated PES into agri-environmental subsidy schemes (Bullock et al., 2011), in developing countries, PES often emerges as a tool for balancing conservation with poverty alleviation and equity (Pascual et al., 2014). For Meghalaya, with its unique system of community and clan-based land tenure, such lessons underline the importance of designing PES models that are context-specific, socially inclusive and institutionally supported.

### 13.1.3 Theoretical Framework

In the context of the GREEN Meghalaya Scheme, the PES initiative serves as an essential tool to encourage local communities to participate actively in environmental conservation while also securing socio-economic benefits. A well-designed PES framework includes clear eligibility criteria, well-defined conservation obligations, financial disbursement mechanisms and a transparent system for monitoring and evaluation.

A key component of PES is the establishment of contractual agreements between landowners and governing bodies, ensuring that conservation commitments are upheld over a long-term period. By linking financial incentives to measurable environmental outcomes, PES fosters a results-based approach to conservation, where payments are contingent on the successful implementation of prescribed activities.

Moreover, PES is instrumental in promoting sustainable land use by addressing economic pressures that often lead to deforestation and ecosystem degradation. It provides a viable alternative to unsustainable practices by offering communities a source of income that aligns with conservation goals. The effectiveness of PES depends on factors such as community engagement, institutional support, policy coherence and the availability of funding sources to sustain the scheme over time.

The implementation of PES under the GREEN Meghalaya initiative emphasises the importance of integrating scientific, economic and social considerations to ensure long-term ecological and livelihood benefits. The approach not only strengthens local environmental stewardship but also contributes to broader climate resilience and biodiversity conservation efforts.

### 13.2 Methodology

The initiative was first started under the funding of the Meghalaya Community-Led Landscape Management Project (MCLLMP, n.d.), with

World Bank support. The aim was to incentivise individuals, clans and communities owning natural forests in the state to participate in conservation efforts through the PES system.

Steps followed under the GM+ Scheme:

- Recruitment of staff: A cadre of dedicated young boys and girls was recruited specifically for the implementation, monitoring and supervision of GREEN Meghalaya. They became known as the Green Field Associates or GFAs. Currently, there are approx 50 GFAs posted across the state.
- Public awareness and engagement: The District Project Management Units of all districts in the state, along with the GFAs and Village Community Facilitators (VCFs), were to work closely together to achieve the target of GREEN Meghalaya by raising awareness among the public about the importance of GREEN Meghalaya. They were to encourage community participation and engagement to apply to the project. The cooperation and support of the local Headman or *Nokma* played a vital role in gaining community involvement and trust.
- Submission of forms: Those interested, including villages, clans, communities and individuals, were urged to submit their application forms to either the District Project Management Unit (DPMU) or the designated GFAs. Submission was to include vital documents like No Objection Certificates (NoC), land *pattas*, sale deeds, etc. This submission procedure would guarantee that participants provide the essential documentation for evaluation by the DPMU/GFAs.
- Verification process: Verification involves demarcating the forest areas using Global Positioning System (GPS) devices to obtain accurate latitude and longitude coordinates of the designated areas. During the verification process, VCFs had to observe the forest areas to determine their classification, such as dense forest, plantation sites, etc.
- GPS data analysis and final data analysis: At the DPMU GIS, where the Keyhole Markup Language (KML) file is extracted, the designated forest area was first evaluated. After an initial analysis, the proposed area would be moved to the State Project Management Unit (SPMU) GIS-Lab for the last round of data analysis and thereafter the calculation of the financial incentive.
- Orientation programme: After the identification of eligible candidates, an invitation would go to them for an orientation session. In this programme, the selected candidates are briefed on the activities to be performed.
- Signing of Memorandum of Understanding: These candidates must sign an agreement with the first party, which is the MBMA.

- Funds transfer: the MBMA calculated the payment amount for each beneficiary based on the size of the forest area and the established base amount of the scheme. The specific formula for these calculations would vary depending on several factors, but it generally involves multiplying the forest area by a predetermined rate. The payment for the 2nd tranche and onwards to the beneficiaries would depend on the MRV scores they had obtained. Once payment amounts had been calculated, funds were disbursed to all beneficiaries via direct account transfer.

This disbursement is a yearly payment for a period of five years.

- Activities carried out under the PES system and collection of acknowledged receipt: The beneficiaries of the GREEN Meghalaya Project are to submit an acknowledged receipt to the MBMA for verification of funds receipt.
- Measurement, reporting and verification: The measurement, reporting and verification (MRV) protocol developed for the PES GREEN Meghalaya initiative was, basically, a set of norms, guidelines, methodological steps and scoring criteria to verify the effective implementation of the prescriptions and guidelines given to forest and landowners registered under the initiative. MRV scores determined the beneficiaries' payments for the second tranche and beyond.
- Monitoring: A team from the GM+ was deployed on a regular basis to monitor the forests of the beneficiaries. This was especially regarding those about whom it had become known that their forests were depleting. This information had been obtained by the MBMA's network in the villages or through GIS.

## 13.3 Results

Response from rural communities has been generally encouraging as they experienced significant benefits from the financial incentives provided. They effectively carried out assigned conservation activities such as assisted natural regeneration, forest boundary fencing and protecting forest flora and fauna through fire line creation. They also implemented community-driven initiatives like repairing community halls, installing streetlights and constructing footpaths.

In contrast, the scheme has seen limited success in urban or semi-urban areas. Here, forest owners are generally more focused on agriculture, land development, etc. They appear less willing to commit to long-term forest conservation for a period of 30 years.

The funds have been distributed equitably across Meghalaya's 12 districts, ensuring widespread participation in forest conservation efforts. Tables 13.2

and 13.3 give a breakdown of beneficiaries, forest area conserved and total financial disbursement across different districts.

The GM+ initiative has successfully onboarded more than 50,000 ha of land across all 12 districts of the state. This wide coverage demonstrates the extensive reach and commitment of the project in promoting environmental sustainability across a significant portion of the state. The inclusion of such a large area involved a mix of forested areas and other natural ecosystems that provide critical services to both the environment and local communities. Initially, there was a challenge in receiving applications under the scheme. However, once the target of 50,000 ha was reached, applications continued to pour in, leading to many remaining unprocessed. This increase in the number of beneficiaries enrolling in the PES project is an indicator of the scheme's growing popularity and success.

**TABLE 13.2**  Details on beneficiaries across the state

| Sl. No. | District | No. of beneficiaries | Verified forest area (in Ha) | Total amount (in Rs.) |
|---|---|---|---|---|
| 1 | East Garo Hills | 474 | 7,791 | 9,10,19,696 |
| 2 | North Garo Hills | 158 | 3,282 | 3,68,04,653 |
| 3 | South Garo Hills | 270 | 5,837 | 3,99,68,472 |
| 4 | South West Garo Hills | 38 | 605 | 78,74,723 |
| 5 | West Garo Hills | 229 | 3,357 | 3,12,10,452 |
| 6 | East Jaintia Hills | 335 | 4,555 | 3,37,79,724 |
| 7 | West Jaintia Hills | 195 | 3,513 | 2,40,04,208 |
| 8 | East Khasi Hills | 382 | 3,883 | 3,13,51,646 |
| 9 | Eastern West Khasi Hills | 139 | 1,327 | 1,18,11,073 |
| 10 | West Khasi Hills | 169 | 6,198 | 3,19,13,327 |
| 11 | Ri Bhoi | 541 | 7,698 | 5,78,33,878 |
| 12 | South West Khasi Hills | 345 | 2,642 | 2,02,05,466 |

*Source:* Internal Report, 2024, MBMA.

**TABLE 13.3**   Break-up of beneficiaries ownership-wise

| District | No. of beneficiaries (Individual) | Area (individual) | Amount (individual) | No. of beneficiaries (community) | Area (commu-nity) | Amount (community) |
|---|---|---|---|---|---|---|
| Khasi Hills | 1,312 | 10,242.74 | 8,08,06,612 | 209 | 7,596.49 | 5,30,76,173 |
| Garo Hills | 708 | 7,941.06 | 8,23,19,844 | 324 | 9,386.83 | 8,42,51,361 |
| Jaintia Hills | 414 | 3,221.31 | 2,48,66,284 | 88 | 3,590.24 | 2,70,05,490 |

*Source:* Internal Report, 2024, MBMA.

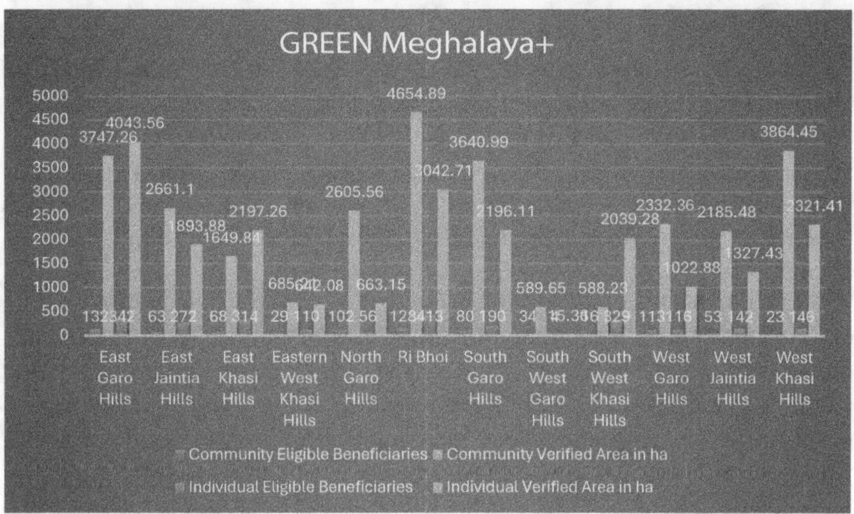


**FIGURE 13.1**    Landholding pattern across districts

*Source:* Internal Report, 2024, MBMA

Based on Table 13.2, we attempted to distinguish the landholding pattern across the state.

Figure 13.1 presents data on community and individual beneficiaries, along with the verified area (in hectares) under the GM+ initiative across different districts of Meghalaya.

**Key takeaways:**

- Ri Bhoi has the highest community-verified area, significantly higher than most Khasi and **Jaintia Hills** districts.
- The Jaintia and Khasi Hills collectively have a large verified area, with the West Khasi Hills, East Khasi Hills and East Jaintia Hills as top contributors.
- The Garo Hills have two leading districts – East Garo Hills and South Garo Hills – which are comparable to Ri Bhoi and parts of the Khasi Hills.
- Ri Bhoi has a high number of community beneficiaries, second only to the East Garo Hills.
- The Jaintia and Khasi Hills show strong community participation, especially in West Khasi Hills, East Khasi Hills and East Jaintia Hills.
- The Garo Hills region shows mixed participation: While the East Garo Hills and South Garo Hills have high numbers, districts like the South West Garo Hills have very low engagement.

## 13.4 Discussion

Following the successful implementation of GREEN Meghalaya in Ganol villages, the GM+ expanded it statewide. The PES system for forest conservation was aimed at the improvement of community forests in the state, which occupy more than 70% of its geographical area. Improved forests will result in higher sequestration of $CO_2$ and provide multiple ecosystem benefits to the communities and, ultimately, aid in adaptation to climate change.

GM+ has been a challenging yet rewarding initiative. Since the MBMA works directly with local, individual forest owners, it has encountered various obstacles while striving to protect the environment. Land disputes and conflicts among beneficiaries have arisen, and forest loss continues to be a problem. All this highlights the complexities of conservation efforts in privately owned forest areas.

To ensure long-term sustainability of the GM+ initiative, the project is closely monitoring the forest areas and preventing deforestation, using state-of-the-art GIS and other technologies. To that effect, we actively engaged with communities, informing them about restricted activities. We will continue to encourage their participation in conservation.

For future sustainable implementation of the GM+, a diversified financing strategy is essential. The funding for this project can be sourced from multiple avenues, ensuring financial stability and continuity over time:

- Government budget allocations: The Meghalaya State Government should allocate a portion of its annual budget specifically for PES-related activities, ensuring consistent funding to communities engaged in forest conservation. This also makes good sense as the state receives tax devolution from the Centre with a weightage of 10% for forest cover according to the 15th Finance Commission's recommendations.
- Externally aided projects: The project can continue leveraging financial support from international organisations like the Asian Development Bank, the World Bank and the International Fund for Agricultural Development. These multilateral agencies often provide long-term funding for sustainable development projects.
- Payment for carbon credits: The MBMA is actively exploring with different agencies to make full use of carbon credits for the forests under the GM+.
- The Government of Meghalaya is actively engaged in negotiations with various financing agencies to explore opportunities for leveraging the carbon market for forests under the GM+.

### 13.4.1  Limitations of the Initiative

The GM+ faces several challenges:

- Meghalaya does not have a cadastral mapping system. This makes it difficult for landowners to identify their land accurately.
- Fears of land or forest confiscation related to governments create hesitation and doubt in the minds of communities to apply for the GM+.
- Green field associates (GFAs) struggle to reach remote areas, and community leaders' reluctance to accept and organise awareness programmes delays successful implementation of the initiative and the spread of information.
- Meghalaya's diverse tribes present an extra difficulty because language barriers complicate communication, especially in districts with varied dialects.
- Geographical distances pose logistical problems, which require innovative transportation and service delivery solutions.
- The monsoon season is problematic for GFAs to reach communities due to landslides, floods and high-velocity winds.
- Community issues, such as ownership disputes and conflicts among committee members, hinder the progress of the GM+.
- There is a lack of responsiveness and ownership from beneficiaries once funds are transferred.
- Because of dense forests, steep slopes, challenging terrains and the presence of wildlife, GFAs and VCFs encounter difficulties in completing verification tasks in these areas. Fortunately, where it is feasible, technology such as the use of drones has been deployed to assist with verification.
- There is a shortage of manpower in GM+ on account of the large number of beneficiaries and villages a single GFA is responsible for, to cover in each block.
- Since Meghalaya is a Sixth Schedule state, most of the land owned lacks proper documentation. In such cases, individuals often submit false applications for land that does not belong to them. This, of course, adds to the problem of identifying the correct land.
- In some remote villages of the state, a lack of electricity, mobile networks and communication infrastructure makes it difficult to reach out to communities.
- Most lands here are owned by the *Syiem or Nokma* (traditional institutions). They are granted to individuals solely for cultivation. This keeps individual persons from applying unless the *Syiem or Nokma* agrees to it.

- Problems come up when a significant portion of the PES plots under observation overlaps with other PES plots, non-PES plots or land managed by other government departments. This leads to confusion, disputes over boundaries and tension related to activity implementation due to changes in the designated area.
- Some applicants do not adhere to the guidelines and have been found engaging in deforestation within the project plots, undermining the objectives of the initiative.
- Lack of cooperation from local headmen causes delays in planning and conducting awareness programmes in the community.

## 13.5 Summary and Conclusion

The project has been tailored for the State of Meghalaya while taking into consideration its geophysical characteristics, the socio-economic features of the population and the environmental conditions, along with a need to address the drivers of forest degradation.

The PES system has turned out to be an effective strategy for motivating people to adopt sustainable land use practices as desired by society at large. They receive monetary incentives for it, based on their adherence to land management prescriptions and performance-based ecosystem improvement works. All these are duly verified.

The PES system offers financial incentives that encourage communities to maintain and enhance their traditional conservation practices, such as the living root bridges and sacred forests, while expanding environmental benefits.

This system can help restore degraded areas and promote biodiversity conservation at a broader landscape level. It goes beyond the capacity of traditional systems alone. By linking forest conservation to economic incentives, the PES system supports alternative livelihoods such as sustainable harvesting and eco-tourism. This reduces reliance on unsustainable practices.

## References

Bladon, A. J., Short, K. M., Mohammed, E. Y., & Milner-Gulland, E. J. (2016). Payments for ecosystem services in developing world fisheries. *Fish & Fisheries*, *17*, 839–859. https://doi.org/10.1111/faf.12095

Blundo-Canto, G., Bax, V., Quintero, M., Cruz-Garcia, G. S., Groeneveld, R. A., & Perez-Marulanda, L. (2018). The different dimensions of livelihood impacts of Payments for Environmental Services (PES) schemes: A systematic review. *Ecological Economics*, *149*, 160–183. https://doi.org/10.1016/j.ecolecon.2018.03.011

Bullock, J. M., Aronson, J., Newton, A. C., Pywell, R. F., & Rey-Benayas, J. M. (2011). Restoration of ecosystem services and biodiversity: conflicts and opportunities. *Trends in Ecology & Evolution, 26,* 541–549. https://doi.org/10.1016/j.tree.2011.06.011

Friess, D. A., Phelps, J., Garmendia, E.. & Gomez-Baggethun, E. (2015). Payments for Ecosystem Services (PES) in the face of external biophysical stressors. *Global Environmental Change, 30,* 31–42. https://doi.org/10.1016/j.gloenvcha.2014.10.013

Le Tuyet-Anh, T., Vodden, K., Wu, J., Bullock, R., & Sabau, G. (2024). Payments for ecosystem services programs: A global review of contributions towards sustainability. *Heliyon,* 10(1), e22361. https://www.sciencedirect.com/science/article/pii/S2405844023095695

Martin, A., Gross-camp, N., Kebede, B., & McGuire, S. (2014). Measuring effectiveness, efficiency and equity in an experimental payment for ecosystem services trial. *Global Environmental Change, 28,* 216–226. https://doi.org/10.1016/j.gloenvcha.2014.07.003

Meghalaya Community-led Landscapes Management Project (MCLLMP). (n.d.). https://projects.worldbank.org/en/projects-operations/project-detail/P157836

Murillo, S. A. M., Castillo, J. P. P., & Ugalde, M. E. H. (2014). Assessment of environmental payments on indigenous territories: The case of Cabecar-Talamanca, Costa Rica. Ecosystem Service, 8, 35–43. https://doi.org/10.1016/j.ecoser.2014.02.003

Pascual, U., Phelps, J., Garmendia, E., Brown, K., Corbera, E., Martin, A., GomezBaggethun, E., & Muradian, R. (2014). Social equity matters in payments for ecosystem services. *Bioscience, 64,* 1027–1036. https://doi.org/10.1093/biosci/biu146

Upadhyay, K. K., Japang, B., Singh, N. S., & Tripathi, S. K. (2019).*Status and socio-ecological dimensions of sacred groves in Northeast India.Journal of Applied and Natural Science,* 11(3), 590–595.https://doi.org/10.31018/jans.v11i3.2121

Yang, H., Yang, W., Zhang, J., Connor, T., & Liu, J. (2018). Revealing pathways from payments for ecosystem services to socioeconomic outcomes. *Science Advances, 4,* 1–8.https://doi.org/10.1126/SCIADV.AAU4932

Zheng Guo, A., Zhang, S., Zhang, Z., & Xie, M. (2024). Benefit evaluation of sloping land conversion program in the upper reaches of the Yellow River. *Heliyon.* https://www.sciencedirect.com/science/article/pii/S2405844024156265

# 14

# FOREST MANAGEMENT PLAN FOR SUSTAINABLE FORESTRY

*Lavinia Mary Dkhar, Jyswill Lyngdoh Nongpiur, Rajaul Karimand H. Liansuanmung*

## 14.1 Introduction

Meghalaya is rich in forests, with 75.6% of its geographical area under forest cover (Indian State Forest Report, 2023). Communities, clans and individuals own over 90% of them. The forest plays a crucial role in supporting local livelihoods and providing essential ecosystem services, including water for the entire population of the State.

Despite their vast extent and importance, limited information is available about these forests. To address this gap, the Centre of Excellence for Natural Resources Management and Sustainable Livelihoods (CoE – NRM&SL), under the Meghalaya Basin Management Agency (MBMA), initiated the preparation of forest management plans (FMPs) in the forests of 400 villages selected for the Community-Led Landscape Management Project (CLLMP) across the State of Meghalaya. These villages, well distributed across all districts of the State, are a representative sample, covering nearly 6% of all of Meghalaya's villages.

Analysis of the forest inventory data from these villages has generated key insights into the extent of dense and open forests, community forest coverage, growing stock, forest carbon, tree numbers and girth-class distribution across different districts and the State as a whole. This information is valuable for policy formulation, planning and implementation of community-led conservation and sustainable management of Meghalaya's forests.

The National Working Plan Code-2023 mandates that all state forest departments develop and implement forest management plans (FMPs) tailored to their unique forest ecosystems and management objectives. The extent and focus of these plans vary, reflecting each state's specific environmental conditions and conservation goals.

DOI: 10.4324/9781003735380-17

### 14.1.1 *Purpose, Scope and Significance of the Case*

- Bring community and private forests in Meghalaya under scientific management.
- Conserve biodiversity.
- Enhance the flow of ecosystem services.
- Create sustainable livelihood opportunities from forests.
- Improve soil and water conservation.
- Mitigate greenhouse gas emissions.
- Enhance resilience to climate change.

The forest management plan in Meghalaya focuses on empowering local communities to sustainably manage and conserve forest resources. It involves identifying community-owned forest areas, assessing resources like timber, non-timber products and water, and promoting sustainable practices such as agroforestry and eco-tourism.

The plan encourages community participation through management structures like joint forest management, ensuring local decision-making and capacity building.

### 14.1.2 *Organisation of the Chapter*

The introduction outlines the context, objectives and background, which is followed by the purpose, scope and significance of the study. A review of literature provides existing research insights, and the identification of gaps highlights areas needing further exploration. A theoretical framework explains guiding models.

The methodology section describes the research design and data collection, with a justification of the methodology spelling out its suitability.

The main case findings present key results, and the discussion relates them to existing literature. The interpretation of findings assesses replicability, while implications for theory, practice and policy explore broader applications.

The chapter concludes with a summary and conclusion, emphasising key findings, restating significance and suggesting future research and practical applications.

### 14.1.3 *Review of Literature*

Forest management has been widely studied across ecological, social and economic dimensions, with a consensus that sustainability requires a holistic and integrated approach. Globally, literature emphasises that forest management plans (FMPs) should balance biodiversity conservation, ecosystem service enhancement and livelihood security (Holling, 1978; Pearce, 2001; Shackleton et al., 2002; Ostrom, 2006; Berkes, 2010; Bawa et al., 2011; Armitage et al., 2020).

A key strand of literature focuses on participatory forest management (PFM). Shackleton et al. (2002) argue that devolving authority to local people enhances conservation outcomes and fosters stewardship, as communities are more likely to protect resources that directly benefit them. Ostrom (2006) similarly highlights the importance of institutions in ensuring successful common-pool resource management.

In India, participatory forest management through joint forest management (JFM) has been relatively successful in states like Odisha and Madhya Pradesh (Saxena, 2000; Sundar, 2000; Agarwal & Gupta, 2005). However, challenges remain in data accuracy, benefit-sharing mechanisms and conflicts with statutory provisions like the Forest Rights Act (Poffenberger, 2006).

From an economic perspective, the literature highlights the growing role of forests in generating financial value through carbon credits, ecotourism and sustainable timber harvesting. Pearce (2001) demonstrates that forest ecosystems provide immense economic services, yet these values are often underestimated in policymaking. At the same time, the uneven distribution of benefits and fluctuating carbon markets are major hurdles to long-term viability.

On the policy and governance front, weak enforcement, illegal logging and fragmented institutional structures hinder effective FMP implementation (Bukhari & Bajwa, 2009; Bawa et al., 2011). Strengthening institutional frameworks and aligning community rights with state policies is crucial to overcoming these barriers.

A recurring theme in the literature is the importance of adaptive management. Holling (1978) introduced the concept of adaptive environmental assessment, emphasising that management plans for dynamic natural systems like forests must evolve with changing ecological and social conditions. This principle is particularly relevant for North-East India, where climate variability, shifting cultivation and sociocultural practices strongly influence forest use.

In the regional context, studies highlight the institutional diversity of community forests in the North-East. Tiwari et al. (2010, 2013) demonstrate how traditional practices continue to shape forest governance and suggest typologies for sustainable management. Khan et al. (2008) emphasise the role of sacred groves across the region in biodiversity conservation.

In the context of Meghalaya, research underlines the significance of traditional forest management systems. Sacred groves, community forests and clan-owned forests exemplify centuries-old practices that have conserved biodiversity while meeting livelihood needs (Sangma & Nampui, 2011; Tiwari et al., 2010; Government of Meghalaya, 2018). These indigenous governance models reflect a strong cultural ethic of conservation, but face mounting pressures from commercial exploitation, population growth and climate change. Recent studies highlight the urgent need to integrate modern tools like Geographic Information System (GIS), forest inventories and carbon assessments with traditional practices to ensure sustainability (Keenan, 2015; Mir et al., 2022; Maru et al., 2023; Laskar, 2024).

Further, literature points to climate change as both a threat and an opportunity for forestry. Keenan (2015) stresses that forests play a dual role: They are vulnerable to climate impacts (fires, pests, storms) yet provide solutions through carbon sequestration and ecosystem-based adaptation strategies. For Meghalaya, whose forests supply water to the state's population and harbour rich biodiversity, integrating climate resilience into FMPs is critical.

In summary, the reviewed literature shows that FMPs must:

- Strengthen community-based governance and participatory structures.
- Internalise the economic value of ecosystem services while ensuring equitable benefit-sharing.
- Incorporate adaptive and climate-resilient management approaches.
- Integrate traditional knowledge with modern scientific tools.

This body of work provides a robust foundation for Meghalaya's pioneering initiative of preparing FMPs for community-owned forests. However, the distinct sociopolitical context of the state, marked by clan ownership and customary rights, requires contextualised frameworks that build upon existing literature while addressing local realities.

### 14.1.4 Identification of Gaps

Since it is the first-ever exercise of this nature, there may be some data gaps. A management plan of dynamic natural resources, like a forest, should also be dynamic itself. So, it is proposed that a forest owner be encouraged to use the FMP as a live document. Every year, there should be a dedicated, time-bound exercise (say, of two weeks) for further enrichment of data in the FMP. This is also to improve **or** modify management practices in the light of fresh data and feedback on previous management interventions. This is in line with the concept of 'adaptive management.'

Further, there is a lack of awareness about the activity at the community level. Local people often have the wrong impression that, through this activity, the government would take over their forest. Human resources pose a challenge to carry out the FMP, as it requires field experts. Consequently, it is difficult to find qualified individuals to carry out the work.

### *14.1.5 Theoretical Framework*

The forest management plan framework in Meghalaya focuses on sustainable forest management through community participation based on their willingness as much as through capacity building and integrated livelihood solutions.

Key components of the framework include the following:

1. **Institutional framework**: Involves active community participation through village councils and local facilitators, Village Community Facilitators (VCFs), with support from government agencies like the Meghalaya Basin Development Authority and the Forest Department.
2. **Mobilisation, capacity building and monitoring**: This means awareness building regarding FMP with the community, particularly forest owners and also training local people, especially youth, in sustainable forest practices and using tools like Geographic Information System (GIS) for monitoring and evaluation.
3. **Planning and implementation**: Includes budgeting, a baseline assessment of forest resources, threats and community needs, followed by the development of a community-led plan with strategies for afforestation, biodiversity conservation and sustainable forest use.

Key activities:

- **Nursery establishment**: To nurture young tree species, especially endangered ones. The seedlings will be used for planting in the village. Surplus seedlings can be sold to generate revenue to maintain the nursery in the long run.
- **Afforestation, reforestation, assisted natural regeneration, gap filling**: To restore degraded lands and plant native species.
- **Forest fire line management**: To create a fire control line along the boundary of the forest.
- **Watershed and biodiversity management**: To protect water sources and create wildlife corridors.
- **Soil and water conservation**: To develop soil and water conservation measures in the forest area where erosion occurs.
- **Climate resilience**: To enhance carbon sequestration and manage climate effects.

**Sustainable forest management and livelihood enhancement:**
Encouraging the cultivation of bamboo, agroforestry and the use of non-timber forest products to promote forest conservation. A strategic approach to sustainable timber harvesting was prescribed based on a thorough

analysis, ensuring responsible resource management. This was provided by the Working Scheme of the Forest approved by the Integrated Regional Office of the Ministry of Environment, Forest and Climate Change (MoEFCC), Government of India, Shillong (GOI, 2023).

By maintaining a sustainable yield, communities can derive long-term economic benefits while preserving forest ecosystems for future generations.

This framework integrates environmental conservation with socio-economic benefits, empowering local communities to manage their natural resources sustainably.

## 14.2 Methodology of the Case

A community-driven approach was adopted for conducting the forest inventory, ensuring active participation from local stakeholders. To assess the forests systematically, plots were identified by stratified random sampling. Forest cover mapping distinguished *dense forests* (canopy density >40%) and *open forests* (canopy density 10–40%). IRS P6 LISS-IV satellite imagery of 2016 was used for it, later updated with the latest available Land Use Land Cover (LULC) and Google Earth imagery.

These two forest categories served as strata for generating sample plots. With GIS software, random points were distributed proportionally across these strata. Due to accessibility problems, some plots could not be surveyed in the field.

To build local capacity, 1,200 Village Community Facilitators (VCFs) were trained by staff from the State Project Management Unit (SPMU), with support from master trainers of the District Project Management Unit (DPMU). These master trainers had previously received training from experts of the Forest & Environment Department, Meghalaya, who had trained approximately 75 of them.

The trained VCFs, along with one or two village youths, conducted the field enumeration under the supervision of the district team. Equipped with Global Positioning System (GPS) devices, the field crews located and surveyed the sample plots, each measuring 0.1 ha (31.62 m × 31.62 m), using prismatic compasses, measuring tapes and nylon ropes.

Within each plot, the girth and height of all tree species above 30 cm girth at breast height were recorded in pre-designed field forms. Tree height measurements were taken using Abney's Level, and where this was not feasible, ocular estimates were recorded. The Quarter Girth Formula was applied to calculate the merchantable timber volume, while the Von Montel Formula was used to estimate potential yield.

This participatory approach not only ensured data accuracy but also empowered communities by enhancing their technical skills, fostering ownership over local forest management and encouraging involvement in decision-making and suggestions.

### 14.2.1 *Justification of Methodology*

The stratified random sampling approach helped ensure a balanced distribution of sample plots across the forests. The number of plots was generated considering the area covered by dense forest and open forest.

Within a certain timeframe and with limited human resources, data collection had to be completed from a huge number of plots. While doing so, the diversity of species in forests had to be kept in mind.

A 0.1 ha plot, which is a national (according to the National Working Plan Code – 2023) and international standard for forest inventories, along

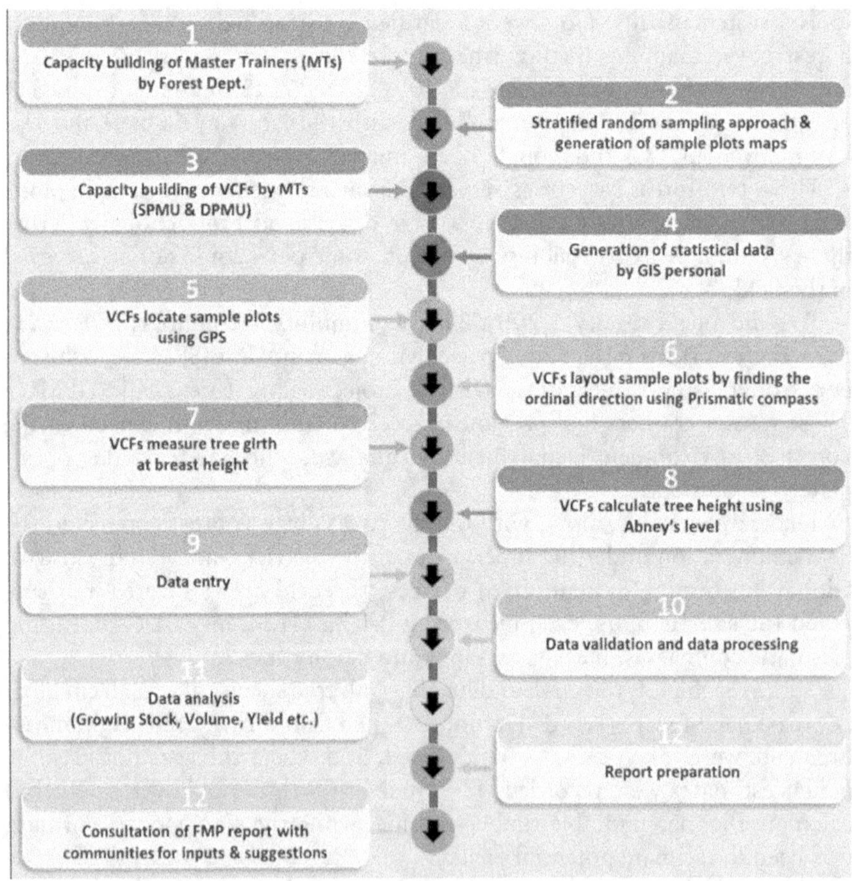


**FIGURE 14.1**   Methodology work flow FMP

*Source:* CoE-NRM&SL, MBMA-MBDA

with biomass estimation and carbon stock assessments, is preferred for forest sampling. It balances representativeness, practicality and statistical accuracy. Also, it is large enough to capture a diverse range of tree sizes and species, providing a good reflection of forest composition.

The size enables efficient data collection across multiple plots within a reasonable timeframe. In addition, with a sufficient number of plots, statistical analysis can yield reliable estimates of forest parameters such as volume, basal area and species composition. This plot size also offers flexibility, allowing slight adjustments to suit different forest ecosystems and dominant vegetation characteristics.

The Quarter Girth Formula was used for the calculation of the volume of timber because it is easier to calculate compared to any other formula.

$$\text{Volume of Timber} = \left(\frac{g}{4}\right)^2 \times h \qquad (14.1)$$

Where:

g = girth of tree

h = clear bole height of Tree

The Von Mantel formula is often considered a preferred method for sustainable yield calculation in forestry on account of its ability to provide a quick estimate of this yield. This is particularly so when dealing with a natural forest with a consistent age distribution, making it readily accessible.

$$Y = \frac{2GS}{R} \qquad (14.2)$$

Where:

Y stands for yield

GS stands for growing stock and

R stands for rotation age

### 14.2.2 Limitations and Challenges

Implementing a forest management plan involves numerous challenges that must be effectively addressed to ensure sustainable forest management. Here are some common implementation challenges and the strategies used to deal with them:

1. **Data collection and accuracy:** Collecting precise and accurate data in difficult terrain and dense forest areas is a challenge. Besides, managing large volumes of data from various sources creates complexity.

   To tackle these problems and challenges, advanced technologies like GPS, prismatic compass and Abney's Level were employed. Mobile apps facilitated real-time data entry and reduced human error. Providing training to field staff and VCFs using such technologies helped improve data collection and management.

2. **Stakeholder engagement**: Different stakeholders, especially local communities and, more so, landowners, encountered various conflicts of interest during the implementation of a forest management plan.

   Involving stakeholders from the beginning of the planning process through regular consultations and meetings helped align their interests and build consensus. Outreach programmes to raise awareness about the benefits of sustainable forest management and encourage community participation.

3. **Resource allocation**: Financial, human and technological resources may be limited, affecting the implementation of an FMP. Deciding how to allocate limited resources effectively can be challenging. So, proper allocation with maximum resources at grassroots level would bring about a good and proper implementation of the plan. Convergence and funding from other organisations and partnerships would supplement resources.

4. **Climate change and natural disasters**: Climate change can lead to unpredictable weather patterns, affecting forest health and management activities. Fires, floods, storms and pests are the cause of natural disasters that bring challenges to the management of forest activities. So it was important to implement adaptive management practices that allowed for flexibility and responsiveness to changing conditions. Also, forest resilience could be improved through practices such as diversifying species and implementing firebreaks.

5. **Monitoring and evaluation**: Ensuring consistent and continuous monitoring of forest conditions can be difficult. Likewise, integrating data from various sources for a comprehensive analysis poses challenges. Using remote sensing and GIS technologies for efficient and continuous monitoring minimised the challenges. This was aided by the development of integrated data management systems that combine data from multiple sources for an integrated analysis.

## 14.3 Results

The implementation of forest management plans (FMPs) significantly affected socio-economic, technical and policy domains.

Socio-economically, FMPs created jobs, improved livelihoods and empowered communities. All along, they preserved cultural heritage – the traditional knowledge, practices, beliefs and customs of indigenous and local communities that have been historically used to manage and conserve forests.

Technically, they promoted sustainable forestry practices, enhanced resource management and integrated advanced technologies like GIS for better monitoring.

On the policy front, FMPs strengthened legal frameworks, built institutional capacities and fostered cross-sectoral coordination, aligning national policies with international sustainability goals.

Overall, FMPs contributed to resilient forest ecosystems, benefiting local communities and the global environment. The Project, covering 400 villages, brought 1,102 km$^2$ under forest management plans, including 4901.43 m$^3$ of growing stock, 19.995 million trees, an annual yield of 152,446.51 m$^3$ and a carbon stock of 1176.87 tonnes C/ha.

The project increased sustainable forest management adoption and community awareness. This included about 75 Master Trainers, while 1,200 VCFs were able to handle the technology during the implementation of the FMP. A working scheme was initiated for the project villages, where the Forest Department and the Autonomous District Council of the State were notified to collaborate with the Meghalaya Basin Management Authority – Meghalaya Basin Development Authority (MBMA-MBDA).

The figures and tables in the following sections present the results of the forest management plans:

### 14.3.1 Growing Stock of Trees

The growing stock of trees in the forest, district-wise, in 400 villages was estimated using the Quarter Girth Formula. It totalled 4,901.43 m$^3$, as shown in Figure 14.2.

From Figure 14.2, it is clear that the East Khasi Hills District has the highest growing stock, with 1,465 m$^3$, contributing about 30% to the total, within the project villages.

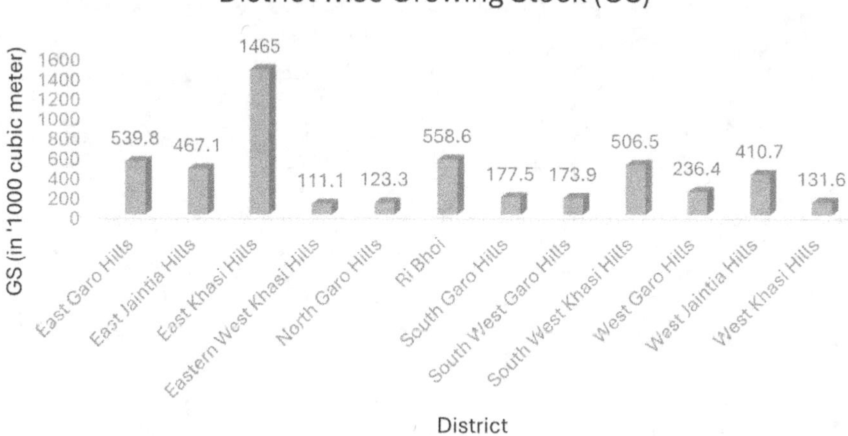


**FIGURE 14.2**  Growing stock of trees

*Source:* CoE-NRM&SL, MBMA-MBDA

### 14.3.2 Distribution of Trees District Wise

Figure 14.3 represents the district-wise number of trees in the project villages for dominant species and miscellaneous species, totalling 19.995 million trees.

Again, the East Khasi Hills District appears to have the highest number of trees, followed by the South West Khasi Hills District, with 3.976 million and 3.934 million trees, respectively. It also shows that the distribution of trees in the North Garo Hills District is the least.

### 14.3.3 Number of Trees in Different Girth Classes

The following graph gives the number of trees in different girth classes for dominant and miscellaneous species (Figure 14.4).

It also shows that trees are present in higher girth classes, but their contribution is relatively small. For this reason, it is recommended that trees from lower girth classes should be protected against moving to higher girth classes.

### 14.3.4 Yield

The following graph represents the district-wise yield in 400 villages for the dominant species and miscellaneous species (grouped together), calculated using the Von Montel Formula, as shown in Figure 14.5.

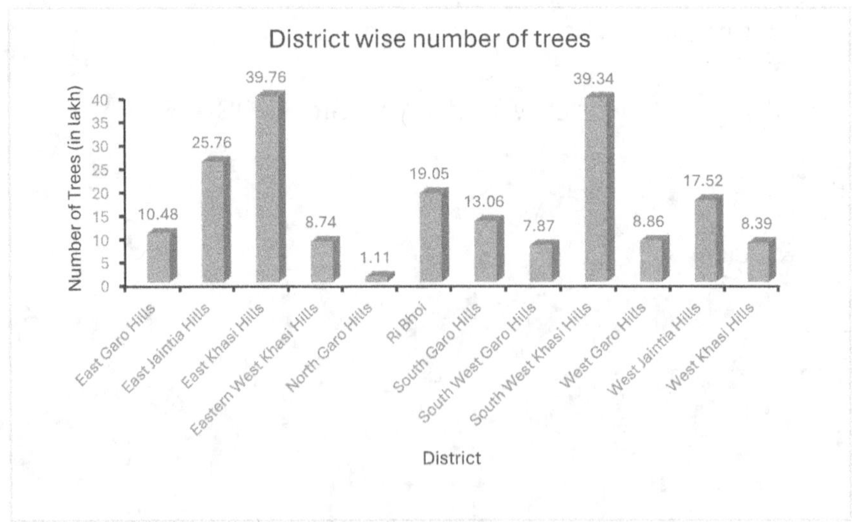


**FIGURE 14.3**   Distribution of trees district-wise

*Source:* CoE-NRM&SL, MBMA-MBDA

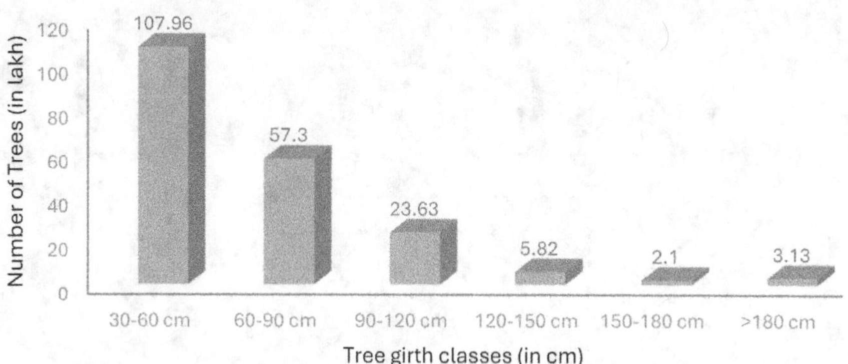



**FIGURE 14.4**  Distribution of trees in different girth classes

*Source:* CoE-NRM&SL, MBMA-MBDA

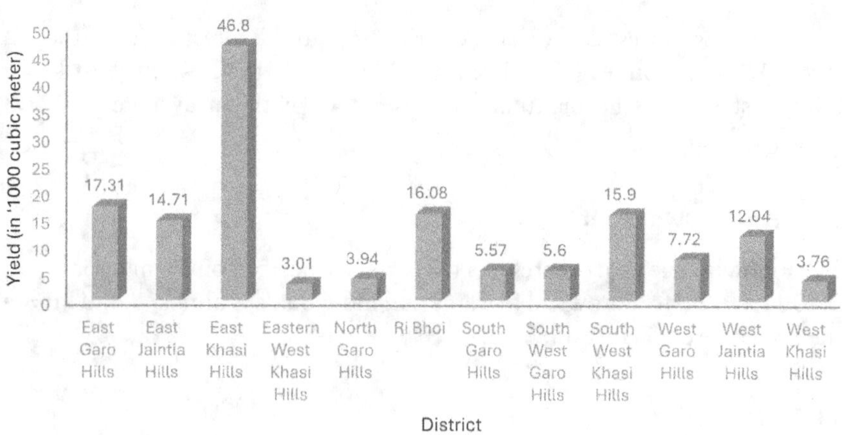



**FIGURE 14.5**  Yield of dominant species and miscellaneous species

*Source:* CoE-NRM&SL, MBMA-MBDA

Clearly, the East Khasi Hills District has the maximum yield of about 46,800 m³ annually.

## 14.4  Forest Carbon

Using the carbon factor given in the India State of Forest Report, 2019 for different forest types in different canopy density classes, the district-wise forest carbon has been estimated, see Figure 14.6.

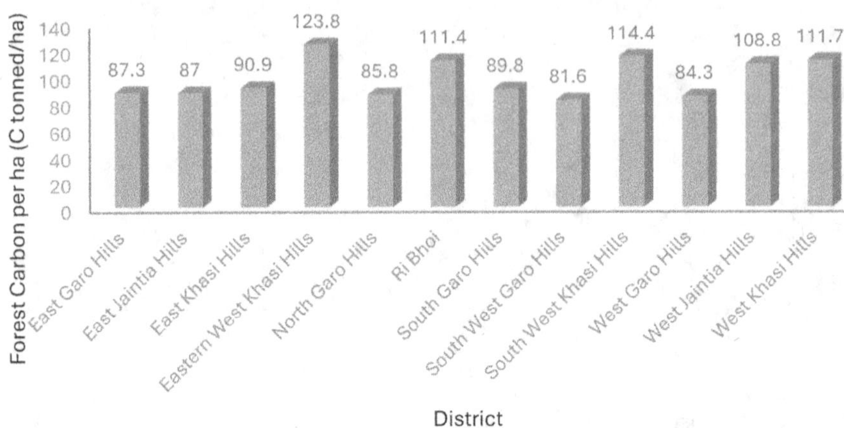


**FIGURE 14.6**  Forest carbon

*Source:* CoE-NRM&SL, MBMA

The average forest carbon stock per hectare in the forest cover of the village = 1,176.87 tonnes C/ha. The graph shows that the Eastern West Khasi Hills District has the maximum carbon stock with an average of 123.8 tonnes C/ha.

## 14.5 Forest Ownership

The following pie chart represents the forest owned by communities is about 2,800 km² and forest owned by clans and individuals is almost 8,470 km² in the project area; Figure 14.7.

### 14.5.1 Forest Cover

Forest cover in **the 400 villages** has been mapped using IRSP6 LISS-IV satellite data on a 1:10,000 scale. The extent of forest cover in the village is 1,102 km² out of which 319.61 km² is dense forest and 783.29 km² is open forest, see Table 14.1 and Figure 14.8. The village clearly has a mixed forest ownership.

### 14.5.2 Common and Dominant Tree Species and Wildlife

Additional findings from the field include common and dominant tree species and wildlife as follows (see Table 14.2).

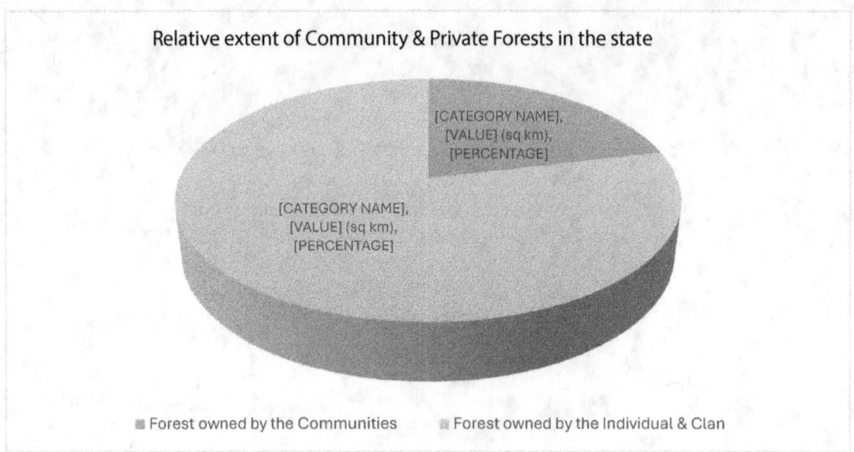


**FIGURE 14.7**    Forest owned by community, clan and individual:

*Source:* CoE-NRM&SL, MBMA-MBDA

**TABLE 14.1**    Forest cover

| Forest cover category | Area (km²) | % of Geographic area of village |
|---|---|---|
| Dense forest | 319.61 | 29 |
| Open forest | 783.29 | 71 |
| Total | 1,102.90 | 100 |

*Source:* CoE-NRM&SL, MBMA-MBDA.

### 14.5.3  Main Case Findings 1

Creating an effective forest management plan is essential for ensuring the sustainable use, conservation and restoration of forest ecosystems.

Here are key learnings and takeaways on developing and implementing a forest management plan:

1. **Forest inventory and assessment:** The Master Trainers and trained VCFs from the project villages were able to conduct a detailed inventory of forest resources, tree species, age classes, wildlife habitats, water resources, soil types and other ecological features. Through such an inventory, the current health and condition of a forest can be assessed. This also means identifying issues such as pests, diseases, invasive species, the status of biodiversity and signs of degradation.
2. **Clear objectives and goals:** A forest management plan creates or sets a long-term goal for sustainability, balancing ecological, economic and social objectives. It defines specific, measurable, attainable, relevant and

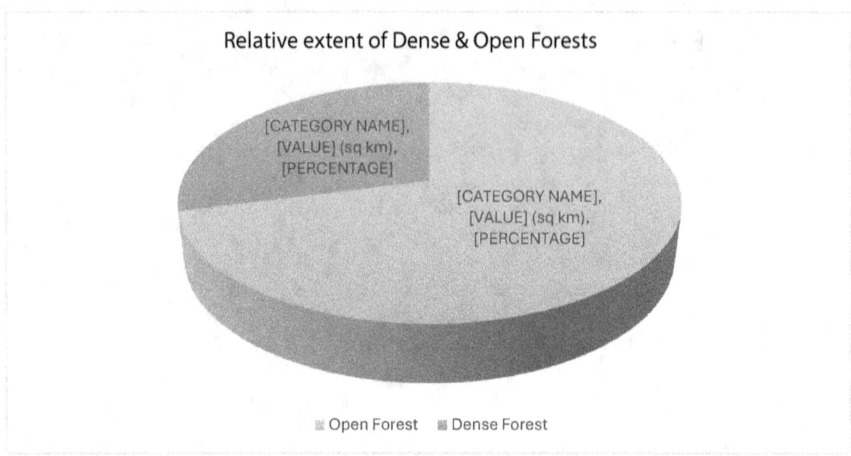



**FIGURE 14.8**    Forest cover in the village

*Source:* CoE-NRM&SL, MBMA-MBDA

**TABLE 14.2**    Common and dominant tree species

| Sl. No. | Vernacular name | Botanical name |
| --- | --- | --- |
| 1 | Dieng Kseh | *Pinus kesiya* |
| 2 | Dieng Ngan | *Schima wallichii* |
| 3 | Dieng Sning | *Casttanopsis indica* |
| 4 | Dieng Blei | *Shorea robusta* |
| 5 | Dieng Sai | *Quercus dealbata* |
| 6 | Dieng Doh | *Exbucklandia populnea* |
| 7 | Dieng Rai | *Michelia champaca* |
| 8 | Dieng Snoin | *Breynia retusa* |
| 9 | Dieng Sohot | *Castanopsis spp* |
| 10 | Dieng Sohstap | *Castanopsis spp* |
| 11 | Dieng Sal | *Shorea robusta* |
| 12 | Dieng Sohphie | *Myrica esculenta* |
| 13 | Arecanut | *Areca catechu* |
| 14 | Boldak | *Schima wallichii* |
| 15 | Bolbret | *Cadrela toona* |
| 16 | Bolsal | *Shorea robusta* |
| 17 | Cashew nut | *Anacardium occidentale* |
| 18 | Bolmatra/Golmatra | *Wrightia arborea* |
| 19 | Masanchi/Makanchi | *Calicarpa arborea* |
| 20 | Matmi | *Croton joufera* |
| 21 | Megong | *Bauhinia purpurea* |
| 22 | Segun | *Tectona grandis* |
| 23 | Tebrong | *Thysanolaena latifolia* |

*Source:* CoE-NRM&SL, MBMA-MBDA.

**TABLE 14.3** Wild life

| Sl. No. | Wild life |
| --- | --- |
| 1 | Elephant |
| 2 | Bear |
| 3 | Clouded leopard |
| 4 | Monkey |
| 5 | Wild boar |
| 6 | Hoolock gibbon |
| 7 | Himalayan marten |
| 8 | Slow loris |
| 9 | Squirrel |
| 10 | Macaque |
| 11 | Jackal |
| 12 | Fox |
| 13 | Snake |

*Source:* CoE-NRM&SL, MBMA-MBDA.

time-bound objectives for various aspects of forest management such as timber production, wildlife conservation and watershed protection.

3. **Stakeholder engagement:** It is important to involve all stakeholders, including government agencies, local communities and private landowners, in the planning to ensure that diverse perspectives and needs are considered. This engagement brings about an effective communication channel to keep all parties informed and involved throughout the implementation and monitoring stages.

4. **Climate change adaptation and mitigation:** By integrating best-practice strategies into a forest management plan, programme designers can significantly enhance the role of forests in carbon sequestration and storage. This not only contributes to climate change mitigation but also promotes healthier, more resilient forest ecosystems, supporting biodiversity and providing long-term benefits to local communities.

5. **Technology:** Scientific technologies used during the preparation of a forest management plan, such as a GPS device, prismatic compass, Abney's-level compass and mobile apps for surveys and information collection, constitute a system of resources that enhances data accuracy, efficiency and effectiveness. This technological synergy enables precise planning, continuous monitoring and adaptive management, ultimately supporting sustainable forest management and conservation efforts.

### 14.5.4 Main Case Findings 2

Implementing a forest management plan brings numerous challenges in its wake that must be effectively taken care of to ensure sustainable forest management.

Here are some common implementation challenges and the strategies used to deal with them:

1.  **Data collection and accuracy:** Collecting precise and accurate data in difficult terrain and dense forest areas is a challenge. Managing large volumes of data from various sources is complex. To remedy this, employing advanced technologies like the GPS, prismatic compass, Abney's-level compass and mobile apps facilitates real-time data entry and reduces human error. Providing training to field staff and VCFs using such technologies ensures better data collection and management.
2.  **Resource allocation:** Financial, human and technological resources may be limited, affecting the implementation of an FMP. Deciding how to allocate limited resources effectively can be challenging. So, proper allocation with maximum resources at the grassroots level will result in good and proper implementation of the project. Convergence and funding from other organisations and partnerships will contribute to supplementing resources.
3.  **Climate change and natural disasters:** Climate change can lead to unpredictable weather patterns, affecting forest health and management activities. Fires, floods, storms and pests are the causes of natural disasters that bring challenges to the management of forest activities. So it is important to implement adaptive management practices that allow for flexibility and responsiveness to changing conditions. Also, to enhance forest resilience through practices such as diversifying species and implementing firebreaks.
4.  **Monitoring and evaluation:** Ensuring consistent and continuous monitoring of forest conditions can be difficult. Likewise, integrating data from various sources for a comprehensive analysis poses challenges. Using remote sensing and GIS technologies for efficient and continuous monitoring and developing integrated data management systems that combine data from multiple sources for a holistic analysis will minimise these challenges.

## 14.5 Discussion

### 14.6.1 Interpretation of Findings in Relation to Their Replicability

As a first step for the preparation of a forest management plan, a two-staged training programme was conducted. First, Master Trainers, who were project officials at the district level, received training. They, in turn, imparted training to village community facilitators (VCF) from each village in their respective district. In addition to the 1,200 VCFs, another 300 VCFs were trained to conduct on-field enumeration for the implementation of FMP in the forests of Payment for Ecosystem Services and the MegLIFE project forests. They carried out a forest inventory survey in their village area under the guidance of district teams. The teams validated the data collected and sent it to the Centre of Excellence for Natural Resource Management & Sustainable Livelihoods (CoE-NRM&SL). There, the information was processed and analysed for all 400 villages, with an additional 100 Payment for Ecosystem Services (PES) of forest owned by the community, clans and individuals.

Replication of this approach has taken place in another 500 villages under the MegLIFE Project; also, it will be implemented in the MegARISE Project.

A forest management plan report was prepared in close consultation with respective villages, incorporating their suggestions regarding the prescriptions of activities such as assisted natural regeneration (ANR), Afforestation, Reforestation, Forest Fire Line Control, Bamboo management and other measures that the community or forest owners should undertake for the sustainable management of the forests.

### 14.6.2 Implications for Theory, Practice and Policy

Forest management plans have key implications:

1. Forest management plans (FMPs) enhance our understanding of ecosystems by providing insights into forest dynamics, species interactions and biodiversity. FMPs support adaptive management by incorporating data from forest management activities (e.g., monitoring biodiversity, tree growth or soil quality) and using these data to adjust management strategies based on real-time information. This aligns with the theory that forest management should be responsive to changing conditions.

   FMPs also bridge ecological and social perspectives by recognising the role of local communities in forest management, integrating both environmental conservation and human activities into a unified approach. This integration helps advance theories that view forests as social-ecological systems, where ecological health and human well-being are interdependent.

2. FMPs promote sustainable forest management and emphasise community involvement in the planning and decision-making process. They help ensure that management practices are not only effective but also fair, considering the social, economic and cultural needs of the people who live in and around the forests. They also encourage adaptive practices by advocating continuous monitoring of forest conditions (such as tree health, soil quality and wildlife populations) and adjusting management strategies based on data.

3.  Several key Acts and regulations guide forest management, support-
    ing the objectives of forest management plans. These include the Forest
    Conservation Act, 1980, which regulates land diversion for non-forest
    use, the Indian Forest Act, 1927, governing forest administration and
    the Wildlife Protection Act, 1972, which protects biodiversity.

The Panchayats (Extension to Scheduled Areas) Act, 1996, empowers local
communities in scheduled areas to manage forests, while the Meghalaya
Forest Regulation Act, 1973, provides state-specific rules for forest use.

The National Afforestation Programme offers funding for restoration,
the Forest Rights Act, 2006, recognises indigenous community rights, and
the Biological Diversity Act, 2002, promotes biodiversity conservation.

Together, these laws ensure sustainable forest management, conservation
and community involvement in Meghalaya.

### 14.6.3 Traditional Forest Management Practices

While there is a huge market demand for high-quality timber in Meghalaya,
large tracts of dense and well-protected patches of community-owned
primary forests can still be seen in many parts of the state. This can be
attributed to the traditional forest management practices followed by tribal
communities since time immemorial.

Under customary law, these forests are classified into different types
depending on their intended use.

Locally, these forests are known as *Law Kyntang* (sacred forest), *Law
Shnong* (village forest), *Law Adong* (village restricted forest), *Law Raid*
(forests belonging to a group of villages), *Law Ri-Sumar* (private forest on
community land), *Law Ri-Kynti* (private forest on private land), *Law Lum
Jingtep* (cemetery forest) and *Law Kur* (clan forest). The sacred forest and
village restricted forests provide ecosystem services such as protection of the
upper catchments of watersheds, conservation of biodiversity and medicinal
plants. These forests are currently called community forests or community
conserved areas. Most villages in the state have one or more types of community forests.

The traditional management practices not only helped in conserving the
resource, as evident from the presence of large patches of well-protected
forests and ensuring its sustainable use but also serve as a common good
and 'safety net' for the communities. This is demonstrated in the village of
Nongpyndeng, where the village council manages a large portion of the forest for the benefit of all inhabitants of the village.

In Mawshun village, three types of forests are managed: Sacred forest,
village forest and agroforest. Each forest type provides different services
to the people in the village. The sacred forest is the home of a deity who,

according to local belief, protects the village from natural calamities, famine and diseases. Village forests provide firewood, wild edible plants and poles for house construction and repairs. Agroforests are the primary source of cash income. (Sangma & Nampui, 2011).

Meghalaya's traditional forest management system remains a strong model of community-led conservation, balancing economic needs with environmental sustainability. These practices highlight the importance of indigenous knowledge in forest governance and ecological preservation.

## 14.7 Summary and Conclusion

The forest management plan has outlined strategies for the sustainable management of forest resources, ensuring biodiversity conservation, ecological restoration and community well-being. It includes detailed forest type classifications, biodiversity assessments and sustainable harvesting practices across various sample plots. A key element of the plan is community involvement, emphasising the importance of local engagement in decision-making to promote sustainable livelihoods.

The plan aims to maintain ecological balance through regular monitoring, adaptive management and continued data collection. Its successful implementation depends on strengthening the capacity of forest personnel and local communities, cultivating robust partnerships and aligning conservation efforts with sustainable resource utilisation.

By following these strategies, the FMP aims to safeguard forest ecosystems for future generations while supporting both environmental and socio-economic goals.

The implementation of FMPs has transformed local communities by enhancing livelihoods, empowering them with knowledge and skills and fostering ecological resilience. By integrating traditional wisdom with modern technologies, the project has laid the foundation for sustainable forest management that benefits both people and the environment.

Community responses to forest management plan initiatives may vary, based on factors such as awareness levels, perceived benefits, trust in implementing agencies and the extent of community participation in decision-making.

Some responses from various community members are as follows:

1. When people can see tangible benefits, such as improved livelihoods and forest regeneration, they actively engage in afforestation, sustainable harvesting and conservation activities.
2. Traditional ecological knowledge is often integrated into an FMP, which leads to greater acceptance and ownership among community members.

3. If the plan includes income-generating activities like producing non-timber forest products, ecotourism or agroforestry, communities show greater enthusiasm.
4. Awareness programmes and incentives encourage community members to adopt sustainable forest management practices, which effectively reduce deforestation and illegal logging.

A forest management plan (FMP) is an essential framework for the sustainable use, conservation and protection of forest resources, while balancing ecological, social and economic needs. It supports biodiversity conservation by safeguarding native species, protecting habitats and reducing deforestation.

FMPs also strengthen community livelihoods by promoting sustainable income opportunities such as agroforestry, non-timber forest products and ecotourism, while encouraging participatory decision-making. From a climate perspective, they enhance resilience by contributing to carbon sequestration, regulating water cycles and reducing the impacts of climate change.

In addition, FMPs serve as practical tools for policy implementation, ensuring alignment with national and international environmental commitments and compliance with legal requirements. The FMP document also serves as a working scheme for communities, practitioners and agencies, providing step-by-step guidance for sustainable forest governance. By combining scientific knowledge, traditional practices and stakeholder participation, it delivers both ecological and socio-economic benefits.

Overall, an FMP acts as a roadmap and working guide for maintaining the health, productivity and sustainability of forests, while addressing global challenges such as climate change, deforestation and rural development.

### 14.7.1 Key Findings Summary

Creating an effective forest management plan involves key components like conducting forest inventories, setting clear goals for sustainability, engaging diverse stakeholders, integrating climate change adaptation strategies and using technology.

Inventorying forest resources helps assess health and identify issues. Clear, measurable objectives are to balance ecological, economic and social goals. Stakeholder engagement ensures diverse perspectives are considered, fostering communication throughout implementation.

Carrying out the FMP involves overcoming challenges such as data collection, resource allocation and effective monitoring. The use of technological tools like GPS and mobile applications enhances data accuracy, facilitates monitoring and supports better management to address these issues.

The implementation of forest management plans has a significant focus on involving local communities in forest care through activities like resource inventory, setting goals for sustainability and using technology. Communities participated in assessing forest health, balancing conservation with income from sustainable harvesting and ecotourism. Training programmes helped locals manage forests better.

The project introduced climate resilience strategies and tools like GPS and mobile apps for better decision-making. It created jobs, increased income, strengthened community leadership and improved forest health, benefiting both the environment and local people through collaboration with authorities and sustainable practices.

## 14.8  Future Research and Practical Applications

Meghalaya, with its rich forest cover (over 76% of its total geographical area), is dealing with both significant challenges and opportunities in sustainable forest management. The state's unique land tenure system, where a large portion of forests is under community and private ownership, necessitates a localised, participatory approach to forest management planning.

Future research and practical applications should continue to focus on community-led, technology-driven biodiversity conservation, climate resilience, sustainable livelihoods and community participation while integrating modern technologies with indigenous knowledge.

## References


Holling. (1978). *Adaptive environmental assessment and management.*

Bawa, K. S., Rai, N. D., & Sodhi, N. S. (2011). Rights, governance, and conservation of biological diversity. *Conservation Biology, 25(3),* 639–641

Pearce. (2001). *The economic value of forest ecosystems.Ecosystem Health, 7(4),* 284–296.

Ostrom. (2006). *PNAS note on linking forests, trees, and people.*

Berkes. (2010). *Devolution of environment and resources governance.*

Armitage, et al. (2020). *Governance principles for community-centred conservation.*

Keenan. (2015). *Forests and climate change* (overview).

Poffenberger, M. (2006). *People in the forest: Community forestry experiences from Southeast Asia.* International Journal of Environment and Sustainable Development, 5(1), 57–69

Sundar. (2000). *Unpacking the 'joint' in joint forest management.*

Saxena. (2000). (JFM assessments).

Agarwal & Gupta. (2005). *Participatory exclusions & gender.*

Bukhari & Bajwa. (2009). *Illegal logging and impacts.*

Forest Survey of India. (2023). *India State of Forest Report 2023* (Vol. 2). Ministry of Environment, Forest and Climate Change. https://fsi.nic.in/uploads/isfr2023/isfr_book_eng-vol-2_2023.pdf

Government of India. (2023). *National Working Plan Code-2023.* Ministry of Environment, Forests, and Climate Change

Tiwari, et al. (2013). *Institutional arrangement and typology (NE India)* Maps traditional, quasi-traditional, and modern forest institutions across NE; Meghalaya stands out for intact customary systems. SpringerLinkResearchGate

Khan, et al. (2008). *Sacred groves & biodiversity*Early pan-India evidence base—widely cited; supports the conservation value of groves common in NE. activeremedy.org

Maru, et al. (2023). *Carbon in sacred forests* Quantifies above-/below-ground carbon in sacred forests—links culture with climate mitigation. Taylor & Francis Online

Government of Meghalaya. (2018, portal). *Sacred Groves (official overview)* Authoritative descriptions/nomenclature (Khlaw Kyntang/Law Lyngdoh/Khloo Blai) and significance—good for baseline framing. megbiodiversity.nic.in

Sangma & Nampui. (2011). *Meghalaya sacred groves* Documents biodiversity/ livelihood linkages in sacred groves; supports integration into formal FMPs. ResearchGate

Shackleton, S. E., Campbell, B., Wollenberg, E., & Edmunds, D. (2002). Devolution and community-based natural resource management: Creating space for local people to participate and benefit. *Natural Resource Perspectives.*

Tiwari, B. K., Tynsong, H., & Lynser, M. B. (2010). Forest management practices of the tribal people of Meghalaya, North-east India. *Journal of Tropical Forest Science, 22*(3), 329–342

Mir, et al. (2022). *Community forests in Khasi Hills—plant diversity & prioritization.*

Laskar. (2024). *Sacred Groves of Meghalaya (book chapter).* Taylor & Francis.

# 15

# RECLAMATION OF MINING-AFFECTED LAND USING AROMATIC GRASS

Hygina Siangbood, Paka I Yo Suja,
Baniaraplang Syiemiong, Rikermeka Kharsahnoh,
Diadema  Nongrum, Markshield Warjri,
Mourdome B Marak and Rishotskhem Kharkongor

## 15.1 Introduction

### 15.1.1 Purpose, Scope and Significance of the Case

The ban on coal mining by the National Green Tribunal in Meghalaya has severely affected local labourers who relied on coal-related activities. With mining halted, many attempted to shift to agriculture, but marginal farmers faced problems due to limited land availability and highly acidic soils. Unscientific rat-hole mining had led to widespread soil contamination, loss of fertility and erosion, turning once-cultivable land into barren wastelands.

Additionally, acid soil-related fertility stress and heavy-metal contamination had worsened crop productivity issues, threatening food security (Lyngdoh & Sanjay-Swami, 2018). To address this, communities and research institutions collaborated to reclaim degraded land using aromatic grasses. These not only improve overall soil health but also offer economic opportunities.

In a community-driven approach to phytoremediation, the use of aromatic grasses was implemented. Among the various bioremediation techniques, phytoremediation stands out as a non-invasive method that enhances soil productivity (Tahir et al., 2016; Tripathi et al., 2017).

The application of aromatic grasses has been shown to offer multiple benefits, including heavy-metal absorption, erosion control and enhancement of microbial activity (Yadav et al., 2023). The introduction of aromatic grasses, such as *Vetiver* (*Chrysopogon zizanioides*), clearly improved water quality significantly and rejuvenated springs by using their deep-rooted characteristics to enhance soil stability, prevent erosion and facilitate groundwater recharge (Truong & Danh, 2015).

DOI: 10.4324/9781003735380-18

Beyond ecological restoration, this approach also presented economic opportunities. The production of aromatic oils from these grasses supported local livelihoods, encouraging long-term community engagement in reclamation efforts. This integration of ecological techniques of livelihood creation promotes a self-sustaining model of community-based land management. This dual-purpose strategy not only stabilised degraded lands and restored biodiversity but also empowered local communities, ensuring sustainable development.

By integrating environmental restoration with economic benefits, this initiative offered a transformative model for mitigating land degradation in Meghalaya.

In this context, the chapter outlines the objectives of employing aromatic grasses for land reclamation, details the methodology used in field interventions, highlights the key findings on soil health, water resources and livelihoods, and discusses their wider implications for sustainability before concluding with lessons and recommendations for future applications.

### 15.1.2 *Review of Literature*

Globally, reclamation of mining-affected lands has involved both engineering and ecological strategies. Physical and chemical methods such as backfilling, soil capping, liming and chemical stabilisation of heavy metals provide rapid improvements but are expensive, labour-intensive and often unsustainable in the long term (Liu & Lal, 2012). As a result, vegetative approaches have gained prominence. These methods use grasses, shrubs and trees to stabilise soils, restore ecological functions and contribute to carbon sequestration while also offering opportunities for local livelihood creation (Tibbett, 2010; Zhou et al., 2017; Madzar et al., 2021). Countries such as Australia, China, South Africa and Brazil have demonstrated success through direct seeding of native legumes, mixed grass-legume plantations and agroforestry systems, all of which promote biodiversity recovery and soil health restoration (Beesley et al., 2011; Pilon-Smits, 2005).

Within this ecological paradigm, aromatic grasses have demonstrated significant potential in rehabilitating degraded and mining-affected soils. Khadija et al. (2021) highlighted the eco-friendly application of medicinal and aromatic plants for soil restoration, noting their ability to improve soil fertility and ecological resilience. Khadija et al. (2021) further showcased how *Cymbopogon citratus* enhances fertility in mining tailings, thereby demonstrating the direct role of aromatic grasses in reclaiming chemically degraded soils.

Research has also emphasised the physical soil stabilisation benefits of grasses. Maiti and Kumar (2016) reported that the deep-root systems of aromatic grasses are highly effective in preventing erosion, stabilising soil structure and preparing mine spoils for ecological succession. Similarly, Pandey et al. (2020) explored the dual role of *Cymbopogon flexuosus* in decontaminating soils and producing essential oils, thereby making the species valuable not only for phytoremediation but also as a source of income through essential oil extraction.

In the Indian context, where coal, bauxite and limestone mining have left vast tracts degraded, both engineering and ecological methods have been attempted. While engineering approaches such as fly ash capping and topsoil replacement have shown promise, they remain expensive and technically demanding (Singh et al., 2014). Vegetative strategies using grasses and legumes are more widely applied, stabilising overburden dumps, reducing erosion and contributing organic matter. Building on this, the introduction of aromatic grasses has created new opportunities to align reclamation with livelihood generation.

Yadav et al. (2023) examined the techno-economic viability of aromatic grasses, demonstrating their effectiveness in mitigating heavy-metal contamination while simultaneously supporting bio-economy development. Their findings suggest that these species can play a pivotal role in creating sustainable, community-led reclamation strategies across India. Ghosh et al. (2017) also documented their contribution to improving acidic and contaminated soils, particularly in mining-affected areas, reinforcing the ecological adaptability of aromatic grasses in diverse conditions.

In Northeast India, and specifically in Meghalaya, the impacts of unscientific rat-hole coal mining have been particularly severe, leading to soil erosion, acidity and heavy-metal contamination (Lyngdoh & Sanjay-Swami, 2018). Conventional reclamation approaches have been costly and less feasible in the hilly terrain, creating the need for low-cost, eco-friendly alternatives. Here, aromatic grasses have emerged as a practical solution. Their deep-rooted systems help control erosion and improve soil stability, while their biomass production enables carbon sequestration. The added advantage of essential oil production links reclamation with livelihood opportunities for local communities.

Sahu et al. (2024) reinforced the role of aromatic grasses in reclaiming contaminated and saline soils, highlighting their potential to improve soil health under challenging edaphic conditions. In Meghalaya, their use has been particularly successful in balancing ecological restoration with economic benefits, offering a replicable model for other mining-affected regions.

### 15.1.3 Identification of Gaps

Mining-induced land degradation in Meghalaya has severely affected local communities and agriculture, leading to the transformation of fertile land into barren wastelands. The current project sought to reclaim degraded lands through the strategic use of aromatic grasses, ensuring cost-effectiveness and community involvement. By prioritising community-owned land, the initiative fostered a sense of ownership and cooperation among locals.

The approach aimed to restore soil health, increase organic matter, reduce soil density and improve nutrient availability. Through the introduction of carefully selected plant species, it aimed to improve ecological conditions and mitigate the adverse environmental effects of human activity. In short, it aimed at promoting the long-term sustainability of affected regions.

### 15.1.4 Theoretical Framework

The reclamation of degraded land using aromatic grasses is grounded in theories of phytoremediation, ecosystem restoration and sustainable land management. Species such as *Cymbopogon flexuosus* and *Cymbopogon martinii* serve as effective phytoremediators thanks to their deep-root systems, high biomass production and heavy metal absorption capacity. These plants stabilise the soil, reduce erosion and remove contaminants, supporting the ecological theory that plant-based interventions can restore soil health.

Beyond soil restoration, they contribute to ecosystem services by improving soil structure, enhancing microbial activity and supporting biological diversity, aligning with ecological restoration principles.

Additionally, their economic viability through aromatic oil production supports integrated development models that balance integrated environmental and socio-economic benefits. The use of by-products like biochar further enhances soil fertility and carbon sequestration. This, in turn, contributes to the promotion of a circular economy.

This theoretical framework underscores the potential of aromatic grasses as a holistic and cost-effective solution for land degradation; at the same time, it strengthens local livelihoods and sustainability.

## 15.2 Methodology of the Case

### 15.2.1 Description of Methodology

The project spanned 46 villages across four districts in Meghalaya, covering 328.5 ha of degraded land. (Refer Tables 15.1, 15.2 and 15.3).

The methodology for rehabilitating mining-affected land and improving livelihoods includes the following:

1. Community mobilisation: Awareness programmes and sensitisation meetings were conducted at block and village levels. Interested villages were selected, and agreements were signed with village councils, self-help groups (SHGs) or integrated village cooperative societies (IVCS), depending on land ownership.
2. Site selection: Severely degraded mining lands were identified, followed by GIS mapping to mark and prioritise intervention areas. Community

**TABLE 15.1**  Area treated under various stakeholders in West Jaintia Hills

| Block | Village | Stakeholder name | Area treated (ha) |
|---|---|---|---|
| Laskein C&RD | Thadmuthlong C Phramer | Phramer VNRMC | 30 |
| | Khlokynrein | Khlookynrein VEC | 5 |
| | Khliehrangnah | Khliehrangnah VNRMC | 21 |
| | Mynska | Mynska VEC | 5 |
| Thadlaskein C&RD | Iongnoh | Iongnoh VNRMC | 3 |
| | Iongnoh | Iongnoh Tourism & Multipurpose Co-Operative Society | 3 |
| Amlarem C&RD | Shkenpyrsit | Shkenpyrsit IVCS | 5 |
| | Umladkur | Umladkur IVCS | 3 |
| Total | | | 75 |

*Source:* CLLMP project data.

and landowner collaboration ensured transparency and mutual agreement.

3. Soil analysis: Baseline soil testing was conducted pre- and post-intervention to assess physicochemical attributes, heavy-metal content and overall soil health. Samples were analysed at the Indian Council for Agricultural Research (ICAR)-North-Eastern Region, Umiam.

4. Selection of aromatic grasses: *Cymbopogon winterianum* (Citronella) was chosen for its phytoremediation capacity and economic value. The Council of Scientific and Industrial Research-Central Institute of Medicinal and Aromatic Plants (CSIR-CIMAP) provided technical support and ensured plant authenticity.

5. Community engagement: Workshops and training programmes were organised to build capacity and encourage participation.

6. Land development and planting: To ensure optimal land rehabilitation, standard protocols for land development, planting and intercultural practices were thoroughly followed, based on the package of practices developed by the Institute of Natural Resources (Meghalaya Basin Development Authority) and CSIR-CIMAP. The local community carried out this phase under the close guidance and supervision of the technical team from the implementing agency.

This collaborative approach not only ensured adherence to best practices but also furthered community ownership and skill development in sustainable land management.

7. Monitoring and evaluation: Periodic assessments measured improvements in soil health, water retention, plant growth and socio-economic impacts.

**TABLE 15.2** Area treated under various stakeholders in East Jaintia Hills

| Block | Village | Stakeholder name | Area treated (ha) |
|---|---|---|---|
| Saipung C&RD | Tluh | Tluh VNRMC | 32 |
| | Lelad | Lelad VNRMC | 12 |
| | Narwan | Narwan VNRMC | 5 |
| | Umkyrpong | Treilang SHG | 2 |
| | Molamylliang | Moolamylliang VNRMC | 7 |
| | Samasi | Samasi vnrmc | 3 |
| | Mynthlu | Ryntihlang SHG | 6 |
| | | Jingkyrmenshkem shkem Mynthlu Multipurpose Co-Operative Society | 3 |
| Wapung C&RD | Sutnga | Togetherness SHG | 3 |
| | | Kyntulang SHG | 1 |
| | | Sutnga VO | 3 |
| | | Shiruplang SHG | 1 |
| | | Sutnga Muliar SHG | 1 |
| | | Nangkiew SHG | 1 |
| | | Kutjingmut SHG | 1 |
| | | Pala Mayalang SHG | 1 |
| | | Pala Lumsashang SHG | 1 |
| | | Shiborlang SHG | 1 |
| | | Tyllilang SHG | 1 |
| | | Maialang SHG | 1 |
| | | Sutnga-2-VO | 1 |
| | | Lyngdoh Family SHG | 1 |
| | | Treishitom SHG | 1 |
| | | Iaikyrshan SHG | 1 |
| | | Kyrsoilang SHG | 1 |
| Khliehriat C&RD | Dein Chynrum | Deinshynrum VNRMC | 12 |
| | Ladwahwapung | Ladwahwapung IVCS | 7 |
| | | Iatyllilang SHG, Ladwahwapung | 1 |
| | | Treilang SHG, Ladwahwapung | 1 |
| | | Mynjurlang SHG, Ladwahwapung | 4 |
| | | Iatreilang SHG, Ladwahwapung | 1 |
| | | Iatreiminot SHG | 6 |
| | Lumshken | Lumshken VNRMC | 5 |
| | Rngad | Rngad IVCS | 2 |
| | lumshyrmit | Lumshyrmit IVCS | 2 |
| | Kairang | Nangiaid Shaphrang SHG | 5 |
| | Iongkaluh | Iongkaluh IVCS | 16 |
| | Mookympad | Iainehshkem SHG | 2 |
| | | Sniothooh na-I pynioo da-I SHG | 6 |
| | Mutong | Mutong IVCS | 2 |
| | Sohkymphor | Sohkymphor IVCS | 5 |
| | Byrwai | Byrwai IVCS | 4 |
| | Suchen Lumiarain | Suchen Lumiarain IVCS | 12 |
| | Suchen Dhana | Suchen Dhana VEC | 5 |
| Total | | | 189 |

*Source:* CLLMP project data.

**TABLE 15.3**  Area treated under various stakeholders in East Khasi Hills and East Garo Hills

| District | Block | Village | Stakeholder name | Area treated (ha) |
| --- | --- | --- | --- | --- |
| East Khasi Hills | Mawphlang C&RD | Umlangmar | Umlangmar VNRMC | 3 |
| | Mawkynrew C&RD | Jatah Lakadong | Jatah Lakadong VNRMC | 10 |
| | | Jatah Nonglyer | Jatah Nonglyer VNRMC | 7.5 |
| | | Syntung | Chiruplang VO | 9 |
| | Shella Bholaganj C&RD | Maraikaphon | Tyllilang Ator PG | 2 |
| | | Laitmawsiang | Laitmawsiang Dorbar shnong | 3 |
| | Mawsynram C&RD | Kmawanrum | Kmawanrum VNRMC | 3 |
| | | Langsymphut | Nuksa Babha VO | 10 |
| | Laitkroh 12 Shnong | Rangtmah | Rangtmah VEC | 1 |
| East Garo Hills | Samanda C&RD | Songmagre | Village Water and Sanitation Committee | 16 |
| Total | | | | 64.5 |

*Source:* CLLMP project data.

## 15.2.2 Justification of Methodology

Aromatic grasses provide a cost-effective and sustainable alternative to chemical soil amendments. Citronella *species* were selected for its dual benefits of phytoremediation and economic viability through essential oil production. GIS-based monitoring facilitated spatial accuracy in evaluations, optimising decision-making. Community involvement ensured long-term project sustainability, bridging environmental restoration and economic empowerment.

## 15.2.3 Limitations and Challenges

Challenges included land disputes, organisational shifts within village natural resource management committees and proximity to active mining sites, necessitating additional interventions. Delays in soil and biomass analysis reports affected the timely implementation. Besides, unpredictable weather and initial community resistance created further obstacles. Despite these hurdles, the initiative has demonstrated significant improvements in soil health and livelihoods, offering a replicable model for future reclamation efforts.

## 15.3 Results

### 15.3.1 Main Case Findings 1: Measurable Impacts on Soil Health, Spring Discharge and Economic Outcomes

The introduction of aromatic grasses with phytoremediation potential brought significant improvements in soil health over an 18 to 24-month period. Before intervention, the soil suffered from severe degradation, characterised by low pH, heavy-metal contamination and low organic-carbon content. Table 15.4 provides a comparison of the soil conditions before and after the intervention.

Pre-intervention findings: Soil pH was critically low at an average of 4.8, indicating high acidity. Soil organic carbon (SOC) was at a meagre 1.7%. Bulk density was high (1.50 g/cm$^3$), leading to soil compaction and reduced water infiltration. Moisture content was low at 15%, affecting plant growth. Heavy-metal contamination was significant, with iron (Fe) at 82,700 µg/g, manganese (Mn) at 140.5 µg/g and lead (Pb) at 23.75 µg/g. Spring water had a pH of 5.2 and a low discharge rate of 4 L per minute.

TABLE 15.4 Comparative analysis of average physico-chemical soil parameters and spring water quality and quantity before and after intervention

| Parameter | Before intervention | After intervention |
| --- | --- | --- |
| Soil pH | 4.8 ± 0.07 | 5.32 ± 0.43 |
| Moisture content (%) | 15 ± 0.88 | 25 ± 1.4 |
| Bulk density (g/cm$^3$) | 1.5 ± 0.02 | 1.3 ± 0.02 |
| SOC (%) | 1.7 ± 0.04 | 2.9 ± 0.092 |
| Fe (µg/g) | 82700 ± 13246 | 42950 ± 6743 |
| Mn (ppm) | 140.5 ± 42.2 | 90.3 ± 31.2 |
| Cu (µg/g) | 18 ± 2.11 | 12.5 ± 1.02 |
| Cr (µg/g) | 57.5 ± 15.8 | 47.5 ± 9.32 |
| Ni (µg/g) | 19.29 ± 4.39 | 18.5 ± 4.04 |
| Pb (µg/g) | 23.75 ± 3.1 | 20.25 ± 3.11 |
| Zn (µg/g) | 53.25 ± 16.2 | 41.5 ± 9.06 |
| Spring water pH | 5.2 ± 0.02 | 5.6 ± 0.02 |
| Spring discharge (L/ min) | 4 ± 0.01 | 11 ± 0.021 |

*Source:* Original unpublished data of INR.

Post-intervention findings: Soil pH improved to 5.32, with reduced acidity and better conditions for plant growth. The SOC content increased to 2.9%, reflecting enhanced soil fertility. Bulk density decreased to 1.30 g/cm$^3$, indicative of improved soil aeration and root penetration. The moisture content rose to 25%, a significant enhancement of water retention. Heavy-metal concentrations declined dramatically, with Fe levels reducing by 48%, Mn by 36% and Pb by 15%. The spring water pH went up to 5.6, while discharge rates improved to 11 L per minute.

These results demonstrate the success of aromatic grasses in improving soil conditions, mitigating heavy-metal contamination and enhancing water availability. This makes a strong case for replicating this approach in other degraded regions.

### 15.3.2 Main Case Findings 2: Community Engagement, Gender Inclusivity and Livelihood Generation

Community engagement and capacity building: A key factor in the project's success was its strong emphasis on community participation. From the beginning, local stakeholders were actively involved, ensuring they played a central role in land reclamation. Over 2,000 farmers from the Jaintia Hills and other districts were engaged in workshops and training sessions focusing on aromatic grass cultivation, essential oil extraction and sustainable farming techniques. These initiatives empowered farmers with knowledge, increasing their sense of ownership and long-term commitment to the project.

Gender inclusivity and empowerment: Women played a pivotal role in the project, making up 62% of direct beneficiaries. Traditionally, women in rural Meghalaya have had limited involvement in decision-making. Now, this initiative actively encourages their participation in planting, maintenance, distillation and sales of essential oils. Women-led cooperatives were established to oversee oil production and marketing, providing a stable income source and reinforcing their socio-economic standing.

As a result, many women gained financial independence, increased decision-making power in their households and took on leadership roles in their communities.

Livelihood generation and economic diversification: The project significantly improved the financial prospects of local farmers by introducing a sustainable alternative to traditional agriculture. Farmers were able to integrate the cultivation of aromatic grasses into their existing farming systems, diversifying their income streams. Essential oil production from crops such as lemongrass and citronella generated annual returns of ₹40,000–₹50,000 per hectare.

Moreover, the initiative created over 10,000 person-days of employment through land preparation, planting, maintenance and marketing. By establishing local markets and supply chains linked to regional and national buyers, farmers gained better access to economic opportunities.

### 15.3.3 Main Case Findings 3: Carbon Sequestration and Soil Carbon Enhancement

Carbon sequestration potential of aromatic grasses: Aromatic grasses such as *Cymbopogon winterianus* and *Cymbopogon flexuosus* have demonstrated high carbon sequestration capabilities, making them valuable in mitigating climate change (Maddhesiya & Singh, 2021; Pandey et al., 2020). These deep-rooted grasses capture atmospheric carbon dioxide and store it in the soil, reducing greenhouse gas emissions while enhancing soil fertility.

Over a three-year period, the project successfully sequestered an estimated 110 metric tons of carbon across 328 ha of reclaimed land. This was achieved through:

Increased soil organic matter.
Enhanced biomass production, leading to a greater carbon uptake.
Improved microbial activity, further promoting soil carbon retention.

Beyond carbon sequestration, the initiative also enhanced soil water retention, leading to a self-sustaining cycle of ecological restoration.

These findings emphasise the role of phytoremediation in tackling both land degradation and climate change simultaneously.

### 15.3.4 Convergence with National Institutions and Agro-technology Exposure

To enhance the effect of land reclamation through aromatic grasses, the project collaborated with national institutions such as CSIR-CIMAP, Council of Scientific and Industrial Research-Indian Institute of Integrative Medicine and Council of Scientific and Industrial Research-Institute of Himalayan Bioresource Technology. Community members participated in exposure visits to these institutions, allowing them to learn about diverse agro-technologies suitable for different agro-climatic conditions. As a result, CSIR-CIMAP established a technology window within MBDA, facilitating access to advanced cultivation techniques and supporting the introduction of high-value crops.

Also, the initiative promoted wild indigenous medicinal and aromatic plants such as wintergreen (*Gaultheria fragrantissima*), which have significant ecological and economic potential.

Through this intervention, farmers have gained practical knowledge of alternative cropping systems, enabling them to experiment with different plant species in their own fields. This collaboration bridges scientific research with community-driven land restoration efforts, ensuring the sustainability and scalability of the initiative.

## 15.4 Discussion

### 15.4.1 Interpretation of Findings in Relation to Replicability

The success of using aromatic grasses to restore degraded mining lands in Meghalaya highlights its potential for replication in similar environments worldwide. Significant improvements in soil health, water availability and economic benefits make this approach promising.

For instance, soil organic carbon (SOC) increased from 1.7% to 2.9%, while heavy-metal concentrations declined, demonstrating the grasses' phytoremediation capabilities. Moreover, spring water discharge improved from 4 to 11 L per minute, indicating enhanced ecosystem functionality. These outcomes suggest that this model can be applied in other regions with acidic, heavy-metal-contaminated soils.

The key to successful replication lies in integrating scientific techniques with community participation. GIS-based mapping enabled precise site selection, while soil testing allowed for customised interventions. The adaptability of the *Cymbopogon* species to different climates makes them viable for various locations.

But strong community engagement is crucial, as demonstrated by the involvement of 2,000 farmers, of which 62% are females. Such inclusivity fosters local ownership and long-term sustainability.

Challenges, such as land disputes and delays in analytical reports, highlight the need for flexible management strategies. Addressing these issues through better governance, policy support and technology will enhance replicability.

Notably, the financial viability of essential oil production, generating annual returns of ₹40,000–₹50,000 per hectare, makes this approach attractive for resource-constrained regions.

With environmental benefits like a 48% reduction in soil iron levels and a 36% decrease in manganese, alongside increased soil pH, this project aligns with global sustainability goals. Its replicable nature makes it a valuable model for land restoration and economic development in mining-affected regions worldwide.

### 15.4.2 Implications for Theory, Practice and Policy

The study has significant implications for environmental restoration theory, practical applications and policymaking. The findings reinforce phytoremediation principles by demonstrating that deep-rooted, high-biomass aromatic grasses restore soil health, enhance biodiversity and contribute to ecosystem services. This supports the ecological theory that plants play a central role in land restoration and sustainable resource management.

In practice, this project offers a scalable and cost-effective model for land reclamation. By integrating GIS mapping with community-driven initiatives, it demonstrated a balanced approach between ecological restoration and economic development. The dual role of aromatic grasses in soil remediation and essential oil production makes them valuable for farmers and local economies. Additionally, the increased spring water discharge enhances regional water security, addressing both environmental and socio-economic concerns.

From a policy perspective, the case highlights the importance of collaboration between scientific institutions, local governments and communities. Technical support from organisations such as the ICAR and CSIR-CIMAP ensured that the project was scientifically sound, while community involvement made it socially sustainable. Policymakers can use this model to design incentive-based programmes for land restoration in mining-affected areas. Strengthening village-level resource management and fostering partnerships can further improve outcomes.

Moreover, the carbon sequestration potential of aromatic grasses aligns with global climate goals. The project sequestered 110 metric tons of carbon over three years, making it a viable climate mitigation strategy. Policymakers could encourage the cultivation of aromatic grasses in degraded areas as part of climate action plans. Further, supporting the use of by-products like biochar could enhance soil fertility while contributing to carbon storage and sustainable agriculture.

### 15.4.3 Comparison with Existing Literature

The study's findings align with previous research on aromatic grasses for environmental restoration. Studies by Khadija et al. (2021) and Pandey et al. (2020) have shown the effectiveness of the *Cymbopogon* species in reducing soil contamination and improving soil fertility. This project further validates those findings, demonstrating a 48% reduction in soil iron levels and a 36% decrease in manganese, among other achievements.

Earlier studies have also recognised the economic potential of aromatic grasses. Khadija et al. (2021) emphasised their ability to rehabilitate chemically degraded soils, while Maiti and Kumar (2016) documented their erosion control benefits. This project expands on those insights by showing how these grasses also improve soil organic carbon and moisture content, making them even more effective for land reclamation.

Research by Ghosh et al. (2017) has pointed out the income-generating potential of essential oil production. This project supports those findings, demonstrating annual returns of ₹40,000–₹50,000 per hectare and the formation of women-led cooperatives. The emphasis on gender inclusivity and community engagement provides a more comprehensive framework for integrating economic and environmental restoration efforts.

A key advancement in this project is the use of GIS-based mapping for site selection and monitoring. Unlike prior interventions, which primarily focused on ecological benefits, this one integrates data-driven decision-making with community participation. Further, its emphasis on circular economy principles, such as valuing by-products like biochar, adds a novel dimension to the sustainable use of aromatic grasses. These contributions position this study as a significant advancement in land reclamation and socio-economic resilience.

## 15.5 Summary and Conclusion

### 15.5.1 Key Findings Summary

This study of the aromatic grasses project demonstrates their potential to rehabilitate degraded soils through phytoremediation. Over an 18 to 24-month intervention, the soil pH improved from 4.8 to 5.32, SOC increased from 1.7% to 2.9% and moisture content rose from 15% to 25%. Heavy-metal concentrations significantly decreased, with Fe levels dropping by 48%. Bulk density reduced from 1.50 g/cm$^3$ to 1.30 g/cm$^3$, improving soil structure. Spring water quality also improved, with the pH increasing from 5.2 to 5.6 and discharge rates rising from 4 to 11 L per minute. These results underscore the effectiveness of aromatic grasses in ecosystem restoration.

### 15.5.2 Significance Restatement

The findings emphasise the value of aromatic grasses as a sustainable solution for restoring degraded soils, particularly in mining-affected regions. Improvements in soil fertility, water retention and reduced heavy-metal contamination demonstrate their environmental benefits. Enhanced spring water availability further supports their role in ecosystem restoration. By addressing land degradation and contamination, this approach offers a scalable, practical solution for improving land productivity and environmental health while supporting global sustainability efforts.

### 15.5.3 Future Research and Practical Applications

Future research should explore the combined use of aromatic grasses and biochar to enhance soil restoration. Biochar, derived from organic waste, has the potential to neutralise acidic soils, improve fertility and increase carbon storage. Investigating the synergy between biochar and aromatic grasses could lead to even more effective phytoremediation strategies.

Practically, integrating aromatic grasses with biochar could provide a powerful solution for post-mining land reclamation. This approach could improve soil health, support plant growth and enhance water quality. By adopting this method, communities could restore degraded lands, promote sustainable agriculture and contribute to climate change mitigation through carbon sequestration. These solutions offer promising opportunities for building healthier ecosystems and improving livelihoods in affected areas.

## References


Beesley, L., Moreno-Jiménez, E., Gomez-Eyles, J. L., Harris, E., Robinson, B., & Sizmur, T. (2011). A review of biochars' potential role in the remediation, revegetation and restoration of contaminated soils. *Environmental Pollution*, *159*(12), 3269–3282.

Ghosh, I., Ghosh, M., & Mukherjee, A. (2017). Remediation of mine tailings and fly ash dumpsites: Role of Poaceae family members and aromatic grasses. In N. A. Anjum, S. S. Gill, & N. Tuteja (Eds.), *Enhancing cleanup of environmental pollutants* (Vol. 1, pp. 67–82). Springer International Publishing.

Khadija, A. E., Mansour, S., & Ali, B. (2021). Chemically degraded soil rehabilitation process using medicinal and aromatic plants. *Environmental Science and Pollution Research*, *28*(1), 73–93.

Liu, R., & Lal, R. (2012). Nano-enhanced materials for reclamation of mine lands and other degraded soils: A review. *Journal of Nanotechnology*, *2012*, Article ID 461468.

Lyngdoh, P. C., & Sanjay-Swami, G. (2018). Impact of soil acidity on agricultural productivity in Northeast India. *Soil Biology and Biochemistry*, *121*(6), 25–32.

Maddhesiya, P. K., & Singh, K. (2021). Effects of perennial aromatic grass species richness and microbial consortium on soil properties of marginal lands and on biomass production. *Land Degradation & Development*, *32*(2), 1008–1021.

Madzar, R., Cloete, T. E., & Bezuidenhout, C. C. (2021). Grass–legume mixtures for the rehabilitation of mine tailings in South Africa. *Journal of Environmental Management*, *280*, 111698.

Maiti, S. K., & Kumar, A. (2016). Energy plantations, medicinal and aromatic plants on contaminated soil. In D. J. Thomas (Ed.), *Managing arsenic in the environment: From soil to human health* (pp. 17–34). CSIRO Publishing.

Pandey, V. C., Rai, A., & Kumari, A. (2020). Cymbopogon flexuosus – an essential oil-bearing aromatic grass for phytoremediation. In V. C. Pandey (Ed.), *Phytomanagement of polluted sites* (pp. 195–209). Elsevier.

Pilon-Smits, E. (2005). Phytoremediation. *Annual Review of Plant Biology*, *56*, 15–39.

Sahu, S. S., Kumar, A., Prasad, M. N. V., & Maiti, S. K. (2024). Aromatic, medicinal, and energy plantations on metalliferous/contaminated soil-Bioremediation and bioeconomy. In *Rehabilitation of arid and semi-arid ecosystems*.

Singh, A. N., Raghubanshi, A. S., & Singh, J. S. (2014). Comparative performance and restoration potential of two Albizia species planted on mine spoil in a dry tropical region, India. *Ecological Engineering*, *71*, 278–285.

Tahir, M. B., Sohail, M., & Javed, M. T. (2016). Bioremediation of heavy metal-contaminated soils using aromatic grasses. *Environmental Science and Pollution Research*, *23*(15), 15553–15565.

Tibbett, M. (2010). Large-scale mine site restoration of Australian eucalypt forests after bauxite mining: Soil management and ecosystem development. *Ecological Management and Restoration, 11*(2), 100–107.

Tripathi, R. D., Singh, S., & Singh, V. P. (2017). Phytoremediation of heavy metal contaminated soils: Advances and opportunities. *Ecological Engineering, 102*(3), 69–78.

Truong, P., & Danh, L. T. (2015). *The Vetiver system for improving water quality.* The Vetiver Network International.

Yadav, D., Yadav, A., Singh, M., & Khare, P. (2023). *Cultivation of aromatic plants for nature-based sustainable solutions for the management of degraded/marginal lands: Techno-economics and carbon dynamics.* Environmental Sustainability.

Zhou, W., Han, G., & Liu, J. (2017). Effects of Robinia pseudoacacia and Populus plantations on soil properties of coal mine spoils in China. *Catena, 148,* 96–103.

# 16
# AGROFORESTRY-CARBON PROJECT IN MEGHALAYA

*Lavinia Mary Dkhar*

## 16.1 Introduction

### 16.1.1 Purpose, Scope, and Significance

Climate change is one of the pressing concerns affecting the livelihoods of communities due to erratic rainfall patterns, primarily impacting rain-fed agriculture. The approach of carbon credits to encourage smallholder farmers in Meghalaya to adopt agroforestry has immense potential as a means to mitigate climate change problems.

Meghalaya has emerged as a trailblazer in adopting environmentally economic approaches and friendly cultural practices to conserve its natural resources. The State initiated a state-level carbon project in 2024 to promote green livelihoods, preserve forest and tree cover, and support sustainable farming. Village natural resource management committees and village employment councils, as well as traditional institutions, played a pivotal role in the successful implementation of projects under the Meghalaya Basin Development Authority (MBDA). They are the primary beneficiaries of the project. Given the multifaceted nature of MBDA's natural resource management activities, this was an opportunity to develop a carbon project that would encompass various aspects of land use-related carbon, including forest plantation and conservation. In short, sustainable forest management.

The State's forests serve as invaluable reservoirs of biodiversity and natural resources. The forestry programmes implemented in India are substantially aimed at the rehabilitation and effective management of over 1 million ha of the country's forests, rendering them more adept at sequestering carbon, augmenting water yields, and advancing the welfare of Indigenous and

DOI: 10.4324/9781003735380-19

tribal communities.The total forest cover in the State is represented in Table 16.1.

The benefits derived from this carbon project were harnessed to bolster Meghalaya's green economy by establishing community-led value chains, supported by robust and transparent institutional frameworks, benefit-sharing mechanisms and safeguards. In this aspect, carbon financing can contribute significantly to forest conservation and reward communities for managing the forests they own.

Carbon finance in Meghalaya involved the generation of financial incentives to enhance carbon sequestration. The State has significant potential for carbon finance projects thanks to its rich biodiversity, vast forest cover, and a growing emphasis on sustainable development. Communities that register would stand to gain benefits from carbon sequestration, climate adaptation and mitigation, agroforestry, water management, and so on. The project sought to protect and revive forests and agricultural lands to cover 2,000 villages in all 12 districts of the State. This was the first phase of a three-phased approach to cover the entire state.

In this context, the Rabobank launched its so-called Acorn programme for the organic restoration of nature. Acorn is a division of Rabobank that developed and manages a platform. This platform supports agroforestry systems for activities that ensure soil health, provide revenue streams (e.g., by the production of fruit trees) and could initiate additional revenue by the generation of Carbon Removal Units (CRUs). Acorn is instrumental in accessing this market by generating CRUs and making it possible to benefit from the potential sales of the CRUs to buyers [1 CRU represents one metric ton (1,000 kg) of carbon dioxide removed from the atmosphere].

The intended beneficiaries of this project were communities, clans, and individuals who have ownership of land and can decide about its use. Since communities were the major stakeholders in this approach, a number of consultation meetings and awareness campaigns were to be conducted to mobilise them and encourage participation.

**TABLE 16.1**  Forest cover in Meghalaya (ISFR, 2023)

| Sl. No. | Class | Area (sq km) | % of total area |
|---|---|---|---|
| 1. | Very dense forest | 594.84 | 2.65 |
| 2. | Moderately dense forest | 9,023.81 | 40.24 |
| 3. | Open forest | 7,348.19 | 32.76 |
| 4. | Total | 16,966.84 | 75.65 |
| 5. | Scrub | 620.41 | 2.77 |

IORA Ecological Solutions and MBDA leveraged this by ensuring the participation of the on-ground IORA team in these meetings. Participants were the residents of a village or a cluster of small villages. The landowners finalised and approved the design for agroforestry in these sessions.

## 16.2 Organisation of the Chapter

This chapter aims to highlight the carbon finance initiatives previously undertaken in other parts of the country based on an examination of relevant literature, and outlines the methodology of the ongoing process and its outcomes.

### 16.2.1 Review of Literature

As the global economy has undergone rapid expansion in recent years, there has been an increasing reliance on fossil fuels for energy production, resulting in a significant increase in atmospheric $CO_2$ pollutants (Kuppan & Chavali, 2017). Agroforestry systems have significant potential to contribute to global climate change mitigation efforts (US Geological Survey, 2008). By integrating trees with traditional agricultural practices, vegetation is increased. So, these systems can effectively sequester carbon dioxide from the atmosphere through photosynthesis (Jose, 2009). The stored carbon helps reduce the concentration of Greenhouse Gases (GHG).

This way, it contributes to the mitigation of global warming and strengthens carbon sinks that offset carbon emissions.

In India, agroforestry is an age-old, traditional resource management and adaptation practice where trees and pastures are intentionally incorporated into existing cropping systems. These agricultural systems are primarily adopted because of the symbiotic relationship between crops, trees, and livestock in addition to food and fibre (Chavan et al., 2015).

Agroforestry in developing countries has attracted increasing attention for both adaptation to climate change and greenhouse gas mitigation. Agroforestry stores more carbon compared to conventional plantations, which is another way of mitigating GHG emissions (Chauhan et al., 2010; Hergoualc'h et al., 2012).

GAgroforestry in India is practised in both irrigated and rain-fed conditions. It produces fuel, fodder, timber, fertiliser, and fibre. Agroforestry has been receiving great attention from researchers, policymakers, and others for its perceived ability to contribute significantly to economic growth, poverty alleviation, and environmental quality (DAC, 2014). It is recognised as an important part of the Green Revolution movement in the country. Moreover, it promotes productive and resilient cropping and farming environments.

Under the Centre of Excellence for NRM and Sustainable Livelihoods, MBDA, forest management plans were created through inventories of forest ecosystems. This exercise covered 400 villages and brought 1,102 km² under forest management plans. It was also estimated that growing stock is approximately 4,901.43 m³, 19.995 million trees, an annual yield of 152,446.51 m³, and a carbon stock of 1,176.87 tonnes C/ha. It was also estimated that the forest area of approximately 2,302 km² is under the ownership of communities while 8,467 km² of forests are owned by individuals and clans.

The Khasi Hills Community REDD+ project, spread across 23,512 ha of forest land, is a pioneering initiative that combines both conservation and restoration strategies to address deforestation, land degradation, and poverty in the region. On the one hand, it emphasises forest protection through the "Reducing Emissions from Deforestation and Forest Degradation" (REDD) approach, aiming to curb the drivers of forest loss and safeguard carbon stocks. On the other hand, it adopts restoration practices such as Assisted Natural Regeneration (ANR), which encourage the recovery of degraded landscapes, enhance biodiversity, and improve ecosystem services like water regulation and soil fertility.

Beyond its environmental goals, the project is deeply rooted in the socio-economic realities of the Khasi Hills. It engages directly with around 4,400 households, the majority of whom – nearly 80 to 90% – live below the poverty line. These households depend heavily on natural resources for fuel, fodder, and income, making forest conservation both a challenge and an opportunity for sustainable development. By integrating livelihood improvement plans with forest management, the initiative seeks to create long-term benefits such as alternative income sources, capacity building, and greater community resilience.

### 16.2.2 Identification of Gaps

Meghalaya possesses a rich diversity in climate, topography, and soil profile that supports the cultivation of a wide range and variety of fruits, vegetables, and spices. In terms of climate change mitigation, there is a need to increase the plantation area, as this will serve to increase the yield and improve carbon sinks.

The Khasi Hills Community REDD+ spans an area of 23,512 ha and engages in strategies for both forest protection (Reducing Emissions from Deforestation and Forest Degradation, or "REDD") and restoration (Assisted Natural Regeneration, or "ANR"). The project also provides detailed and long-term plans for improving the livelihoods of 4,400 households, 80 to 90% of which live below the poverty line. The region experienced a rapid 28% loss in forest cover between 2000 and 2005. Further, India ranks

136th of 186 countries in the UNDP's 2012 Human Development Index, and annual incomes in the region are $490 for a family of five to six.

There is a need for interventions such as the Khasi Hills Community REDD+ as the present situation is underscored by the alarming loss of forest cover in the State (approximately 79 km², ISFR 2023), a rate of deforestation that threatens not only ecological balance but also the cultural and economic lifelines of Indigenous communities. This crisis unfolded against the backdrop of India's broader development challenges: In 2012, the country ranked 136th out of 186 countries in the United Nations Development Programme's Human Development Index (HDI), reflecting gaps in health, education, and living standards. Within the Khasi Hills specifically, annual household incomes average just $490 for families of five to six members, highlighting the scale of poverty and vulnerability that conservation efforts must account for.

It is important to note the existence of several gaps that need to be addressed to leverage carbon finance in the state fully. Many projects there do not have the financial resources or incentives to participate in carbon finance schemes. The upfront costs to establish carbon credit projects are an impediment for local communities. By linking climate action, along with community welfare, a holistic model of sustainable development is represented. It not only works to reduce carbon emissions and regenerate ecosystems but also seeks to empower marginalised communities, improve livelihoods, and demonstrate how global climate goals can align with local development priorities.

## 16.3 Methodology

The project is being implemented by IORA Ecological Solutions together with Meghalaya Basin Development Authority (MBDA). It acts on behalf of the State government to facilitate community actions and participation, while IORA Ecological Solutions was brought in as the technical partner for the development of this carbon project and to enable carbon investments and benefit sharing of carbon credit revenue. It also deals with monitoring and engagement with carbon investors.

The main activities included are the following:

1. Facilitation of conservation and sustainable management of community forests by involving local communities themselves.
2. Reforestation and afforestation, focused on enhancing the green cover of the state.
3. Livelihood development by creating green income-generating opportunities for communities to reduce forest exploitation.

4. Capacity building of communities for implementing aforementioned project activities, as well as monitoring, reporting and verification, and green enterprise development.

    The key steps for conducting awareness among community members and the farmer onboarding process are indicated below:
5. Members designated as agroforestry-carbon associates (ACAs) received training on the onboarding process and data collection.
6. These ACAs assumed responsibility for the conduct of awareness programmes along with respective village heads regarding the project, highlighting its benefits and detailing criteria.
7. Post these meetings, smaller cluster meetings were held at a village with Headmen/Nokmas and farmers from neighbouring villages also participating.
8. The ACAs are also responsible for carrying out verification of plantations of interested people along with providing information and verifying necessary documents.
9. The profiles of the plantation owners were uploaded onto an app especially designed for this purpose.
10. If farmers meet the eligibility criteria, the ACAs would then:
    a. Discuss the farmers' agreement in detail, ensuring the farmers are aware of all particulars and clauses
    b. Record farmer details using digital data collection apps.
    c. Digitise the land parcel boundary using an inbuilt GPS mapping tool in the app

Plot Eligibility Criteria:

1. Plantation Age: Trees planted must be less than 5 years old.
2. Land Ownership: Farmers must present valid land documents or a No Objection Certificate (NOC) from the Nokma/Headman.
3. Land Size:
    a. Private lands: Between 0.1 and 10 ha
    b. Community lands: No size limit.
4. Tree Density: Minimum of 300 trees per hectare, preferably with mixed species. Areas with monoculture or lower tree density are categorised for gap filling.
5. Canopy Cover: Existing tree cover should not exceed 40%.

## 16.4 Result and Discussion

The project's primary objective was to improve the management and restoration of community forest areas. After the initiation of the project in 2024, significant progress was observed as shown in Table 16.2.

**TABLE 16.2** District-wise farmer onboarding – number of farmers and area onboarded till September 2025

| Sl. No. | District | Farmers | Area |
| --- | --- | --- | --- |
| 1 | North Garo Hills | 2,526 | 3,194.20 |
| 2 | West Garo Hills | 4,278 | 4,989.35 |
| 3 | South Garo Hills | 1,953 | 2,489.76 |
| 4 | East Garo Hills | 1,789 | 2,598.16 |
| 5 | South West Garo Hills | 1,073 | 924.285 |
| 6 | West Jaintia Hills | 493 | 3,751.88 |
| 7 | East Jaintia Hills | 322 | 1,314.59 |
| 8 | East Khasi Hills | 244 | 1,600.48 |
| 9 | Eastern West Khasi Hills | 182 | 928.182 |
| 10 | Ri Bhoi | 229 | 746.428 |
| 11 | South West Khasi Hills | 245 | 667.314 |
| 12 | West Khasi Hills | 76 | 205.153 |
| | Total | 13,410 | 23,409.78 |

*Source:* IORA Ecological Solutions.

The number of farmers who have enrolled in the carbon finance programme serves as a key indicator of positive outcomes, whereby community members are actively participating in climate change efforts by adopting practices that contribute to carbon sequestration.

The enrolment also signals a growing awareness and engagement from farmers in environmental stewardship, especially in maintaining and improving the ecosystem. As part of the carbon finance programme, these farmers were and are expected to follow guidelines related to agroforestry systems.

The payment cycle is structured in such a way that it ensures the accuracy and credibility of the carbon offsets. This involves monitoring and verifying the carbon sequestration processes over a period of a year. Once verification is over and carbon credits are certified, the payment process can begin.

### 16.4.1 District-Wise Carbon Payments

As part of the Rabobank Acorn process, post quality checks were carried out and 11,839 ha belonging to 6,099 farmers were enrolled on their platform. These plots were eligible for carbon measurements for CRU generation in this payment cycle. In the first round of measurements (growth from February 2024 to January 2025), 233 farmers have generated 468 CRUs.

### 16.4.2 District-Wise Benefits Provided

This year the project has carried out plantations on 3,500 ha for 2,260 farmers, providing saplings and support with fencing and fertilisers, where required. Project farmers will also be trained on management of agroforestry

**TABLE 16.3**  Number of farmers receiving carbon payments in 2025

| Sl. No. | District | Number of farmers |
|---|---|---|
| 1 | North Garo Hills | 1 |
| 2 | West Garo Hills | 57 |
| 3 | South Garo Hills | 15 |
| 4 | East Garo Hills | 8 |
| 5 | South West Garo Hills | 66 |
| 6 | West Jaintia Hills | 37 |
| 7 | East Jaintia Hills | 6 |
| 8 | East Khasi Hills | 7 |
| 9 | Eastern West Khasi Hills | 11 |
| 10 | Ri Bhoi | 1 |
| 11 | South West Khasi Hills | 8 |
| 12 | West Khasi Hills | 0 |

*Source:* IORA Ecological Solutions.

**TABLE 16.4**  Number of farmers receiving different benefits

| Sl. No. | District | Saplings | Fencing | Fertiliser |
|---|---|---|---|---|
| 1 | North Garo Hills | 326 | 52 | 2 |
| 2 | West Garo Hills | 409 | 778 | 224 |
| 3 | South Garo Hills | 113 | 2 | 0 |
| 4 | East Garo Hills | 98 | 1 | 2 |
| 5 | South West Garo Hills | 374 | 486 | 41 |
| 6 | West Jaintia Hills | 112 | 13 | 1 |
| 7 | East Jaintia Hills | 112 | 13 | 1 |
| 8 | East Khasi Hills | 10 | 6 | 0 |
| 9 | Eastern West Khasi Hills | 10 | 3 | 0 |
| 10 | Ri Bhoi | 131 | 71 | 45 |
| 11 | South West Khasi Hills | 10 | 3 | 0 |
| 12 | West Khasi Hills | 43 | 6 | 0 |

*Source:* IORA Ecological Solutions.

plantations, pest and nutrient management, fire prevention and control, etc. These plantations were carried out using India's first carbon-backed loan.

## 16.5 Summary and Conclusion

The project is an innovative and sustainable approach to incorporating agroforestry systems to acquire carbon credit benefits. It encouraged farmers and communities to adopt agroforestry practices that would involve integrating tree species with agricultural systems to support biodiversity, improve soil health, and generate income. By promoting indigenous tree species, the project sought to create more sustainable farming, reduce soil erosion, improve water retention, and increase resilience against climate change.

A key component to ensure progress and success of the project was community mobilisation, capacity building and local leadership. At the same time, it ensured that farmers were empowered and actively involved in the transition to more sustainable agricultural practices. The reward system based on carbon sequestration further incentivised the adoption of these practices, fostering a sense of ownership and responsibility within the community. Ultimately, this project offers a model for integrating environmental sustainability with socio-economic development.

There lies an immense opportunity for carbon sequestration in agroforestry and other sectors in Meghalaya, and it is essential to ensure that the benefits are shared equitably among the various stakeholders. This would lead to sustained additional investments in the natural resources in Meghalaya.

Notably, the growing enrolment of farmers in the programme is a clear sign of positive community engagement in ecosystem management. Once the final steps of the verification cycle are completed, the farmers will be compensated for their important role in carbon sequestration.

## References


Chauhan, S. K., Sharma, S. C., Beri, V., Ritu Yadav, S., & Gupta, N. (2010). Yield and carbon sequestration potential of wheat (*Triticum* aestivum) and poplar (*Populus* deltoides) based agri-silvicultural system. *The Indian Journal of Agricultural Sciences, 80*(2), 129–135.

Chavan, S. B., Keerthika, A., Dhyani, S. K., Handa, A. K., Newaj, R and Rajarajan, K. (2015). National agroforestry policy in India: A low hanging fruit. *Current Science, 108*(10), 1826–1834.

DAC. (2014). *National agroforestry policy–2014*. DAC, Ministry of Agriculture, GOI. http://agricoop.nic.in/imagedefault/whatsnew/Agroforestry.pdf

Forest Survey of India. (2023). India state of forest report 2023 (Vol. 2). Ministry of Environment, Forest and Climate Change, Government of India.

Hergoualc'h, K., Blanchart, E., Skiba, U., Henault, C., & Harmand, J. M. (2012). Changes in carbon stock and greenhouse gas balance in a coffee (*Coffea arabica*) monoculture versus an agroforestry system with *Inga densiflora*, in Costa Rica. *Agriculture, Ecosystems & Environment, 148*, 102–110.

Jose, S. (2009). Agroforestry for ecosystem services and environmental benefits: an overview. *Agroforestry Systems, 76*, 1–10

Kuppan, C. S., & Chavali, M. (2019). $CO_2$ sequestration: Processes and methodologies. In L. M. T. Martínez, KharissovaO. V. , & B. I. Kharisov (Eds.), *Handbook of ecomaterials* (pp. 619–668). Springer International Publishing.

U.S. Geological Survey. (2008). Carbon sequestration to mitigate climate change. *Technical Report U.S. Geological Survey.* https://sites.middlebury.edu/biol0490a -s11/files/2011/01/ C_sequestration_USGSFactSheet.pdf

# 17
# LEVERAGING BAMBOO RESOURCE

*Nangteibor Shabong and Lavinia Mary Dkhar*

## 17.1 Introduction

### 17.1.1 Purpose, Scope and Significance of the Case

Meghalaya, one of India's most biodiverse states, is known for its vast bamboo resources. With approximately **3,100** km² of bamboo forests, the forest cover accounts for about 12.5% of the state's total geographical area. It is home to **nearly** 40 species across 11 genera (Tripathi, 2007). Bamboo plays a critical role in Meghalaya's ecosystem, economy and cultural heritage. Still, despite its abundance, the sector remains **underused** due to poor market integration, unsustainable harvesting and lack of industrial demand.

Leveraging bamboo effectively requires an integrated approach that combines sustainable plantation, organised harvesting, industrialisation and community engagement. We emphasise in this chapter bamboo's potential as a sustainable livelihood source while outlining how Meghalaya can balance supply and demand through industry-led initiatives. We will show how communities look at the gap of how local people have not utilised the potential of the bamboo resource in the State.

We explore how the Community-Led Landscape Management Project (CLLMP) has expanded its effect through the establishment of a Centre of Excellence for Natural Resource Management & Sustainable Livelihoods (COE NRM & SL), involving the community itself to assess locally available bamboo resources under the guidance of the COE NRM & SL.

DOI: 10.4324/9781003735380-20

By fostering community participation, the CLLMP has demonstrated an ability to work with local people for an understanding of the potential sources of bamboo resources within Meghalaya.

We also examine its significance as a sustainable resource and contrast this with the current scenario in Meghalaya, where a lack of structured data and resource mapping has hindered its full potential. Recognising this gap, the CLLMP initiated efforts to generate baseline data on bamboo resources as a foundational step to better planning and use. Through its community-based approach, particularly by engaging village community facilitators, the project has facilitated bamboo resource assessments, contributing to a broader understanding of the sector's potential.

In this chapter, we underscore how the project's intervention serves as a model for integrating community engagement in natural resource management. It demonstrates how localised efforts can lead to broader policy insights and sustainable development strategies for Meghalaya's bamboo sector.

### 17.1.2 Organisation of the Chapter

The chapter opens with an introduction that outlines the study's objectives, scope and significance, setting the stage for the presentation. A literature review situates the study within existing academic and policy discussions from other areas. It brings out how important it is to have strong policies for a State to leverage the resources to their maximum potential. Also, as an initial step, a baseline study needs to be carried out using a community-based assessment approach.

The theoretical framework explores the state's strategic direction in promoting community participation in bamboo resource management, ensuring an inclusive approach to decision-making using data. The methodology section details the research design, data collection processes and analytical techniques used to assess bamboo resources effectively. The findings and discussion present key results, offering a critical examination of the bamboo resource assessment.

The chapter concludes by summarising major insights, discussing broader implications and proposing recommendations for future initiatives. The references provide a source of supporting information.

### 17.1.3 Review of Literature

Meghalaya has yet to formulate a dedicated bamboo policy, unlike states like Tripura and Assam, where well-structured policies have played a pivotal role in advancing the bamboo sector. Establishing a comprehensive policy framework would enable systematic planning, coordinated stakeholder engagement and optimised resource allocation, minimising redundant efforts across various agencies while maximising sectoral impact.

### 17.1.3.1 Bamboo Policies by States

A few states in India have developed bamboo policies, which have helped them leverage their bamboo resource in a better state in comparison to states that have not. Let us look at the state of Assam: It introduced the Assam Bamboo and Cane Policy, 2019, to promote the sustainable development and utilisation of bamboo and cane resources through scientific management and active stakeholder participation.

Their policy aims to harness the economic, social and environmental potential of bamboo and cane in the state. It has carved out key elements such as protection and conservation, which help protect biodiversity. It promotes resource enhancement, which helps encourage the cultivation of bamboo and cane in the forest, with a focus on industrial and infrastructure development.

The policy is a key consideration ensuring the state is forward in improving the sector.

We can learn from the state of Tripura. It has been proactive in recognising and harnessing the potential of its abundant bamboo resources since 2001. In doing so, it became the first Indian state to adopt a dedicated state bamboo policy. The state has understood the under-utilisation of its resource right from an early stage. It was clearly observed that only 1–2% of the total extracted bamboo had received value addition. This indicated a significant untapped potential, as mentioned in its State Bamboo Policy in 2001.

The State, in its **NTFP** policy, 2020 (NTFP: Non-Timber Forest Product), is looking to strengthen the bamboo sector by:

1) Enhancing turnover and livelihood by promoting value addition.
2) Encouraging the production of bamboo-based industrial products.
3) Facilitating market linkages.

Without a clear policy direction, Meghalaya risks underutilising its bamboo resources and missing opportunities for economic and environmental gains. Through its concerted efforts, Tripura has fully realised the ecological and economic benefits of its bamboo resources. Meghalaya could and should learn from Tripura's decade of experience in this field.

Bamboo has vast applications across industries such as construction, furniture, paper, textiles and bioenergy. In Meghalaya, over 60% of rural households depend on bamboo for income (Nongkynrih & Bisht, 2019). But market inefficiencies and limited value addition have restricted its economic potential. By 2030, it is estimated that bamboo cultivation and processing can create over 100,000 jobs in the state (Tiwari, 2015).

On the whole, bamboo-based industries in India are projected to grow significantly, with engineered bamboo, bio-based packaging and eco-friendly products in high demand (Gogoi & Singh, 2021).

As a start, the CLLMP gave attention to the need for integrating accurate baseline data on bamboo resources, including species distribution across Meghalaya. This, obviously, is essential for effective policy formulation and strategic planning. Comprehensive resource inventories provide a foundation for informed decision-making, enabling the development of targeted management and conservation strategies.

A study assessing bamboo management sustainability in central Indian forests highlighted the importance of such data. It noted that 'graduating to sustainable bamboo management will require better protection, resource augmentation, sustainable harvest, enhancing livelihood benefits and creating new bulk markets' (Tambe et al., 2021). A review on bamboo resources in Africa emphasised this same need (Bahru et al., 2021).

Clearly, establishing a comprehensive database of Meghalaya's bamboo resources is a critical step to developing an effective bamboo policy and realising the sector's full potential.

### 17.1.4 Identification of Gaps

Despite this resource wealth, the use of bamboo remains largely local, with limited large-scale external adoption. The aforementioned lack of comprehensive, real-time data on bamboo resources is a critical barrier to external adoption. To be sure, Meghalaya has initiated geospatial mapping and UAV (Unmanned Aerial Vehicle)-based surveys for bamboo assessment, but the coverage remains limited. This makes it difficult for industries to make investment decisions based on reliable yield projections. With unclear data, it becomes a hindrance to map out the entire state resource, which keeps large-scale investors from coming in (NESAC, 2020).

In this context, the observation in a comprehensive analysis of the National Bamboo Mission is noteworthy: 'Research institutions have developed new technologies for treatment, preservation and processing of bamboo, making it more suitable for industrial applications' (Bhatt & Joshi Associates, 2025). This underscores the critical need for accurate baseline data to inform policy and attract industrial investment.

The initiative is grounded in the principles of community-based resource management, emphasising collaborative data collection and local empowerment. Meghalaya's unique land tenure system, where forests and land are mainly community-owned, presents challenges in acquiring comprehensive resource data. Traditional top-down approaches to data collection often face resistance due to concerns over land rights and external interventions.

### 17.1.5 Theoretical Framework

Through the CLLMP, the state took up a bottom-up, participatory approach. It linked village community facilitators along with project staff to bridge the gap between state agencies and local communities.

This community-driven model brought about trust and transparency in data collection. This, in turn, contributed to mitigating resistance to external data gathering. Similar to a large-scale PRA methodology, this approach fosters local stewardship, enhances resource governance and provides a sustainable mechanism for long-term environmental planning.

## 17.2 Description of Methodology

The bamboo resource assessment at the district and block levels applied stratified random sampling. For an even distribution of sample plots, a 10 km × 10 km grid was created for the entire state. This grid was overlaid on a land-use land-cover map, which categorised the area into six strata. Within each grid, 30 sample points were spatially distributed in proportion to the stratum's area using the probability proportional to size method.

The geographic coordinates of the selected sample plots were extracted with GIS software. Field teams, guided by GPS, established 30 m × 30 m plots at each designated point. For this, they followed a standardised procedure, explained in a field training.

The bamboo inventory was in partnership with the community: It involved local youth as village community facilitators.

Fieldwork included laying out sample plots with predefined coordinates, measuring clump size and culm circumference, identifying bamboo species and preparing herbarium samples for unidentified species through leaves and sheaths. Then green weight was recorded while dry weight was measured after 45 days for 8 to 10 clumps per species in each district.

## 17.3 Result

The assessment revealed that the total number of bamboo culms across the state exceeds 2,100 million (Table 17.1). They are spread over 23% of the total geographical area of Meghalaya.

Regions with the highest concentration of bamboo are the West and Eastern West Khasi Hills, while the West Garo Hills, South Garo Hills and East Khasi Hills also show a notable bamboo distribution.

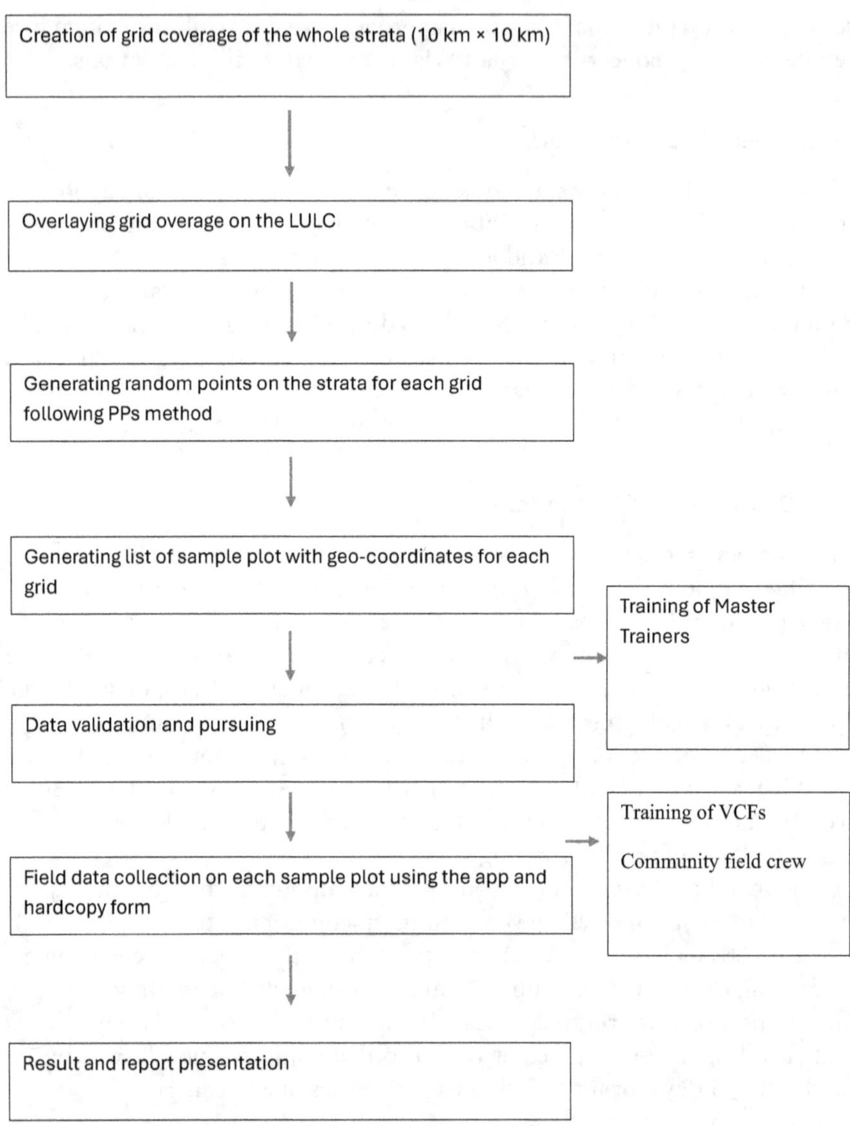


**FIGURE 17.1**　Workflow methodology

*Note:* Developed by Centre of Excellence for NRM & Sustainable Livelihoods

This large-scale availability of bamboo resources presents valuable opportunities for income generation and improved livelihoods for local residents. It encourages entrepreneurial initiatives.

With these resources, people can establish small businesses, develop value-added bamboo products, increase employment opportunities and

**TABLE 17.1**  District-wise number of culms with green weight and dry weight

| Sl. No. | District | Number of culms (in million) | Green weight (in tonnes) | Dry weight (in tonnes) |
|---|---|---|---|---|
| 1 | East Garo Hills | 160 | 1,375,245 | 652,084 |
| 2 | East Jaintia Hills | 193 | 1,497,226 | 805,712 |
| 3 | East Khasi Hills | 224 | 1,441,988 | 752,625 |
| 4 | North Garo Hills | 137 | 1,054,195 | 500,721 |
| 5 | Ri-Bhoi | 163 | 2,295,240 | 1,195,380 |
| 6 | South Garo Hills | 247 | 2,270,625 | 1,040,224 |
| 7 | South West Garo Hills | 117 | 1,081,222 | 496,562 |
| 8 | South West Khasi Hills | 63 | 360,848 | 176,998 |
| 9 | West Garo Hills | 284 | 1,620,842 | 706,316 |
| 10 | West Jaintia Hills | 102 | 892,987 | 445,614 |
| 11 | West Khasi Hills and Eastern West Khasi Hills | 499 | 7,047,742 | 3,700,021 |
| Total | | 2,189 | 20,938,161 | 10,472,257 |

*Source:* Bamboo Resource Inventory of Meghalaya, MBDA (2025).

contribute to the general economic well-being and sustainable development of the region.

With help from local communities, the assessment yielded the number of culms district-wise, along with species-wise estimates of their number and weight for each district. Moreover, it resulted in strata-wise estimates of culms across districts.

### 17.3.1  Success: Detailed Baseline Data Forms the Basis for Investors to Plan

The initiative has significantly contributed to Meghalaya's development, particularly by optimising its abundant bamboo resources. With comprehensive data now available, the state has a clear understanding of how these resources can be strategically used.

The dataset provided precise information on the location and species distribution, offering valuable insights for stakeholders. Notably, this information proved useful to investors to formulate their investment strategies under the Meghalaya Industrial and Investment Promotion Policy.

As a result, the state has received proposals from over five investors, collectively amounting to more than Rs.800 crore. This development has been made possible thanks to both the state's investor-friendly policies and the improved visibility of baseline data on bamboo resources.

The next critical step for investors is conducting ground truthing. This will enable more precise and efficient planning, further strengthening the effect of the bamboo resource assessment findings.

## 17.4 Discussion

The assessment identified a total of 26 bamboo species across the region, out of which 10 were recognised as economically viable. This identification was a critical step in helping local communities strategically select bamboo species that offer the greatest potential for generating income and improving livelihoods. By focusing on species that are economically valuable, communities can optimise the cultivation of bamboo to meet both local needs and broader market demands. Identified key species such as *Bambusa balcooa*, *Bambusa tulda*, *Bambusa bambos*, *Dendrocalamus hamiltonii* and *Melocanna baccifera* are valuable for their strength, durability and versatility. They are commonly used in building structures, making furniture, creating flooring materials and even in the production of bamboo-based products like mats and baskets.

By selectively cultivating these economically viable species, communities can not only meet local demand for construction materials but also tap into regional markets. Moreover, the cultivation and sustainable harvesting of these species open up opportunities for value-added products such as furniture, handicrafts and paper. This further diversifies income sources and supports sustainable economic development.

The ability to grow high-demand species offers a pathway for communities to strengthen their economic resilience and improve their overall standard of living, while also promoting the sustainable use of natural resources.

### 17.4.1 Importance of Policies for Meghalaya in Order to Leverage Its Resources

Lessons for Meghalaya from other state bamboo policies and recommendations for an enhanced policy by learning from the weaknesses of others; a bamboo policy that is progressive, FRA-compliant and sustainable.

The State can focus on key pointers to improve Meghalaya's bamboo policy:

1.  Strengthening the legal and institutional framework: To ensure compliance with the existing framework and centralise community-led

bamboo resource management for their full-fledged participation in decision-making.

2. Sustainable bamboo cultivation and conservation: To develop an agro-forestry-based bamboo policy (learning from Manipur), stimulate private and community-based bamboo plantations and encourage the provision of incentives to farmers, SHGs and cooperatives.
3. Reclamation of wasteland and degraded land: To be looked at for the potential of increasing state green coverage.
4. Promoting scientific bamboo management (learning from Kerala and Assam): To be done through the adoption of scientific methods of bamboo harvesting, to prevent over-extraction and degradation. This will also help improve productivity by introducing high-yield bamboo varieties.
5. Industrial development and value addition: To set up bamboo industrial clusters (learning from Tripura and Assam). In this context, the state Hub and Spoke Model can learn from the Meghalaya Livelihoods and Access to Markets Project.
6. Going from a PPP (Public Private Partnership) model to a CPPP (Community Private Partnership) model, a core community-centric approach. This goes with a shift in focus on value-added bamboo products beyond handicrafts. That is, we need to start looking at bamboo-based construction materials, furniture, textiles, charcoal, bioenergy and composites.
7. Encouraging local entrepreneurs: through the CPP model, we can open up a window to provide financial support, subsidies and training for entrepreneurs wishing to set up bamboo-based businesses.
8. Building up market linkages and export potential for an organised bamboo supply chain. Two important elements in this are to improve bamboo procurement, storage and transportation networks, and to establish primary processing centres at the community level using the IVCS (Integrated Village Coorporative Societies) model that states are focusing on.

For the state to have a strong policy to leverage its bamboo resource, it has to make sure these factors are addressed:

1. Resource mapping and inventory assessment.
2. Socio-economic and livelihood assessment.
3. Industrial and value addition potential.
4. Policy, governance and institutional framework.
5. Financial and infrastructure planning.

## 17.5 Summary and Conclusion

A key strength of the assessment was the active involvement of local communities, particularly through the engagement of village community facilitators. This approach not only leveraged community-based resource management but also helped in building local capacity for scientific data collection. The integration of traditional knowledge with modern scientific methodologies enhanced the accuracy and relevance of the findings. Community participation also fostered a bottom-up approach, ensuring that field data collection was inclusive and representative of ground realities.

The project resulted in the creation of a strong baseline dataset on bamboo resources, categorised by species. This dataset serves as a crucial reference for resource planning, conservation and sustainable management.

The initiative has also led to the formation of a local data-collecting cadre, empowering local people with skills in field inventory and scientific observation. The successful implementation of community-driven data collection demonstrates the community's ability to adopt and integrate scientific methodologies while retaining their traditional knowledge.

Overall, the state has established a robust, species-wise, ground database for bamboo resources. This gives a strong foundation for future research, policymaking and sustainable bamboo management, ensuring both environmental and economic benefits for local communities.

By integrating local knowledge and participatory mapping, the CLLMP framework exemplifies a scalable model for community-based resource assessment, setting a precedent for decentralised data-driven policymaking in Meghalaya.

### 17.5.1 Future Research and Practical Application

The community-based bamboo resource management approach has demonstrated flexibility, time efficiency and optimal resource utilisation. This model can be scaled for larger programmes, illustrating that activities once considered complex can be effectively supported by community participation. It has also brought out the critical role of communities in data collection and validation, reinforcing their contribution to evidence-based resource management.

As a result of this initiative, the state now has access to a comprehensive bamboo resource map with precise coordinates, enabling improved visibility and informed decision-making regarding bamboo resources. This approach can be extended to other forest products and resource-based activities that require community-driven data collection.

The successful application of this method has also facilitated the CLLMP Forest Management Plan. There can be a future for trained community members as data collectors for the state. These individuals can serve as frontline contributors to data digitisation and act as key points of contact for future resource assessments.

The findings from this initiative have also shaped the bamboo industry's participation in the state by providing sufficient evidence to support strategic decisions. Besides, the collaboration with the Centre of Excellence, Natural Resource Management & Sustainable Livelihoods has streamlined governmental advisory processes, ensuring that departments and institutions formulate their policies on accurate baseline data.

This achievement has been made possible through the extended reach of the CLLMP project, which has strengthened community engagement in resource management.

## References


Bahru, Tinsae, Ding, Yulong, A Review on Bamboo Resource in the African Region: A Call for Special Focus and Action, International Journal of Forestry Research, 2021, 8835673, 23 pages, 2021. https://doi.org/10.1155/2021/8835673

Bhatt & Joshi Associates. (2025). *National Bamboo mission: A comprehensive analysis of Bamboo development.* https://bhattandjoshiassociates.com/national -bamboo-mission-nbm-a-comprehensive-analysis-of-bamboo-development/

Tambe, S., Patnaik, S., Upadhyay, A. P., Edgaonkar, A., Singhal, R., Bisaria, J., Srivastava, P., Dahake, K., Hiralal, M. H., Tofa, D., Telharkar, S., Edlabadkar, V., Dethe, V., & Shekhar, K. (2021). Assessing the sustainability of bamboo management in central Indian forests. *Forests, Trees and Livelihoods*, 30(1), 28–46. https://doi.org/10.1080/14728028.2020.1852975

Gogoi, B., & Singh, R. J. (2021). Utilisation pattern of bamboo in North Eastern Region of India. *Indian Journal of Extension Education*, 57(1), 1–5.

Forest Department, Government of Tripura. (2020). *Operational policy for Non-Timber Forest Produce (NTFP) in Tripura.* Government of Tripura. https:// risingnortheast.in/wp-content/uploads/2025/04/NTFP-Policy-1.pdf

Nongkynrih, W. S., & Bisht, N. S. (2019). Bamboos: Diversity and its utilisation in Meghalaya, Northeast India. *Plant Archives*, 19(2), 3106–3110.

Industries and Commerce Department, Government of Assam. (2019). The Assam bamboo and cane policy, 2019 (Government Notice No. Ci. 227/2019/61). Government of Assam. https://industriescom.assam.gov.in/sites/default/files/ swf_utility_folder/departments/industries_com_oid_4/portlet/level_2/assam _bamboo_and_cane_policy_2019.pdf

Tiwari, B. K. (2015). Bamboo mat making and its contribution to the rural livelihood of women in South Meghalaya, India. *Journal of Bamboo and Rattan*, 14(3–4), 65–72.

Tripathi, Y. C. (2007). *Utilisation of bamboo resources in Meghalaya: Opportunities and challenges.* Presentation: State Level Workshop on Development of Suitable Strategy for Promotion of Bamboo Sector in Meghalaya.

# PART IV

# Governance and Policy

# 18

# THE POLICY AND GOVERNANCE IMPACT OF THE MBDA

## A Case Study

*Wankit K. Swer*

## 18.1 Introduction

### 18.1.1 Purpose, Scope and Significance of the Case

Meghalaya has been actively implementing several climate change and sustainable natural resource management actions over the last decade.

The unique challenges arising out of the complicated land tenure system, difficult terrain, dispersed villages, land-use pattern and socio-economic context have led the government to reimagine its developmental approach from the ground up.

This resulted in the creation of unconventional institutions such as the Meghalaya Basin Development Authority (MBDA). It came into being specifically to overcome hurdles of the typical government machinery and to enable the implementation of various pilot initiatives. It made use of funding from externally aided projects. Initiatives that are co-financed through grants or loans from multilateral institutions such as the World Bank, International Fund for Agricultural Development (IFAD), Japanese International Cooperation Agency (JICA) and Kreditanstalt für Wiederaufbau (KfW) bank – among others.

In the MBDA bottom-up approach, communities are the prime movers and the institution plays the role of a facilitator and enabler. This approach gives greater importance to the empowerment of communities and local actors. This is achieved through purpose-driven capacity building, frequent handholding, technical backstopping and funding. This equips community actors for grassroots mobilisation, planning and execution of community-identified and relevant, localised solutions.

DOI: 10.4324/9781003735380-22

Lessons drawn from the various projects then inform the government on key policy decisions and enable the adoption and scaling up of the most effective ones.

An outcome of this approach can be seen in the several policy measures that the state government adopted during the implementation of various initiatives of the MBDA.

Notable among these are the following:

1. The launch of a state policy for establishing a natural resource management committee in each and every village.
2. The creation of a state GIS (Geographic Information System) and UAV (Unmanned Aerial Vehicle) Centre.
3. The creation of a State Geo Portal.
4. The strengthening of the Meghalaya State Council for Climate Change and Sustainable Development. This is an apex body headed by the Chief Minister of Meghalaya. It supervises and provides strategic and policy direction to various climate change initiatives.
5. The development of a climate emergency framework, a legal framework for strengthening climate resilience through focused measures related to health, energy, agriculture, economy and social security, among others.

This case study is a reflection on the journey that led to various policy decisions, documenting critical elements that led to this outcome. It will provide deeper insight to enable replication by other **practitioners** and policymakers.

### 18.1.2 Review of Literature and Comparative Analysis

Several states in India have established climate change councils or cells in compliance with the National Action Plan on Climate Change (Government of India, Ministry of Environment, Forest and Climate Change, n.d.). Meghalaya has expanded the scope and depth of its climate change agenda through the creation of an apex-level council headed by the Chief Minister, with other ministers and all relevant departments as members, aiming for better coordination and institutional convergence (Centre for Policy Research, 2023).

The landscape approach to natural resource management adopted by the MBDA centres on balancing the multiple and often competing uses of land – such as agriculture, forestry, conservation, settlement, infrastructure and ecosystem services (Pimbert, 2009). Unlike traditional sectoral or watershed management, the landscape approach considers environmental, social and economic objectives holistically and promotes participatory decision-making across diverse stakeholder groups (Pimbert, 2009). International research recognises the landscape approach as a way to reconcile conservation priorities with local livelihood needs, climate resilience and sustainable development (Minang et al., 2015).

Critical elements of successful landscape management include engaging communities, government agencies, private sector actors and civil society in joint planning, implementation and monitoring processes (Pimbert, 2009). Participatory mapping, multi-stakeholder forums and integrated land-use planning are used to identify synergies – while recognising trade-offs between short-term economic needs and long-term ecosystem health (India Development Review, n.d.). The landscape approach draws on adaptive management principles, recognising changing conditions and the need to continually negotiate priorities and trade-offs (Minang et al., 2015).

Evidence from published studies shows that where the landscape approach has been implemented – in countries such as Ethiopia, Indonesia, Mexico and in Indian states such as West Bengal and Sikkim – it has led to improved conservation outcomes, increased agricultural productivity and strengthened community resilience to climate risk (Minang et al., 2015). Additionally, research affirms the importance of empowering indigenous and marginalised communities within the landscape decision process, enhancing both equity and sustainability (Pimbert, 2009; India Development Review, n.d.).

Policies and institutional frameworks that support landscape-scale governance are now viewed as critical for fighting climate change, biodiversity loss and resource degradation globally (Pimbert, 2009). The approach aligns with emerging international standards like the UN FAO's "Sustainable Land Management" and the CBD's "Socio-Ecological Production Landscapes" initiatives (Minang et al., 2015). Adopting landscape approaches can safeguard ecosystem services while supporting local economies and social well-being (Minang et al., 2015).

Studies of basin-based governance in river systems such as the Ganga, Brahmaputra and Indus confirm that participatory, decentralised management – when combined with strong institutional support – leads to measurable improvements in resource stewardship, climate adaptation and poverty reduction (India Water Portal, 2024; Government of Meghalaya & Meghalaya Basin Development Authority, 2024). These lessons are echoed in international examples from the Murray-Darling and São Francisco Basins (Quinn et al., 2007).

The World Bank-supported Community-Led Landscape Management Project (CLLMP) exemplifies the practical application of a basin/landscape approach, integrating community mobilisation with landscape restoration, social inclusion and policy reform. Project documents, evaluations and monitoring records underline how women's representation, village planning and knowledge management led to sustainable models for both governance and policy innovation in Meghalaya (Government of Meghalaya & Meghalaya Basin Development Authority, 2024; World Bank, 2025).

The landscape approach as operationalised by the MBDA, promotes the convergence of government schemes, local empowerment and data-driven decision-making – offering a replicable template for other regions with similarly complex social and ecological systems (Government of Meghalaya & Meghalaya Basin Development Authority, 2024; World Bank, 2025).

### 18.1.3 Identification of Gaps

The state is taking several actions to mitigate climate change effects and build resilience. Multiple actors, including government departments, specialised agencies such as the Meghalaya Basin Development Authority (MBDA), local NGOs and others are engaged in this sector.

Yet, there has not been a single body to coordinate the activities of these various stakeholders. This has led to inefficiency because of duplication of work, inaccessibility to resources, knowledge available within other agencies and lack of visibility of each other's work.

For the beneficiaries, there are too many agencies working on the same or similar issues. This leads to confusion, which undermines the focus and understanding of local communities on the criticality of climate change actions. As it is, there is no institution at the village level to take up this subject and enable the planning of effective actions. Besides, there is generally poor awareness of climate change, even though people feel its effects. Convergence of resources and initiatives from various agencies is difficult.

These issues are compounded by the general lack of baseline data and other basic information. This renders decision-making extremely challenging.

### 18.1.4 Theoretical Framework

In the course of the implementation of various MBBDA initiatives, several critical lessons have been learned. These have contributed to policy decisions and guided actions on the ground. Policy decisions and measures materialised in conjunction with MBDA's actions through its various projects and initiatives. They are significant contributions to the overall policy approach of the state.

### 18.2 Methodology of the Case

### 18.2.1 Description of Methodology

The case study is a documentation of the initiatives and their subsequent effect on policy and governance decisions of the state through the experiences of the practitioners of the MBDA.

### 18.2.2 Justification of Methodology

As the author of this case study, I am a practitioner with first-hand knowledge of the events and their subsequent policy and governance consequences and effects.

### 18.2.3 Limitations and Challenges

To be sure, the case study is not intended to establish an empirical correlation between MBDA's actions and subsequent corresponding policy and governance decisions. It merely documents facts as I found them as a practitioner.

### 18.3 Results

#### 18.3.1 Main Case Findings 1: State Policy for Village Natural Resource Management Committees

The World Bank-financed Community-Led Landscape Management Project (CLLMP) is one of the successful natural resource management projects the MBDA has implemented (Government of Meghalaya & Meghalaya Basin Development Authority, 2024). The project began in 2018 and formally completed its implementation in June 2024.

As we have seen, Meghalaya does not have any institutional arrangement at the village level capable of natural resource management (NRM). Conventionally, a body created for implementing MGNREGA was the village employment council (VEC). But, in the absence of a Panchayat Raj system in the state, it did not have the specific capacity to plan for NRM, despite being responsible for planning NRM, since 60% of MGNREGA funds were to be used.

Institutions such as joint forest management committees, which in other states contribute to the planning of NRM under the MGNREGA, are non-existent due to the complex land tenure system in the state.

To overcome this, the MBDA established so-called village natural resource management committees (VNRMCs) in all CLLMP villages to enable it to execute the project. As a community-driven project, the VNRMCs were capacitated for NRM and empowered to plan and implement interventions that were most meaningful for them to address their NRM priorities.

The MBDA provided technical facilitation, in terms of training of community participants on processes, safeguards, accounting and record keeping and financial support. Under this project, 400 villages created NRM plans and implemented interventions across 50,000 ha of land. More than 24,000 ha were treated with soil and water conservation measures. Nine thousand seven hundred forty-six hectares were brought under afforestation, while another 12,000 ha were brought under agriculture or agroforestry.

Village community facilitators (VCFs) actively supported the village committees. These VCFs were boys and girls that the VNRMCs had selected and appointed from within their own villages to be trained and become village functionaries for the project.

The VNRMCs proved to be pivotal in the successful implementation of the project. They supervised the implementation and took care of record keeping for validation of success.

With respect to composition, the VNRMCs were mandated to have women serve as secretaries. Besides, women were to constitute 50% of the membership. Notably, all VNRMCs adhered to this second requirement.

The VNRMC has demonstrated a viable model for decentralised decision-making and the implementation of unconventional subjects such as NRM. Its success resulted in the state replicating the model for implementation of the NRM component of the MGNREGA programme in the state through a policy to create natural resource management committees (NRMCs) that was officially notified in 2021. Since then, the NRMCs have been empowered to plan for the 60% of MGNREGA funds meant for their respective villages. The VCFs were made mandatory members of the VECs.

By then, the CLLMP had already trained more than 13,000 VCFs across all villages of the state. This way, all NRMCs could leverage the services of the VCFs. The expenditure share for NRM under the MGNREGA in Meghalaya reached a peak of 46% in 2022–2023.

### 18.3.2 Main Case Findings 2: Creation of State GIS and UAV Centre

Notably, technology, particularly Geographic Information Systems (GIS) and Unmanned Aerial Vehicles (UAVs), has played a crucial role in the implementation of various initiatives of the state. What started out as a small GIS and UAV unit under the MBDA was gradually able to take up significant initiatives, such as the mapping of village boundaries for project implementation. This was the first ever in Meghalaya, which otherwise does not have cadastral maps. This was important for the preparation of state-wide land-use land-cover maps, support during natural disasters and calamities, support in infrastructure planning, traffic management, pilot training and documentation – among other activities. With a growing demand for these services, the state government formally inducted the unit and upgraded it into a full-fledged independent State GIS and UAV Centre in December 2022.

### 18.3.3 Main Case Findings 3: Creation of State Geo Portal

The Meghalaya State Geo Portal is an online geospatial platform. It was developed to streamline the use of geospatial data and facilitate better planning, monitoring and decision-making for various government programmes and projects in the state. This initiative was part of broader efforts to leverage technology and digital tools for enhanced governance and transparency in Meghalaya.

GIS technology is not new; various departments of the state have used it in the past. But most of their work was done in silos; there was no awareness of what was available in which department. This has resulted in duplication of effort and wastage of time and financial resources. The portal has been able to bring together, for the first time, and consolidate all the data layers available with each department of the state. As a result, it has enabled access to all information for all developmental actors on a single platform.

The portal is powered by advanced GIS technology, allowing the government to combine various layers of data, including topography, land use, demographics and infrastructure. This ensures that the platform is comprehensive and can be used for a wide range of planning activities.

It also integrates satellite imagery for real-time updates on environmental conditions, land cover changes and urban expansion. All these are crucial for sustainable development and natural resource management in Meghalaya.

### Key objectives

1. *It is a centralised geospatial data repository*: The portal serves as a centralised hub for all geospatial data. It enables the state government and other stakeholders to access, visualise and use maps, satellite imagery and geospatial information easily, for informed decision-making.
2. *It supports rural development initiatives*: The Geo Portal is particularly valuable for the implementation of rural development programmes like MGNREGA, as well as natural resource management and infrastructure planning. It allows the government to monitor the progress of various projects, such as afforestation, water conservation and road construction, by linking geographical data with real-time updates.
3. *It enhances transparency*: By making geospatial data available to the public and local bodies, the portal promotes transparency in governance. Citizens and local communities can access relevant information. This helps them engage more effectively in planning and hold officials accountable for project execution.
4. *It assists in disaster management and planning*: The Geo Portal also plays a crucial role in disaster management. It enables authorities to map out *vulnerable areas* – those that are prone to natural disasters like floods, landslides and droughts and plan mitigation measures accordingly.

### 18.3.3.1 Significance for Meghalaya

Given the hilly terrain and scattered population of Meghalaya, this Geo Portal enables more effective resource management and implementation of government schemes across the state's rural and remote areas. It addresses the challenges posed by geographical isolation and provides data-based insights for efficient service delivery in critical sectors like agriculture, water resources and infrastructure development.

### 18.3.4 *Main Case Findings 4: Strengthening of the Meghalaya State Council for Climate Change and Sustainable Development*

The state government officially notified the Meghalaya State Council for Climate Change (MSCCC) in 2012. The aim was to lead the state's efforts in tackling the effects of climate change, developing adaptation strategies and promoting sustainable development in alignment with national and global climate commitments.

The MSCCC was created in response to the growing recognition that climate change poses significant risks to the environment, livelihoods and natural resources in the state. This was particularly relevant on account of the state's hilly terrain, fragile ecosystems and reliance on rain-fed agriculture.

Since its establishment, the council has played a key role in formulating the Meghalaya State Action Plan on Climate Change. It was also instrumental in coordinating the state's efforts to mainstream climate action across various sectors, including forestry, water resources, agriculture and disaster management. Nevertheless, its full potential as the apex-level strategic coordinator could not be fully realised in the absence of quality data and visibility of individual actions of departments and other actors.

The MBDA added potency to the council by leveraging the GIS and MIS technology already in use. Take the example of the creation of a water budget for every village.

The agency used a combination of rainfall data from the Meteorological Department, its own real-time weather data from its 80 basic and 22 advanced automatic weather stations, Census data and various GIS layers available with them. These included village boundary maps, land-use land-cover maps, contour maps, slope maps and drainage maps – among others.

Working this way, they could assess the quantity of water available to a village from various sources. They could also estimate both domestic and agricultural demand, which, in turn, enabled the estimation of deficit or surplus for every village.

Using the same system, it was also possible to identify potential interventions to address the gaps best. Then various departments could adopt them under the coordination of the council.

A second example was the mapping of the entire Shillong catchment. It enabled the MBDA to simulate interventions using check dams to meet the water demand of Shillong City. The agency was able to calculate accurately the number of check dams required, their locations and the amount of water that would be collected.

These capabilities provide state-of-the-art insights. These have greatly enhanced the quality of decisions being made at the highest level. In 2024, the MBDA was formally notified as a Secretariat to the council.

### 18.3.5  Main Case Findings 5: Development of a Climate Emergency Framework

The state government drafted the Meghalaya Climate Emergency & Green Growth Framework, a comprehensive initiative to address the growing climate crisis facing the state. This framework, which is in the process of becoming law, is one of the notable outcomes. The MBDA facilitates entirely under the CLLMP. The objective is to institutionalise and strengthen the state's ongoing efforts to deal with climate change.

#### 18.3.5.1  What the Framework Is about

The framework aims to provide a robust enabling structure to unify existing climate interventions and integrate them into a coherent state-level strategy. It is grounded in extensive research and consultations with stakeholders from communities, administrative bodies and political leadership.

Key areas of focus include the following:

- **Water conservation and management,** including the use of a real-time water data portal and village-level water budgeting.
- **Payment for ecosystem services (PES),** to incentivise communities to preserve forests. This will enable the adoption and scaling up of the MBDA's PES by the state government.
- **Health sector integration,** ensuring that climate-related health risks are addressed.
- **Coordinating across government departments,** strengthening existing interventions and introducing new laws and policies for climate mitigation and adaptation.

The framework also reflects a forward-thinking approach by including youth engagement. Since 37% of the state's population is **youths,** making them key stakeholders in its climate future is a path-breaking step.

For Meghalaya, the framework is crucial as it helps the state respond to climate effects that threaten its fragile environment, such as landslides, flooding and soil erosion. By framing climate change as an emergency, the state signals its commitment to addressing immediate risks and challenges, while ensuring long-term sustainable development. It consolidates its previous efforts in sustainable development, including a Water Policy of 2019 and the CLLMP.

The framework is also health-sensitive, integrating climate risks with public health planning. This is especially relevant in the face of emerging climate-related diseases and health challenges. Further, it places an emphasis on community-led action, recognising local knowledge and involvement as vital for effective climate action.

Nationally, Meghalaya's adoption of this framework places it at the forefront of climate action in India. This is especially so among the northeastern states, where the vulnerabilities to climate change are often more pronounced because of challenging terrain and socio-economic factors. The framework also aligns with India's national climate goals. Accordingly, it can serve as a model for other states seeking to implement coordinated climate strategies.

Internationally, Meghalaya's declaration of a climate emergency is significant as the state joins over 40 countries that have made similar declarations, underscoring the urgency of the climate crisis.

It marks Meghalaya's entry into the global movement for climate resilience and green growth. This could attract international climate finance, technology partnerships and collaboration with global organisations working on sustainable development and climate mitigation.

By adopting this framework, Meghalaya demonstrated its commitment to both undertaking local action and contributing to global climate goals. In doing so, it has positioned itself as a leader in climate adaptation and green growth in India's northeastern region.

## 18.4 Discussion

### 18.4.1 Interpretation of Findings in Relation to Replicability

#### 18.4.1.1 Local Replicability

The initiatives designed by the MBDA, especially those centred around community-led natural resource management, have proven to be highly adaptable to local contexts. The success of the village natural resource management committees (VNRMCs) under the Community-Led Landscape Management Project, which empowered local communities to manage and implement NRM projects, can be replicated on a smaller scale in other regions within Meghalaya or in similarly structured states.

The village community facilitators model, wherein youth from villages themselves are trained to facilitate community programmes, also offers a locally applicable and adaptable model that strengthens community participation and ownership.

The Geo Portal and GIS/UAV technologies have clear local benefits. These tools can be used to address local governance challenges, such as resource mapping, disaster management and monitoring of development projects, particularly in remote or difficult terrains. Communities, especially in rural areas, can use this platform for localised decision-making, resource allocation and emergency management.

### 18.4.1.2 Regional Replicability

On a regional level, the MBDA's approach to climate change and NRM offers a model for other states in Northeast India and similar regions with complex land tenure systems and mountainous terrain.

For example:

- The VNRMCs model could be adapted to states like Arunachal Pradesh, Nagaland and Assam, where land use and community participation are crucial yet often underserved in development planning.
- The GIS and UAV Centre, especially for mapping village boundaries and land-use patterns, could be applied across other states facing similar issues related to fragmented data and inefficient spatial planning. The integration of real-time data, like weather patterns and infrastructure monitoring, could greatly enhance regional disaster management efforts.

The Meghalaya State Council for Climate Change could serve as a model for other states looking to institutionalise climate action and coordinate various departmental efforts. By replicating the framework of a centralised, apex-level climate body, other regions could foster better institutional convergence and address climate-related risks more effectively.

### 18.4.1.3 National Replicability

Nationally, the findings underscore the importance of cross-departmental coordination and data-driven governance in addressing climate change challenges. The Meghalaya Climate Emergency & Green Growth Framework could inspire other states, especially in regions vulnerable to extreme climate events, to adopt similar frameworks for enhancing climate resilience and sustainable growth.

The focus on integrating health and climate within governance is particularly valuable, considering the rising climate-related health risks in India.

Importantly, the natural resource management committees established under the MGNREGA programme in Meghalaya, leveraging the VNRMC structure, could be scaled up nationally. In states without Panchayat Raj systems, such models can bridge the gap in decentralised governance. Here, local bodies are empowered to manage their resources, enabling more effective implementation of national schemes like the MGNREGA.

### 18.4.1.4 International Replicability

Internationally, the MBDA's integrated, community-centric approach to climate change can be seen as a viable model for regions in developing countries that face similar ecological and socio-economic challenges. The establishment of a climate emergency framework and payment for ecosystem services (PES) could inspire global adoption of policies that align local economic development with environmental sustainability, particularly in countries with a high reliance on natural resource-dependent sectors.

The Geo Portal, which integrates and centralises geospatial data from various departments, is also a practice that could be replicated in other countries with significant challenges in cross-agency data sharing and spatial planning.

## 18.5 Summary and Conclusion

The Meghalaya Basin Development Authority's approach to climate governance, natural resource management and community empowerment offers a highly replicable model at local, regional and national levels.

The key lessons and initiatives, from community-based NRM planning to advanced GIS technologies, can be adopted and scaled in other regions with similar socio-economic and environmental contexts. Additionally, the MBDA's innovations in governance and institutional frameworks have the potential to inform national policies, especially in the areas of climate adaptation, disaster management and sustainable development.

## References

Centre for Policy Research. (2023). *A comprehensive overview of departmental convergence under MGNREGS.* https://cprindia.org/wp-content/uploads /2023/04/A-Comprehensive-Overview-of-Departmental-Convergence-Under -MGNREGS.pdf

Government of India, Ministry of Environment, Forest and Climate Change. (n.d.). *State action plan on climate change.* http://moef.gov.in/state-action-plan-on -climate-change

Government of Meghalaya & Meghalaya Basin Development Authority. (2024). *The Meghalaya Community Led Landscape Management Project (MCLLMP).* https://mbda.gov.in/meghalaya-community-led-landscape-management-project -mcllmp

India Development Review. (n.d.). *Taking a community-led approach to climate resilience.* https://idronline.org/article/climate-emergency/taking-a-community-led-approach-to-climate-resilience/

India Water Portal. (2024). *Collaborative approaches to water governance in Indus, the Ganga and Brahmaputra Basin.* https://www.indiawaterportal.org/governance-and-policy/governance/collaborative-approaches-water-governance-indus-ganga-and-brahmaputra-basin

Minang, P. A., van Noordwijk, M., Freeman, O. E., Mbow, C., de Leeuw, J., & Catacutan, D. (Eds.). (2015). *Climate-smart landscapes: Multifunctionality in practice.* Nairobi, Kenya: World Agroforestry Centre (ICRAF).

Pimbert, M. P. (2009). *Natural resources, people and participation.* International Institute for Environment and Development. https://www.iied.org/sites/default/files/pdfs/migrate/G02104.pdf

Quinn, G.P. et al. (2007). *River basin development: A framework for case studies.* IWMI. http://www.iwmi.cgiar.org/publications/working-papers/iwmi-working-paper-118/

World Bank. (2025). *Implementation completion report for the Meghalaya community led landscape management project.* https://www.cllmp.com/wp-content/uploads/2025/07/Meghalaya-CLLMP-IEG-ICRR-.pdf

# 19

# BRIDGING SKILL GAPS IN MEGHALAYA THROUGH THE APPRENTICESHIP TRAINING PROGRAMME

*T. Niang Suan Ching and Sucielia Mylliemngap*

## 19.1 Introduction
### 19.1.1 Purpose, Scope and Significance of the Case

Meghalaya has emerged as a key beneficiary of international development funding, receiving support from organisations including the International Fund for Agricultural Development, the World Bank, the Japan International Cooperation Agency and the German *Kreditanstalt für Wiederaufbau*. This is evident in the rise of externally aided project (EAP) funding: From ₹25 billion (approximately USD 300 million) in 2019 to ₹125 billion (around USD 1.5 billion) by 2023. MBMA plays a critical role in executing these projects, with the World Bank-funded MCLLMP as a flagship initiative. Launched in 2018 and running until June 2024, the MCLLMP was the first of its kind in the state's natural resource management sector (Meghalaya Basin Management Agency & World Bank, n.d.). Given its scale and complexity, advanced technologies such as GIS and UAV remote sensing have proven indispensable for effective planning, monitoring and data collection. But the MBMA struggled to source local professionals with the necessary expertise, affecting project execution. In response, it launched the Apprenticeship Training Programme in March 2022 to build technical capacity and stimulate workforce readiness – with encouraging results. This case study examines the programme's importance for filling the skills gap through trainee development, employment outcomes and EAP contributions (MBDA 2022).

DOI: 10.4324/9781003735380-23

### 19.1.2 *Organisation of the Chapter*

The chapter is structured to provide a comprehensive analysis of the MBMA's Apprenticeship Training Programme and its effect. It begins with an introduction, outlining Meghalaya's evolving role in international development funding and the significance of EAPs. The literature review explores existing research on capacity-building initiatives. The methodology section details the research approach, data collection and analysis methods. Findings and a discussion assess the programme's effectiveness, challenges and opportunities. Finally, the conclusion and recommendations summarise key insights and propose strategies to improve the technical workforce development in Meghalaya.

### 19.1.3 *Review of Literature*

Apprenticeship programmes are widely recognised as essential for workforce development, offering a practical blend of theoretical education and hands-on training to improve both employability and productivity. Fuller and Unwin (2017) emphasise that 'good quality apprenticeships' enhance employment opportunities, raise earnings potential and support long-term career progression. The International Labour Organisation (ILO, 2024) echoes this, stating that quality apprenticeships help learners acquire relevant skills, respond quickly to labour market changes, enhance workplace culture and often lead to better job placement outcomes than traditional education systems. Globally, Germany's apprenticeship programme, the Dual Vocational Education and Training (Dual VET), serves as a leading example of effective skill development, and is frequently cited as a benchmark of success. The programme plays a key role in the country's economic development and low rate of youth unemployment, and it has been adopted by both developed and developing nations (Euler, 2013).

Apprenticeship programmes are especially critical for bridging the education-to-employment gap in developing nations. As Almeida et al. (2012) observe, the majority of the labour force in these nations possesses limited formal education. Even individuals with advanced academic qualifications often lack essential job-ready skills – particularly those related to analytical thinking, problem-solving and interpersonal communication – which are increasingly valued in today's dynamic labour markets. The Asian Development Bank (2014) reinforces this concern, identifying the shortage of skilled human resources as 'one of the major binding constraints' to sustaining high economic growth in developing economies, especially in South Asia. In response, developing nations have turned to apprenticeship programmes as a strategic solution to address skills mismatches and promote workforce readiness.

But several challenges hinder the effectiveness of these programmes. These include inadequate funding, weak industry participation and outdated training curricula that fail to keep pace with technological advancements and market demands (ILO, 2022). Overcoming these barriers requires comprehensive policy reform, enhanced collaboration between educational institutions and industries and regular curriculum updates aligned with labour market trends. By adopting globally informed, context-specific apprenticeship models, developing nations can significantly improve employment outcomes, build a more resilient workforce and support long-term economic development.

## 19.2 Identification of Gaps

While the MBMA's Apprenticeship Training Programme effectively builds technical skills in GIS and UAV remote sensing, several gaps remain. The programme could benefit from expanded industry collaborations to provide trainees with broader exposure to real-world applications. Limited access to advanced UAV technology and high-end GIS software may hinder hands-on learning. Besides, there is no structured job placement support. This reduces direct employment opportunities post-training. The programme could also increase soft skills training, such as communication and project management, to improve overall employability. Lastly, long-term impact assessments are needed to track trainees' career progress and programme effectiveness.

## 19.3 Methodology of the Case

The evaluation of the Apprenticeship Training Programme used the case-study approach for a detailed analysis of its structure, implementation and overall impact. By integrating both qualitative and quantitative research methods, the study captures key statistical outcomes alongside the personal experiences and perspectives of participants, offering a well-rounded understanding of the programme's effectiveness.

### 19.3.1 Qualitative Methodology

Qualitative methodology focuses on gathering insights from apprentices, trainers and other stakeholders. Document review is an essential part of this process, where programme guidelines, training materials and policy documents are analysed to assess the programme's objectives and structure. Observations of training sessions help evaluate instructional quality, engagement levels and the integration of theoretical and practical learning. Also, structured surveys and questionnaires collect feedback from apprentices regarding their experiences and challenges, as well as suggestions for improvement.

### 19.3.2  Quantitative Methodology

Quantitative methodology emphasises measuring key performance indicators to assess the programme's effectiveness. Enrolment figures help track participation levels. Retention and completion rates serve to determine the programme's ability to sustain engagement and ensure successful training outcomes. Job placement statistics are analysed to assess how many apprentices secure relevant employment after completing the programme. Further, pre- and post-training assessments measure skill development. This allows for an evaluation of apprentices' progress. Demographic data, including age, gender and educational background, contribute to an understanding of participant diversity and inclusivity.

### 19.3.3  Description of Methodology

This study adopts a case study approach within a mixed-methods research design to evaluate the effectiveness of the Apprenticeship Training Programme. The quantitative component is based on secondary data analysis of MBMA's internal records. Key indicators – such as enrolment figures, demographic characteristics, retention and completion rates and post-training employment outcomes – were examined using descriptive statistical methods to identify patterns and trends. Complementing this, the qualitative component involved the analysis of open-ended survey responses from programme participants. These responses provided insights into personal experiences, particularly regarding the quality of mentorship, relevance of assigned tasks, opportunities for skill development, alignment with career aspirations and challenges encountered during the programme. Thematic analysis was employed to identify recurring themes and meaningful insights. Additionally, participants were asked about their likelihood of recommending the programme to others, offering a broader perspective on overall satisfaction and perceived value.

### 19.3.4  Justification of Methodology

The case study approach is appropriate for evaluating the Apprenticeship Training Programme as it allows for an in-depth, context-specific analysis of its structure, implementation and outcomes. This methodology is particularly useful for examining complex programmes where both numerical data and human experiences are important. By combining quantitative and qualitative methods, the study captures key performance indicators – such as enrolment, completion and employment outcomes – alongside participants' perspectives on mentorship, skill development and overall satisfaction. This integrated approach provides a more holistic understanding of the programme's effectiveness and helps identify strengths, challenges and opportunities for improvement that might not emerge through a single-method analysis.

## 19.4 Limitations and Challenges

The case study approach also has certain limitations and challenges. A key limitation is that the findings are specific to this programme; they may not be fully generalisable to other apprenticeship models or industries. Since case studies focus on a particular context (in this case, Meghalaya and the MBMA), applying the results to different settings may require adjustments. Data collection problems may arise, particularly in gathering consistent and accurate responses from participants. Apprentices and trainers may have varying levels of willingness to share feedback, leading to potential gaps in qualitative insights. Besides, variations in record-keeping may affect the completeness of quantitative data, such as job placement rates and skill assessments. Time constraints and resource limitations could influence the depth of the study. Despite these challenges, careful data validation and diversified data sources help produce a reliable and well-rounded evaluation of the Training Programme's effectiveness.

## 19.5 Results

### 19.5.1 Main Case Findings 1: Proven Model for Workforce Development

Launched in March 2022, the Apprenticeship Training Programme began with two courses: GIS and UAV remote sensing. Five apprentices enrolled in each. After the success of the initial batch, the programme expanded in August 2022. It added Finance & Accounts, Procurement, Environment Management and Monitoring & Evaluation. Enrolment increased to 39. Since then, the programme has grown significantly, offering a diverse array of courses, including GIS, UAV remote sensing, Knowledge Management, Finance, IT Development and Artificial Intelligence (AI), among others.

Each course integrates theoretical learning with hands-on application. This way, apprentices not only gain academic knowledge but also practical experience. Real-world projects, coupled with mentorship, bridge the gap between classroom learning and industry requirements. So, apprentices can acquire the skills needed to thrive in their careers.

### 19.5.1.1 Selection Process

The selection process is highly competitive; it is designed to admit only the most qualified candidates. It begins with widespread advertisements through newspapers, online job portals and the MBMA's official website, inviting applications from eligible candidates. Applicants must be residents of Meghalaya, aged 18–26 and meet specific educational criteria, such as a degree in GIS, Remote Sensing or Geography for GIS applicants, or a Mass Communication qualification for Knowledge Management. Applications undergo a thorough review to confirm eligibility; shortlisted candidates take a written exam. Successful candidates proceed to an interview, while final selections are based on performance in both the exam and the interview. Once selected, candidates are enrolled in training programmes ranging from three to twelve months. The MBMA's Apprenticeship Training Programme has successfully admitted 222 apprentices across nine batches, spanning from March 2022 to January 2025. Of these, 164 apprentices (74%) have completed their training, while 58 (26%) are currently attending their respective courses.

### 19.5.1.2 Course-Wise Distribution of Apprentices

Among the various courses, Finance & Accounts is the most sought-after, with 47 apprentices (21% of total intake). GIS with 42 apprentices (19%) follows closely, and UAV Remote Sensing with 29 apprentices (13%) comes third. Field Engineering training has attracted 19 apprentices (9%), while Procurement has 18 (8%). Environment Management accounts for 13 apprentices (6%), and IT Developer has 10 (5%). Monitoring & Evaluation and Data Analyst training each have eight apprentices (4%). IT, AI and Spring Shed Management each have 5 (2%). And, finally, IT GIS has four apprentices (2%), Seed Secure six (3%) and Knowledge Management three (1%). This distribution reflects a strong demand for financial, geospatial and technical skills, which are highly valued in today's workforce. Simultaneously, the diverse spread across other fields clearly shows the Training Programme's comprehensive approach to workforce development: Apprentices are enabled to acquire a versatile skill set that is applicable across multiple industries.

### 19.5.1.3 Apprentice's Response

To assess the effectiveness of the Training Programme further in workforce development, all 222 apprentices received a structured survey. These were the ones who had either completed or were currently undergoing training. An impressive 94% response rate was achieved from 209 apprentices – 108 female (52%) and 101 male (48%) respondents. The survey examined key aspects such as overall satisfaction, skill development and employment

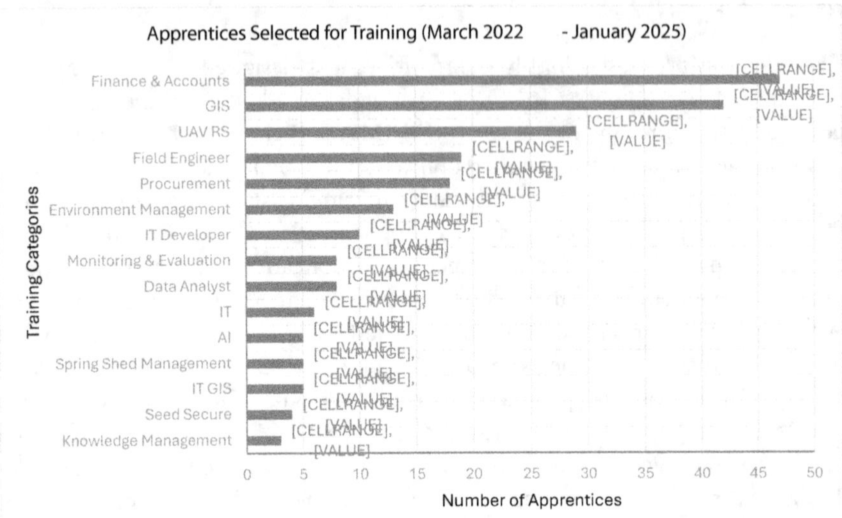


**FIGURE 19.1**   Apprentices selected for training

*Source:* MBMA documentation and programme records.

outcomes. They provided valuable insights into the programme's effect on career readiness. The findings not only highlighted the Training Programme's strengths but also identified opportunities for improvement, ensuring continued success in preparing apprentices for the workforce.

**Overall experience:** A majority of apprentices expressed high satisfaction with the programme: 118 rated their experience as 'Good' and 69 as 'Excellent.' Meanwhile, 22 apprentices (11%) found it 'Average,' indicating areas for potential strengthening. These results point to the Training Programme's effectiveness in meeting apprentices' expectations while also highlighting opportunities for ongoing improvement.

**Guidance and mentorship:** Guidance and mentorship are crucial to the apprenticeship experience, and the survey results reflect a strong support system within the programme.

A majority of apprentices, 141, reported 'Always' receiving sufficient guidance, while 50 stated they 'Often' received support. A further 14 indicated they 'Sometimes' had access to mentorship, with only three reporting 'Never' and one saying 'Rarely.' These findings pinpoint the Apprenticeship Training Programme's robust mentorship set-up. It also underscores a need for greater consistency to ensure every apprentice receives the support necessary for her or his professional growth.

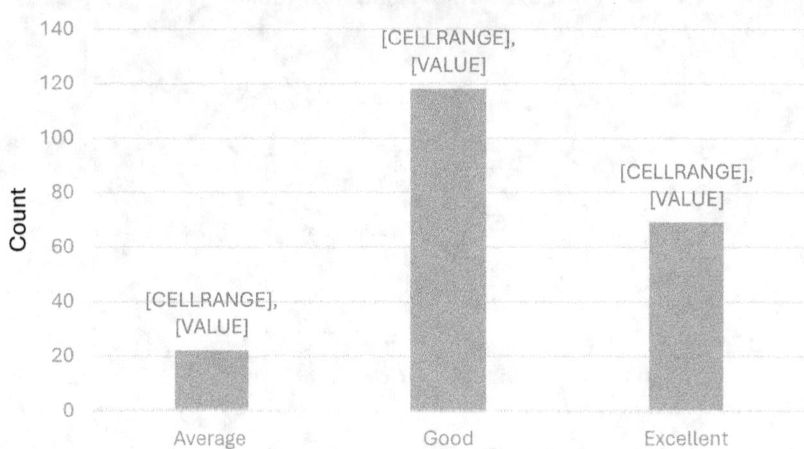



**FIGURE 19.2**   Overall experience

*Source:* MBMA Survey.

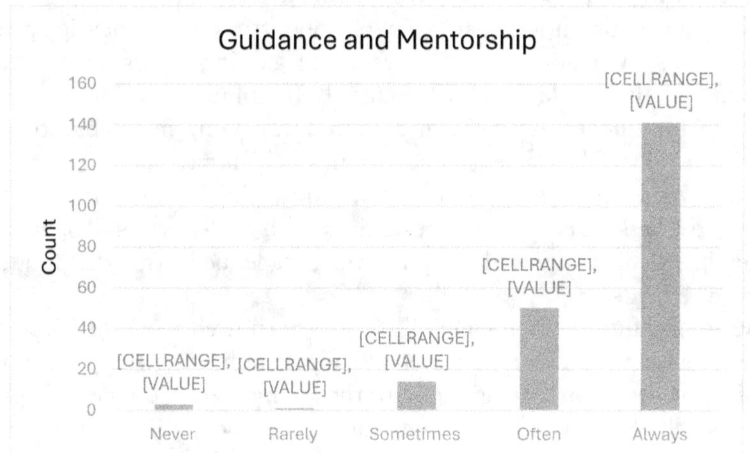



**FIGURE 19.3**   Guidance and mentorship:

*Source:* MBMA Survey.

**Relevance of assigned tasks:** The majority of apprentices found their tasks beneficial to their learning. A significant 108 rated them 'Highly relevant,' while 86 considered them 'Moderately relevant.' Notably, 11 found their tasks 'Slightly relevant,' and four (2%) deemed them 'Not relevant.' These results reflect the Training Programme's effectiveness, with room for improvement in task allocation.

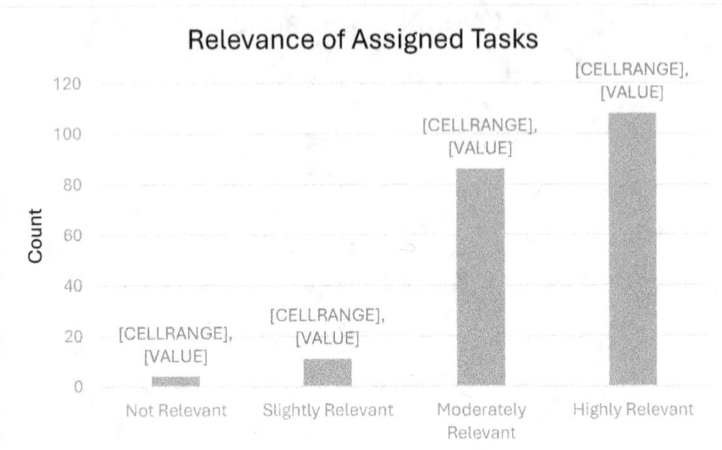


**FIGURE 19.4**  Relevance of assigned tasks

*Source:* MBMA Survey.

**Opportunities for practical skill development:** The survey results indicate that most apprentices had valuable opportunities to develop practical skills. A majority – 133 – reported gaining hands-on experience 'Extensively,' while 71 had benefitted 'To some extent.' This suggests room for further exposure in certain areas. Only five apprentices felt they had lacked sufficient opportunities, which means a minor area for improvement. These findings demonstrate the Training Programme's effectiveness in equipping apprentices with real-world skills, reinforcing its role in bridging the gap between academic learning and industry requirements.

**Relevance for future employment:** The survey showed that 173 apprentices felt well-equipped for future employment, indicating the programme's effectiveness in preparing them for the job market. A further 36 apprentices felt 'Somewhat' prepared, suggesting room for improvement in training or support. These results highlight the Apprenticeship Training Programme's success in career readiness while identifying opportunities to strengthen job preparation efforts.

**Willingness to recommend the programme:** A strong voice of 202 apprentices would recommend the programme, showing high satisfaction. Still, seven apprentices expressed concerns about the lack of job placement after training, suggesting the programme should focus more on preparing for alternative career options. This feedback highlights the Apprenticeship Training Programme's success while pointing to areas for improvement, particularly in job placement support and career development.

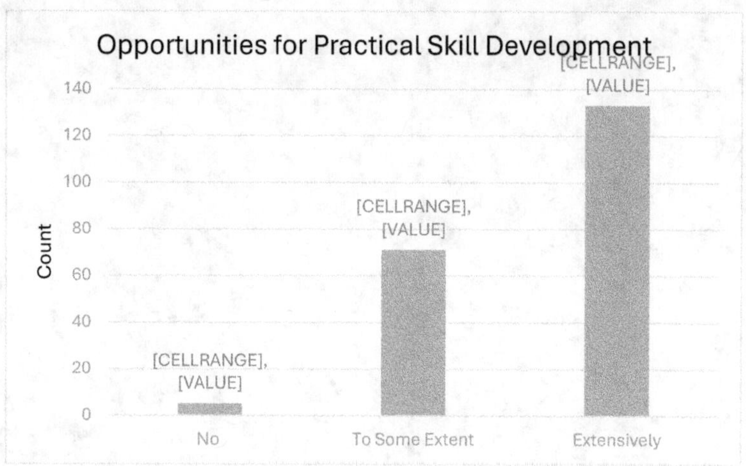



**FIGURE 19.5**  Opportunities for practical skill development

*Source:* MBMA Survey.

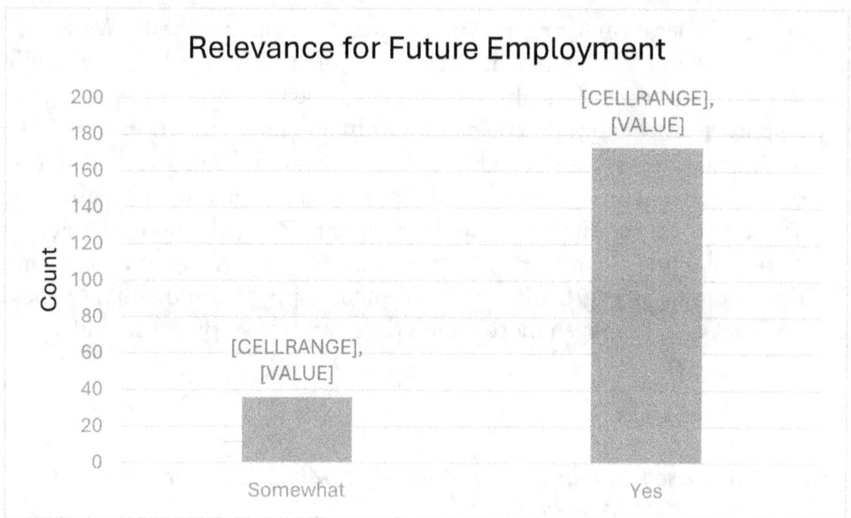



**FIGURE 19.6**  Relevance for future employment

*Source:* MBMA Survey.

**Employment outcomes:** The survey results revealed that 119 apprentices secured employment, while 68 remained unemployed and 22 were still completing their apprenticeship. Many of those employed had joined the MBMA; others had found opportunities in various organisations, including the Meghalaya Public Works Department, Meghalaya

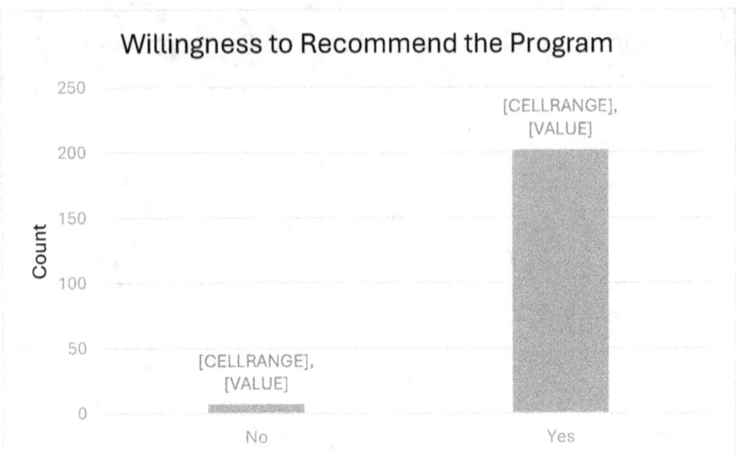


**FIGURE 19.7**  Willingness to recommend the programme

*Source:* MBMA Survey.

Power Generation Corporation Limited, Meghalaya State Watershed & Wasteland Development Agency, North Eastern Indira Gandhi Regional Institute of Health and Medical Sciences, North Eastern Space Applications Centre, Earthtree, Sauramandala Foundation and Shah Technical Consultants (P) Ltd., among others. This broad distribution of employment is a pointer to the effectiveness of the Apprenticeship Training Programme in equipping apprentices with industry-relevant skills, enabling them to secure diverse career opportunities across multiple sectors. Strengthening job placement support could enhance these outcomes, so that even more apprentices can transition successfully into the workforce.

### 19.5.2 Main Case Findings 2: Strengthening of MBMA's Talent Pipeline and Employment In Meghalaya

Of the 222 apprentices who participated in the Apprenticeship Training Programme, 164 have successfully completed their training. Of these, 106 had taken up full-time positions within the MBMA. This 65% absorption rate is an impressive achievement. Importantly, this figure not only shows the programme's effectiveness in equipping apprentices with industry-relevant skills but also underscores its success in seamlessly transitioning them into the workforce. The strong alignment between the Apprenticeship Training Programme's training objectives and MBMA's staffing needs reinforces the relevance of the skills imparted. This apparently ensures that graduates are

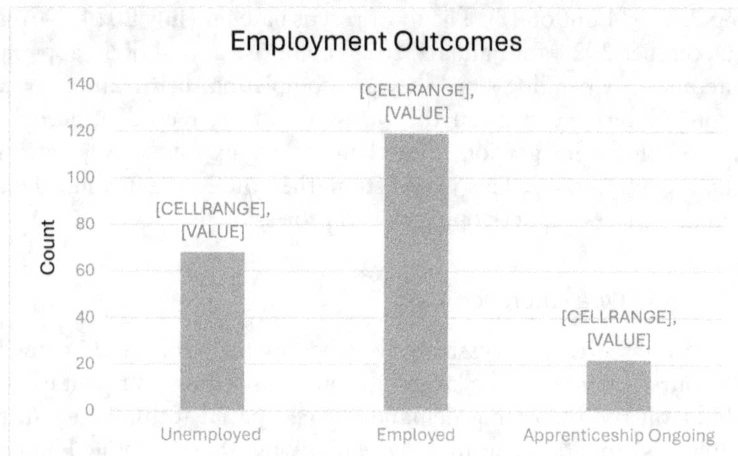


**FIGURE 19.8**  Employment outcomes

*Source:* MBMA Survey.

well prepared for long-term employment. Moreover, with more than half of the programme's graduates securing permanent positions, the initiative has had a tangible effect on career progression: It gives apprentices valuable opportunities for professional growth. This success benefits both the individuals and the MBMA, since it fosters a skilled, dedicated workforce and reinforces organisational continuity.

### 19.5.2.1 Batch-Wise Absorption Rates

The inaugural March 2022 batch, consisting of 10 apprentices, received training in GIS and UAV remote sensing. On completion, seven apprentices secured placements within the MBMA, a strong 70% absorption rate. This early success established a solid foundation for the Apprenticeship Training Programme. Building on this achievement, the second batch, launched in August 2022, expanded its curriculum and broadened the programme's scope. New courses were introduced, as already outlined. This expansion significantly increased enrolment, up to 39 apprentices. Of these, 22 secured a place within the MBMA, a 56% absorption rate. Although slightly lower than the first batch, this is a strong result, particularly given the broader range of disciplines introduced.

The programme's continuous evolution has led to consistently impressive placement results across successive batches. The March 2023 batch saw a 100% absorption rate, with all four apprentices securing positions within the MBMA. The April 2023 batch followed with a solid 67% placement rate (39 out of 58 apprentices); the March 2024 batch continued the upward trajectory, reaching 72% (34 out of 47). The most recent batches (July 2024, November 2024, December 2024 and January 2025) comprise a total of 64 apprentices. The outcomes are pending since they are completing their training or await evaluation. Nevertheless, given the MBMA's strong record of accomplishment in workforce integration, these batches are expected to achieve similar high absorption rates. This would strengthen the Apprenticeship Training Programme's success in developing skilled professionals.

### 19.5.2.2 Course-Wise Absorption Rates

Of the 106 apprentices successfully absorbed by the MBMA, 36 come from the GIS course, making it the largest group. This strong representation puts a spotlight on the increasing demand for geospatial technologies in planning, analysis and decision-making. Following the GIS, the Finance & Accounts course secured 23 placements, making it the second-largest group. This strong placement rate highlights the critical need for financial expertise in any organisation. By equipping apprentices with skills in budgeting, financial reporting and fiscal responsibility, the MBMA is actively working so that they can contribute effectively to both organisational efficiency and industrial standards. Meanwhile, the UAV Remote Sensing accounted for 18 placements, encouraging MBMA's commitment to technological innovation and its application in real-world scenarios. The Procurement course accounted for 12 placements, underscoring the critical role of procurement for efficient sourcing of goods and services. This strong placement rate reflects the growing demand for expertise in vendor management, logistics

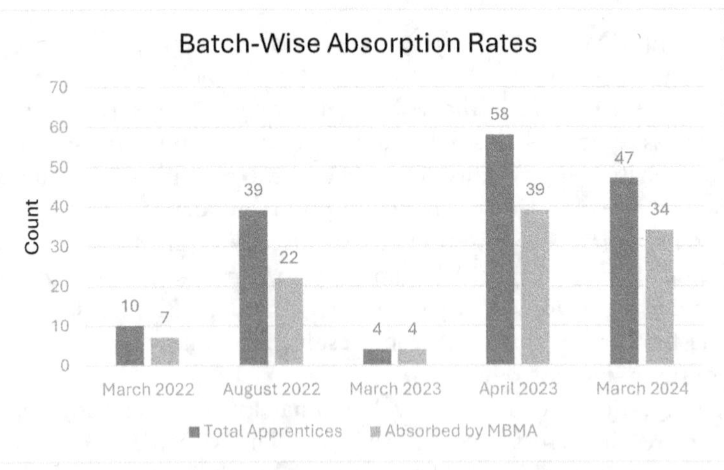


**FIGURE 19.9** Batch-wise absorption rates

*Source:* MBMA documentation and programme records.

and supply chain operations – key areas essential to MBMA's continued efficiency and broader industry trends.

Seven apprentices secured IT Developer positions, highlighting the organisation's dedication to digital transformation and the increasing need for tech-driven solutions. Smaller but noteworthy placements were achieved in Environment Management and Data Analysis, with three apprentices in each field. These apprentices are poised to contribute to MBMA's sustainability efforts and data-driven decision-making. Further diversification is evident in Knowledge Management (2 apprentices), as well as Monitoring & Evaluation and IT GIS (each with 1 apprentice). While these represent smaller segments, they add to the programme's broad scope, preparing apprentices for specialised and niche roles that are integral to MBMA's operations.

### 19.5.2.3 Apprentice's Response

Survey results from 209 apprentices further showed the Apprenticeship Training Programme's role in building a strong and sustainable talent supply for the MBMA. A significant 186 apprentices expressed interest in future job opportunities with the organisation. This is a clear sign that the programme has successfully nurtured a skilled workforce aligned with MBMA's operational needs. In addition, 23 apprentices responded with 'Maybe.' One may understand this as an indication that, while they found the programme valuable, their long-term employment decisions could be influenced by job

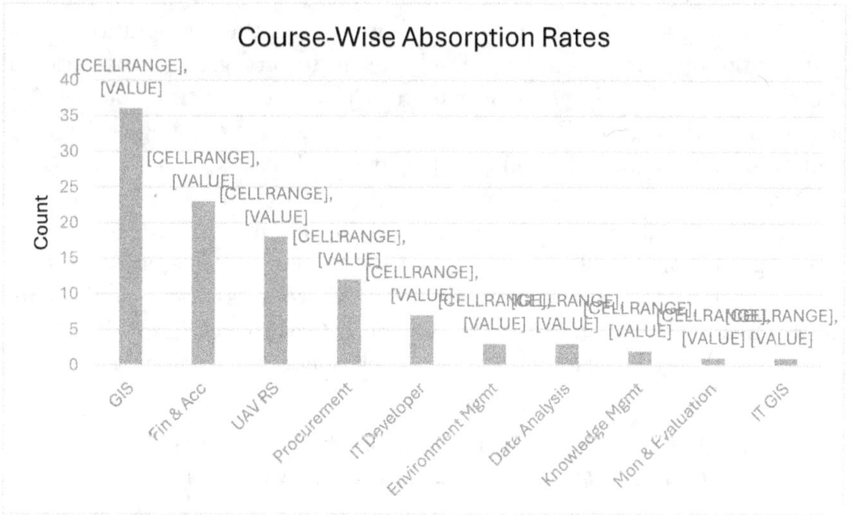


**FIGURE 19.10**   Course-wise absorption rates

*Source:* MBMA documentation and programme records.

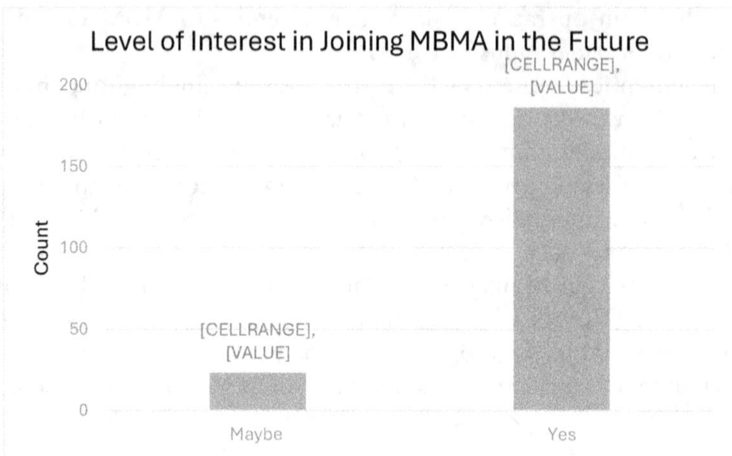


**FIGURE 19.11**   Interest in joining MBMA in the future

*Source:* MBMA survey.

role availability, career growth prospects or external market opportunities – among other factors.

Beyond talent development, the MBMA's Apprenticeship Training Programme has played a crucial role in reducing the agency's reliance on external hires. By equipping apprentices with specialised skills tailored to its needs, the programme ensured a steady supply of trained professionals, already familiar with the organisation's objectives, methodologies and operational practices. This has reduced the time and resources typically required for onboarding and training new hires from outside the organisation.

Additionally, internal talent development fosters greater institutional knowledge retention. Apprentices who transition into full-time roles bring a deep understanding of MBMA's landscape management strategies, environmental initiatives and community engagement efforts.

To strengthen this internal reservoir, the agency can promote retention strategies by introducing structured career pathways, mentorship programmes and targeted skill development initiatives. Establishing clearer progression routes – such as defined career ladders, internal promotion opportunities and competitive benefits – would encourage apprentices to view the MBMA as a long-term career destination rather than seek opportunities elsewhere.

### 19.5.3 Main Case Findings 3: Vital Role in Successful Implementation of Meghalaya Community-Led Landscapes Management Project (MCLLMP)

The MCLLMP was implemented from 2018 to June 2024, with a total funding of USD 43 million from the World Bank, USD 10 million from the Government of Meghalaya and USD 2 million from community contributions and other sources. As a pioneering initiative in Meghalaya's natural resource management sector, the project focused on restoring degraded landscapes, promoting sustainable land-use practices and empowering local communities to manage their resources effectively. It was implemented across 400 villages across 46 blocks in 11 districts, prioritising the restoration of degraded and highly degraded landscapes.

Initially, the project was tasked with bringing 31,510 ha under sustainable landscape management. It surpassed its target, successfully managing 35,764 ha within the 400 project villages and an additional 10,940 ha in non-project villages. This was an achievement of a total of 46,704 ha and an overall success rate of 148%. The World Bank recognised the MCLLMP as a successful and replicable model for community-led resource management. Some of its key achievements include the following:

1. **Nurseries:** 443 nurseries established, producing over 2.6 million saplings. Of these, 430,625 were cultivated for commercial sale, while the rest were used for plantation.
2. **Spring chambers:** 1,080 spring chambers constructed to improve groundwater recharge.
3. **Spring mapping:** Over 55,000 springs mapped, with 3,000 monitored monthly for discharge and water quality to enable timely interventions.
4. **Afforestation:** 868 afforestation sites developed, covering approximately 5,690 ha of land.
5. **Composting:** 539 composting units established to promote organic waste management.
6. **Water harvesting structures:** 280 structures built to support water conservation and sustainable agriculture.
7. **Agroforestry:** Introduced in 276 villages to integrate tree planting with sustainable farming practices.
8. **Soil and water conservation:** 24,948 conservation measures implemented to prevent erosion and enhance land productivity.
9. **Treatment of mining-affected areas:** Since 2020, 672 ha of land across 73 villages have been rehabilitated, benefiting 6,262 households.

### 19.5.3.1 Role of the Apprenticeship Training Programme

The Training Programme has played a pivotal role in the success of the MCLLMP by cultivating a skilled local workforce. Through specialised training, the programme reduced reliance on external professionals while creating employment opportunities within the agency. Many apprentices transitioned into full-time positions, contributing significantly to landscape management and environmental conservation. This has improved project

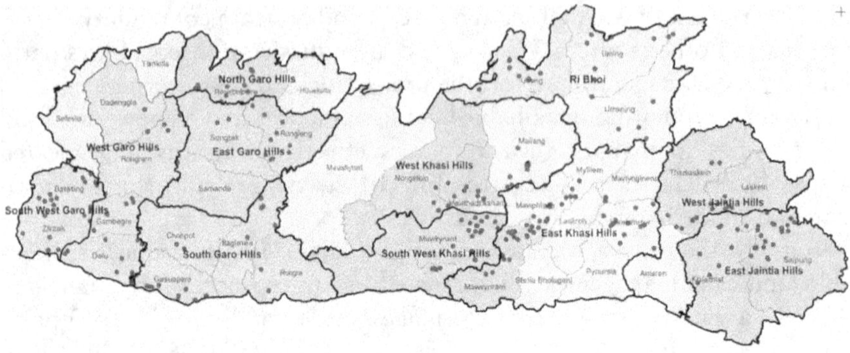

**FIGURE 19.12** MCLLMP coverage

*Source:* Centre of Excellence for NRM and Sustainable Livelihood. Government of Meghalaya.

efficiency and ensured that critical expertise remains within local communities, contributing to long-term resilience and self-sufficiency.

One of the programme's key achievements is its significance for GIS and UAV Remote Sensing. Apprentices trained in these fields have achieved the highest absorption rate within the MBMA, with many securing jobs as programme associates. Their expertise has been instrumental in monitoring reforestation efforts, analysing soil erosion patterns and optimising land-use planning.

By developing comprehensive geospatial databases, these professionals have empowered policymakers and project teams to make informed, data-driven decisions. Leveraging advanced remote sensing techniques, apprentices have enhanced the precision and efficiency of data collection, enabling timely interventions in environmentally sensitive areas. Their contributions have significantly strengthened watershed management, climate adaptation strategies and sustainable agricultural practices, amplifying the long-term effect of MBMA's initiatives.

Recognising the growing need for in-house geospatial expertise, the Government of Meghalaya established the Meghalaya State GIS & UAV Centre on 22 December 2022. This milestone positioned Meghalaya as the first state in Northeast India to have a dedicated GIS and UAV facility, bringing advanced geospatial capabilities with wide-ranging applications. Previously, Meghalaya had relied primarily on the North East Space Application Centre (NESAC) for geospatial data. But, since the NESAC serves the entire Northeast region, it often struggles to meet the state's immediate and large-scale requirements. The establishment of the GIS and UAV Centre is dealing with this problem by providing Meghalaya with greater autonomy in geospatial data collection, analysis and application. This state-of-the-art facility is set to enhance developmental planning across multiple sectors by delivering precise geospatial insights. Various government departments can leverage its services to improve spatial awareness, optimise resource management and strengthen evidence-based decision-making. From infrastructure development and environmental conservation to disaster management and agriculture, the Centre is poised to play a transformative role in streamlining planning processes and driving sustainable growth across the state.

The MBMA's Apprenticeship Training Programme has also significantly strengthened the financial and administrative capacities of the MCLLMP. By training apprentices in finance and procurement, the programme has improved financial management, ensured adherence to public procurement principles and enhanced transparency in resource allocation. These efforts have bolstered the project's accountability and efficiency, leading to smoother operations and a more effective utilisation of funds.

Also, the Training Programme has played a crucial role in community engagement and knowledge dissemination. Apprentices specialising in Knowledge Management have contributed to documentation, communication and outreach efforts, ensuring that best practices and success stories from the MCLLMP would be systematically recorded and shared. Their expertise in content writing, photography and videography has helped amplify awareness and support for sustainable landscape management initiatives. By capturing real-world impacts and community-driven efforts, they have facilitated broader public engagement and encouraged knowledge sharing among stakeholders.

## 19.6 Discussion

### 19.6.1 Interpretation of Findings in Relation to Their Replicability

The findings from the Training Programme bring to the fore its effectiveness in bridging the gap between academic learning and professional employment. The programme's ability to train and absorb apprentices into the workforce demonstrates a replicable model that can be adapted in other regions, particularly in areas with similar socio-economic and environmental challenges.

A key indicator of replicability is the high absorption rate of apprentices. This suggests that targeted skill development in high-demand sectors increases employability and can be replicated in other development-focused organisations. By aligning training with sectoral needs, similar programmes can ensure that apprentices are equipped with relevant and in-demand skills.

Another factor contributing to the programme's success is its multi-disciplinary approach. By integrating training in fields such as finance, procurement, environmental management and IT development, the programme prepares apprentices for a range of professional opportunities. This flexible, demand-driven model can be adapted in other regions by identifying key sectors that require skilled professionals and tailoring training programmes accordingly.

Additionally, the programme's role in supporting large-scale development projects, such as the MCLLMP, highlights its scalability. Apprentices have contributed to critical activities, including GIS mapping, financial planning, monitoring and evaluation and environmental conservation. Similar apprenticeship initiatives could be integrated into other government (and donor) funded projects to enhance project implementation while simultaneously building a skilled workforce.

But successful replication requires tackling challenges such as ensuring adequate employment opportunities, providing financial support to apprentices and fostering industry partnerships. Establishing collaborations with private sector companies and government agencies can expand job placement opportunities and create a sustainable employment reservoir.

Overall, the MBMA's Apprenticeship Training Programme serves as a strong example of how structured, hands-on training can improve workforce readiness and project efficiency. With careful adaptation to local needs, similar models can be replicated in other states and regions, enhancing employment outcomes and strengthening institutional capacities for sustainable development.

## 19.7 Implications for Theory, Practice and Policy

The MBMA Apprenticeship Training Programme has significant implications across theory, practice and policy, particularly in workforce development, skill-based education and sustainable livelihoods. Its success provides valuable insights into how structured training programmes can enhance employability, support development projects and strengthen institutional capacities.

From a theoretical standpoint, the programme reinforces a human capital development model that emphasises the importance of education and skills training in improving employment outcomes and economic productivity. It also aligns with experiential learning theories, highlighting the role of hands-on, practical experience in reinforcing theoretical knowledge.

The programme's success in absorbing apprentices suggests that structured, sector-specific training models can bridge the gap between education and employment, a concept that can be further explored in skill development research.

In practice, the programme demonstrates the importance of industry-aligned, demand-driven training. By offering the various courses mentioned, the programme ensures that apprentices acquire skills that are directly applicable to real-world projects. This approach can serve as a model for other training programmes seeking to enhance workforce readiness. Additionally, the programme's integration into large-scale projects like the MCLMP shows how apprenticeship models can support sustainable development initiatives. The apprentices' contributions to activities such as GIS mapping, financial planning and environmental conservation emphasise the value of trained personnel in project execution and monitoring.

At the policy level, the programme underscores the need for greater investment in apprenticeship and skill-based training initiatives. Governments and development organisations can use this model to design policies that promote vocational training and align education with industry needs. Further, the programme's high absorption rates suggest that well-structured apprenticeship models can address unemployment challenges, particularly among the youth.

To maximise results, policies should encourage public-private partnerships, ensuring that apprenticeship programmes are co-developed with industries to meet workforce demands. Moreover, scaling such programmes at the national level can contribute to broader economic growth and sustainable development.

## 19.8  Comparison with Existing Literature

The Training Programme aligns with existing literature on skill development, workforce readiness and vocational training. It reinforces established theories while also presenting new insights into region-specific implementation. Research on apprenticeships and vocational training is bringing out their role in reducing youth unemployment and bridging the gap between education and employment.

Studies by the ILO suggest that apprenticeship programmes contribute to economic growth by creating a skilled labour force tailored to industry needs. Similar findings are echoed in reports on Germany's dual vocational training system, which has been widely recognised for its success in workforce development. The MBMA programme reflects these principles by offering hands-on training in high-demand sectors such as GIS, UAV Remote sensing and environmental management, which have already led to high absorption rates.

Existing literature on experiential learning also supports the MBMA model. Scholars such as Kolb (1984) argue that learning is most effective when individuals gain direct experience in real-world settings. The programme's structured training, combined with on-the-job exposure, aligns with this perspective by equipping apprentices with practical skills applicable to large-scale projects like the MCLLMP.

Additionally, literature on rural and sustainable development highlights the need for localised capacity-building initiatives. Studies on community-led resource management stress the importance of skilled personnel in implementing conservation and livelihood programmes. The MBMA's apprenticeship initiative contributes to this discourse by demonstrating how targeted skill development can improve project implementation in rural landscapes.

Even so, while the MBMA programme shares similarities with global models, it is distinct in its integration with community-driven environmental initiatives. Unlike traditional apprenticeships that focus solely on technical skills, this programme embeds apprentices within development projects, emphasising the role of vocational training in sustainable development. This unique aspect adds to the existing literature by outlining an adaptable, region-specific approach to workforce development.

## 19.9  Summary and Conclusion

The Training Programme has proven to be a significant initiative in bridging the gap between academic learning and practical application, particularly in sectors critical to Meghalaya's sustainable development. Through structured training in GIS, UAV Remote Sensing, Finance, Procurement, Environmental Management and other disciplines, the programme has successfully equipped apprentices with industry-relevant skills. The high absorption rates of graduates into MBMA projects highlight the programme's effectiveness in workforce development and capacity building.

A key strength of the programme is its alignment with large-scale initiatives like the MCLLMP. By integrating apprentices into real-world projects, the programme is ensuring that trainees gain firsthand experience in implementing conservation, resource management and infrastructure development activities. This approach not only reinforces their employability but also strengthens the effect of community-driven projects by ensuring the availability of skilled personnel.

The findings from the programme demonstrate its potential for replicability. Similar apprenticeship models could be implemented in other regions facing workforce shortages in specialised fields. The emphasis on experiential learning, coupled with structured mentorship, serves as a model for vocational training programmes seeking to improve employment outcomes. The programme also contributes to broader policy discussions on skill development, showcasing how targeted apprenticeship initiatives can address sector-specific labour demands.

Despite its successes, there are areas for improvement. Expanding the reach of the programme to more remote villages, strengthening partnerships with private sector employers and incorporating emerging technologies into training modules could add to its impact. Besides, continuous monitoring and evaluation will be crucial in refining the curriculum to meet evolving industry needs.

In conclusion, the MBMA's Apprenticeship Training Programme has played a pivotal role in workforce development, skill enhancement and community-driven resource management. Its success underscores the importance of integrating vocational training into development projects, ensuring that local communities benefit from both skill acquisition and employment opportunities. As the programme continues to evolve, its model can serve as a benchmark for similar initiatives aimed at building a skilled, self-sufficient workforce capable of driving sustainable development in diverse contexts.

## 19.10  Key Findings Summary

The MBMA's Apprenticeship Training Programme has successfully enhanced workforce development by equipping apprentices with industry-relevant skills in GIS, UAV Remote Sensing, Finance, Procurement, Environmental Management and other disciplines. The programme's high absorption rates indicate a strong demand for trained professionals, particularly in large-scale projects like the MCLLMP. Apprentices gained hands-on experience, improving project implementation and sustainability. The programme also demonstrated its potential for replicability in other regions facing similar workforce challenges. Key areas for improvement include expanding outreach, strengthening industry partnerships and updating training modules to align with emerging technologies and evolving sectoral needs.

## 19.11  Significance Restatement

The MBMA's Apprenticeship Training Programme plays a crucial role in bridging the skill gap and fostering workforce readiness in Meghalaya. By providing hands-on training in specialised fields, the programme is enhancing employability while directly contributing to the successful implementation of community-led projects. Its integration with large-scale initiatives like the MCLLMP underpins its role in sustainable development. The programme's high absorption rates underscore its effectiveness, making it a model for vocational training in resource management and rural development. Its success demonstrates the importance of structured apprenticeship programmes in strengthening local economies and driving long-term community resilience .

## 19.12  Future Research and Practical Applications

Future research on the MBMA's Apprenticeship Training Programme could focus on long-term career trajectories of apprentices to assess sustained employment impacts. Comparative studies with similar programmes in other regions could provide insights into best practices and areas for improvement. Additionally, research on the scalability of the programme, particularly its integration with private sector industries, could increase its effectiveness. From a practical perspective, expanding the programme to include emerging technologies such as AI-driven GIS analysis, advanced UAV applications and climate resilience strategies could further strengthen its relevance. Reinforcing industry partnerships and creating pathways for apprentices to transition into leadership roles would also contribute to long-term benefits. Replicating this model in other states or regions with similar socio-economic and environmental challenges could maximise its impact. By continuously refining training methodologies and expanding its reach, the programme has the potential to serve as a blueprint for workforce development in sustainable resource management.

## References


Almeida, R., Behrman, J., & Robalino, D. (Eds.). (2012). *The right skills for the job? Rethinking training policies for workers.* International Bank for Reconstruction and Development/The World Bank. https://documents.worldbank.org/en/publication/documents-reports/documentdetail/535251468156871924/the-right-skills-for-the-job-rethinking-training-policies-for-workers\

Asian Development Bank. (2014). *Technical and vocational education and training for accelerated human resource development in South Asia.* https://www.adb.org/sites/default/files/publication/416/innovative-strategies-technical-vocational-education-training.pdf

Center of Excellence for NRM and Sustainable Livelihood. (n.d.). *Meghalaya Community-Led Landscape Management Project (MCLLMP).* Government of Meghalaya. https://coenrm.megplanning.gov.in/mcllmp/

Euler, D. (2013). *Germany's dual vocational training system: A model for other countries?* Bertelsmann Stiftung. https://www.bertelsmann-stiftung.de/fileadmin/files/BSt/Publikationen/GrauePublikationen/GP_Germanys_dual_vocational_training_system.pdf

Fuller, A., & Unwin, L. (2017). *Better apprenticeship: Access, quality and labour market outcomes in the English apprenticeship system.* The Sutton Trust. November. https://www.suttontrust.com/our-research/better-apprenticeships-quality-access-social-mobility/

International Labour Organisation (ILO). (2022). *Good practices in apprenticeships in India: Challenges and opportunities. Apprenticeships Development for Universal Lifelong Learning and Training (ADULT).* Geneva. https://www.ilo.org/publications/good-practices-apprenticeships-india-challenges-and-opportunities#:~:text=The%20country%2Dlevel%20report%20%E2%80%9CGood,promote%20apprenticeship%20in%20the%20country

International Labour Organisation (ILO). (2024). *Quality Apprenticeship Recommendation, 2023 (No. 208). Guide for policymakers.* https://www.ilo.org/sites/default/files/2024-12/1684_SKILLS_Quality_Apprenticeships_WEB.pdf

Kolb, D. A. (1984). *Experiential learning: Experience as the source of learning and development.* Prentice Hall. https://www.fullerton.edu/cice/_resources/pdfs/sl_documents/Experiential%20Learning%20-%20Experience%20As%20The%20Source%20Of%20Learning%20and%20Development.pdf

Meghalaya Basin Development Authority. (2022). *MBDA apprenticeship training programme*. https://mbda.gov.in/mbda-apprenticeship-training-programme

Meghalaya Basin Management Agency & the World Bank. (n.d.). *Meghalaya Community-Led Landscape Management Project (MCLLMP): Implementation completion report*. Government of Meghalaya. www.cllmp.com/wp-content/uploads/2024/11/Borrowers-Implementation-Completion-Report.pdf

# 20

# CONCLUSION

## Policy Convergence and the Blueprint for Regenerative Governance in Meghalaya

*Anjal Prakash, Sampath Kumar, and Gunanka D. B.*

The volume culminates by analysing a decade of transformative shifts in Natural Resource Management (NRM) within Meghalaya, India. The core challenges discussed across these chapters arise from the state's distinctive governance structure under the Sixth Schedule of the Indian Constitution, wherein communities, clans, and individuals collectively own and manage more than 90% of the land. While this deeply decentralised system safeguards traditional autonomy and local stewardship, it often falls short of achieving genuine empowerment. In the absence of cohesive planning, reliable geospatial data, and coherent policy frameworks, decentralisation without adequate institutional capacity has, at times, exacerbated resource degradation rather than mitigating it.

The initiatives spearheaded by the Meghalaya Basin Development Authority (MBDA) and Meghalaya Basin Management Agency (MBMA), notably the Community-Led Landscape Management Project (CLLMP), systematically tackled these impediments. They pioneered an approach centred on Integrated Adaptive Co-Management (IACM), which strategically integrated traditional ecological knowledge (TEK) with modern scientific methodologies and market-based incentives.

This concluding chapter aims to map the critical issues examined throughout this volume, interpret the significance of the resulting policy convergence, and articulate the model's replicable components. The core thesis confirmed by this volume is that sustainable and equitable resource governance in indigenous regions is achieved by transforming local communities into empowered, technically proficient, and financially incentivised stewards of their natural capital.

DOI: 10.4324/9781003735380-24

## 20.1 Key Issues in Natural Resource Management and Governance

Table 20.1 summarises the core challenges, interventions, and resulting ecological and institutional outcomes discussed in this volume, cross-referencing the chapters where these issues were primarily investigated. This detailed cross-referencing highlights the interconnected, multidisciplinary nature of the Meghalaya model, illustrating how foundational issues in governance were resolved through targeted technical and financial solutions.

## 20.2 The Convergence of Bottom-Up Governance and Modern Ecology

The critical insight derived from Table 20.1 is the systematic convergence of grassroots capacity, technological application, and policy adaptation. The interventions moved beyond mere project delivery to establish durable, institutionalised structures that address the state's inherent governance challenges.

### 20.2.1 The Institutional Pivot: From VEC to NRMC

Prior to these initiatives, the Village Employment Council (VEC), established under MGNREGA, lacked the specific capacity or mandate to effectively plan and execute NRM interventions, despite 60% of MGNREGA funds being allocated for this purpose. The MBDA's creation of Village Natural Resource Management Committees (VNRMCs) within the project villages solved this structural deficit. By demonstrating success in planning and executing interventions across 50,000 ha of land, the VNRMC model provided the necessary evidence base for a fundamental policy shift. In 2021, the state officially notified a policy to replicate this model by creating Natural Resource Management Committees (NRMCs) in *every* village, enabling them to leverage the large NRM component of the MGNREGA budget. This action closes the structural loop on decentralised governance fragmentation, validating the bottom-up approach (Chapters 2 and 16).

### 20.2.2 Technology Democratisation and Data Ownership

A defining feature of the Meghalaya model is the transformation of advanced technology (GIS, RS, UAVs) from an external imposition to a tool of community empowerment. The Apprenticeship Training Programme (ATP) and the training of VCFs successfully created an internal human resource pipeline, directly addressing the local skills gap. The high absorption rate of GIS and UAV trainees into the MBMA (65% absorbed overall, GIS being the largest group) ensured that geospatial expertise – essential for monitoring

**TABLE 20.1**  Core challenges, interventions, and institutional outcomes

| Category | Key issue/challenge | Intervention/solution implemented | Quantitative/qualitative outcome |
|---|---|---|---|
| I. Governance & Capacity (chapters 2, 11, 16, 18) | Lack of local institutional capacity for NRM planning (Pre-2021) | Establishment and capacity building of village natural resource management committees (VNRMCs) | VNRMCs demonstrated a viable model for decentralised decision-making; established in 400 villages under CLLMP; prepared over 400 NRM plans |
| | Skills gap in advanced technical fields (GIS, UAVs, Finance) for EAPs | Apprenticeship training programme (ATP) | 222 apprentices admitted across nine batches; 164 completed; 65% absorption rate (106 staff) into MBMA/MBDA. GIS and Finance had the highest absorption rates |
| | Lack of government reach and community trust in remote villages | Deployment of trained local youth as village community facilitators (VCFs) | Over 1,161 VCFs trained in technical and participatory skills; over 13,000 community professionals trained cumulatively; facilitated seamless bridging between the State and communities |
| | Fragmentation of geospatial data and duplication of effort across state departments | Creation of the State GIS and UAV Centre and State Geo Portal | Consolidated all existing data layers; enabled access for all developmental actors on a single platform; facilitated data-driven decision-making in sectors like agriculture and water resources |
| | Community fear of land confiscation due to government intervention | Community-led participatory mapping of village boundaries using GIS/RS | Maps reflected community-recognised land divisions, strengthening local ownership, and reducing resource conflicts between villages |
| II. Financial Mechanisms (chapters 15, 17) | Unrecognised value of natural forest cover and lack of incentive for preservation | Implementation of payment for ecosystem services (PES) under GREEN Meghalaya and GM+ | India's first large-scale PES initiative; enrolled 3,300 beneficiaries; protected over 51,000 ha of natural forest; disbursed Rs.48 crores in financial incentives |

| Category | Key issue/challenge | Intervention/solution implemented | Quantitative/qualitative outcome |
| --- | --- | --- | --- |
| | PES Model Exclusion of existing conservation efforts | PES specifically compensates for conserving existing natural forests, viewing this as more cost-effective than rejuvenation | PES-based incentive raised up to ₹20,000 per hectare/year under GM+; minimum commitment of 30 years |
| | Lack of financial resources/ incentives for smallholder farmers to participate in climate mitigation | An Agroforestry-Carbon financing project under Rabobank's Acorn platform | Over 13,410 farmers enrolled across 12 districts; managing 23,409 ha of land; baseline carbon stock estimated at 1,176.87 tonnes C/ha in 400 villages surveyed |
| | Dependence on EAP funding and lack of long-term revenue for NRM | Exploring diversified financing strategy, including Government budget allocations and Carbon Credit leveraging | MBMA is actively negotiating to leverage the carbon market for forests under GM+ |
| III. Ecological Restoration (chapters 3, 7, 8,9,14) | Severe soil acidity and heavy metal contamination from rat-hole mining | Phytoremediation using aromatic grasses (*Cymbopogon winterianum*) | Soil pH improved from 4.8 to 5.32; Soil organic carbon (SOC) increased from 1.7% to 2.9%; Iron (Fe) levels reduced by 48%; Manganese (Mn) reduced by 36% |
| | Acid mine drainage (AMD) contamination of water sources (pH 2–3) | Passive treatment using open limestone channels (OLCs), integrated with Vetiver grass phytoremediation | Chambered OLCs demonstrated 90% success rate in neutralising acidic water compared to 60% in drainage ditches; treated pH consistently remains within BIS standards |

(*Continued*)

**TABLE 20.1** Continued

| Category | Key issue/challenge | Intervention/solution implemented | Quantitative/qualitative outcome |
|---|---|---|---|
| | Acute water scarcity during the lean season despite high rainfall | Community-led hydrological spring mapping and springshed management | Over 55,000 springs mapped statewide; 3,000 critical springs monitored monthly; intervention models successfully led to improvements, e.g., in Mawkyrdep, enhancing water security |
| | Soil erosion and nutrient depletion from shifting cultivation (Jhum) in hilly terrain | Implementation of sloping agricultural land technology (SALT) using nitrogen-fixing plants (NFPs) as hedgerows | On managed farms (e.g., East Jaintia Hills), SALT led to improvement in soil pH (from 4.5–5.5 to 5.6) and a significant increase in Nitrogen (426 ppm) |
| | Lack of quality planting materials (QPMs) suited to microclimates and challenges in transporting saplings | Establishment of decentralised community nurseries managed by communities/ SHGs | 739 community managed tree nurseries established; produced over 3.5 million seedlings; contributed to restoring 37,137 ha of degraded land; achieved over 90% survival due to reduced transportation shock |
| IV. Data, Technology & Monitoring (chapters 5, 10, 11, 12, 16,17) | Lack of accurate baseline data on key natural resources (e.g., bamboo, forest stock) | Creation of community-led village resource inventory, Forest management plans (FMPs) and statewide bamboo resource assessment | 1,102 km² brought under FMPs; growing stock estimated at 4,901.43 m³; carbon stock estimated at 1,176.87 tonnes C/ha; bamboo culms mapped (over 2,100 million culms). Comprehensive statewide bamboo assessment estimating 2,649 million culms (22.9 million tonnes) across Meghalaya, identifying key species and a sustainable annual yield of 165.7 million culms |

| Category | Key issue/challenge | Intervention/solution implemented | Quantitative/qualitative outcome |
| --- | --- | --- | --- |
|  | Need for transparent verification of conservation activities for performance-based payments (PES) | Development of measurement, reporting, and verification (MRV) protocol utilising GIS, RS, and the PES Forest Watcher Dashboard | Dashboard provides real-time spatial and temporal data on forest cover changes and fund disbursement status; VCFs/GFAs conduct verification using the transect line methodology |
|  | Difficulty in large-scale afforestation (cost, logistics, community engagement) | Mass community engagement through the seed ball initiative using school students and VCFs | 3 million seed balls prepared and dispersed by 75,000 students from 1,840 schools; post-dispersal survey showed an average 55% germination and survival rate |
|  | Lack of documented indigenous knowledge and heritage threat | Research, documentation, and conservation of the Jingkieng Jri/Lyu Chrai Cultural Landscape | Conservation of 74 living root bridges; landscape included in the UNESCO World Heritage tentative list (2022) and final dossier submitted to ASI (2025); formalised collaborations with IISc, Geological Survey of India, etc |

*Source:* Compiled from the chapters of this volume.

and planning – became an institutional capability, secured by the new State GIS and UAV Centre.

This decentralised technical capacity enabled community-led GPS mapping to systematically delineate village and forest boundaries, providing objective, verifiable data where traditional cadastral maps are often lacking in Meghalaya and thereby supporting conflict resolution. It also supported performance transparency: The PES Forest Watcher Dashboard and the MRV protocol, driven by GIS, link the disbursement of substantial financial incentives (up to ₹20,000 per hectare annually under GM+) and (base price is 10,000) to quantifiable ecological outcomes, with results cross-checked using satellite imagery and field surveys (Chapters 11 and 18).

### 20.2.3 Market Alignment and Ecological Viability

The volume consistently highlights the necessity of providing economic alternatives to destructive practices like unscientific mining and high-impact shifting cultivation. The integration of market mechanisms provided a credible solution.

The pioneering GREEN Meghalaya PES scheme directly incentivised custodians for the ecological services provided by their natural forests – carbon sequestration, water regulation, and biodiversity – recognising that conserving existing forests is far more cost-effective than restoration. At the same time, the agroforestry – carbon financing project integrated smallholder farmers into global climate finance, enabling them to generate Carbon Removal Units (CRUs) from sustainable land use and thereby establish a green livelihood. Restoration efforts were designed for dual purposes: Ecological interventions also generated revenue. For example, phytoremediation of acid-mine spoils using aromatic grasses raised soil pH and reduced heavy metal concentrations while providing an annual return of ₹40,000–₹50,000 per hectare through essential oil production. This confirms the success of linking conservation commitment (e.g., the 30-year requirement for PES) directly to immediate, performance-verified financial reward (Chapters 17 and 15).

### 20.2.4 Nature-Based Solutions and Resilience Building

The book documents the successful deployment of context-specific, nature-based solutions to reverse severe environmental damage. Passive treatment systems effectively addressed geological contamination: Acid Mine Drainage (AMD) was neutralised using Open Limestone Channels (OLCs), with an enhanced chambered design achieving about a 90% success rate, while Vetiver grass integrated into OLCs and aromatic grasses used in phytoremediation absorbed heavy metals and improved water quality.

To bolster water security – critical where 80% of the rural population depends primarily on springs – a large-scale spring-mapping project (over 55,000 springs mapped) produced the hydrogeological baseline needed for targeted catchment restoration.

For slope stability, widespread adoption of Sloping Agricultural Land Technology (SALT) gave farmers a systematic agroforestry-based approach to curb soil erosion and nutrient loss from shifting cultivation, yielding measurable soil health improvements in targeted areas. These ecological solutions were all executed through the deployed VCF/VNRMC structure, reinforcing the principle that local capacity is indispensable for large-scale environmental intervention (Chapters 8, 9, and 7).

## 20.3 Policy and Practice Convergence, Significance, and Limitations

### 20.3.1 Implications for Policy and Practice

The findings have profound implications for policy formulation and practice, both within Meghalaya and globally, particularly for regions operating under similar devolved governance structures or facing large-scale land degradation.

#### 20.3.1.1 Policy Implications

The project activities have successfully influenced state policy, demonstrating that ground-level experience can guide governance reform. Firstly, the project activities have successfully influenced state policy, demonstrating that ground-level experience can guide governance reform. Secondly, future policy must formalise geospatial governance by institutionalising geospatial training within local governance structures so that VCFs and VNRMCs have continuous access to data, tools, and decision-support systems, leveraging the State Geo Portal and GIS Centre as crucial assets. Thirdly, policymakers should scale incentive-based conservation by expanding PES and Carbon Finance frameworks, integrating community-led MRV protocols to ensure transparency and performance-based compensation, and sustaining financial incentives to secure the 30-year conservation commitments required under PES.

Fourthly, restoration models that are successful and low-cost – such as the chambered OLC design for AMD neutralisation, aromatic grasses for heavy metal phytoremediation, and the SALT model for sloping terrains – should be mainstreamed into state NRM and rural development schemes to ensure convergence and adequate funding. Finally, strengthening local institutions through continued support for VNRMCs to manage funds, resolve local conflicts, and leverage schemes like MGNREGA NRM funds is essential for decentralised management, and the success of this model makes it a viable approach for states lacking Panchayat Raj systems.

The case studies in this volume also show that the implementation of EAPs functioned as an effective policy laboratory, producing robust evidence that informed statewide reforms. Three clear convergences emerged: First, the 2021 state notification to establish Natural Resource Management Committees (NRMCs) in every village – replacing VECs for NRM planning – directly stems from the demonstrable success of VNRMCs across 400 CLLMP villages and formally institutionalises grassroots authority over NRM budgeting and planning (MGNREGA); second, the creation of the State GIS and UAV Centre and the Geo Portal, driven by needs identified during spring mapping and MRV activities, centralises geospatial governance to ensure sustained, data-driven decision-making beyond project timelines; and third, the drafting of the Meghalaya Climate Emergency & Green Growth Framework aims to embed successful MBDA initiatives – especially the PES model – into state law, securing long-term financial and policy commitments to climate resilience and green growth.

The Meghalaya model holds significant implications for global sustainable development, especially in regions with high biodiversity and complex land tenure systems. Firstly, the governance approach – training a local cadre of technical facilitators (VCFs/VDVs) who are accountable to community institutions (VNRMCs/NRMCs) – is highly replicable, notably in areas such as Northeast India where Panchayat Raj systems are absent. Secondly, the financing approach demonstrated by PES and carbon finance provides a pragmatic pathway for governments to justify and sustain conservation funding; by showing financial viability (₹40,000–₹50,000 per hectare revenue from reclaimed land), the model supports integrating conservation into economic planning in resource-constrained regions. Thirdly, the technical solutions – specialised, low-cost interventions like chambered OLCs, aromatic grass phytoremediation, and community nurseries for QPM supply – offer proven, transferable blueprints for addressing specific forms of ecological degradation worldwide.

### 20.3.1.2 Practical Implications

For practitioners engaged in NRM and sustainable development, the Meghalaya experience offers a critical roadmap. Firstly, prioritising community engagement and ownership is essential: Local participation – from initial awareness through implementation and long-term monitoring – must be secured, as seen where up to 16% of the population joined the Cham Cham seed-saving initiative. Secondly, adopt adaptive management: NRM plans should evolve with changing ecological and social conditions, using continuous inputs from the GIS/MRV system and VCF field reports to refine practices (for example, optimising hedgerow maintenance in SALT areas). Thirdly, diversify restoration funding by pursuing sources beyond government budgets, including carbon credits, and community–private partnerships for resource-based industries such as bamboo. Finally, emphasise capacity building with practical, hands-on, multidisciplinary training (GIS, finance, procurement, agroforestry) to develop a versatile local workforce capable of implementing complex EAPs.

### 20.3.2 Limitations and Open Loops

Despite the successes, the volume identifies several systemic limitations that require continued attention. Firstly, land tenure ambiguity remains a fundamental challenge: The absence of a cadastral mapping system means that VCF-produced GPS village maps are functional rather than legal, complicating long-term land security, afforestation projects with extended commitments (such as the 30-year PES mandate), and resolution of resource conflicts. Secondly, technical literacy and geographic isolation persist as obstacles: Despite strong capacity building, ongoing gaps in technical skills – combined with difficult terrain, dense forests, steep slopes, and monsoon disruptions – hinder field monitoring and verification and contribute to chronic manpower shortages in remote blocks. Finally, sociocultural barriers continue to impede programme uptake, with land ownership disputes among beneficiaries and reluctance from traditional leaders (Syiem or Nokma) limiting full participation in initiatives like GM+.

## 20.4  Conclusion and the Way Forward

The comprehensive evidence provided across these chapters demonstrates a highly successful operational response to the core dilemma facing Meghalaya: How to foster large-scale environmental sustainability when land governance is profoundly decentralised and locally owned. The answer lies in transforming the custodians of the land into the architects of its future, supported by rigorous data and financial incentives.

The initial challenge – the absence of effective institutional arrangements at the village level to plan for NRM – has been decisively addressed by the institutionalisation of the VNRMC/NRMC model. The structural foundations for regenerative governance are now in place, underpinned by a capable, locally trained workforce (VCFs/VDVs) and centralised, transparent geospatial data management.

The lasting impact of the Meghalaya model is its dual commitment to the ecology and the economy. It proves that forest health is a superior economic asset when its ecosystem services are monetised (via PES and Carbon CRUs) and when degraded lands are simultaneously restored and converted into profitable green livelihoods (e.g., essential oils from phytoremediation).

The core principles demonstrated in Meghalaya are highly replicable in other ecologically sensitive regions, particularly in the Indian Himalayan Region and in developing countries confronting similar challenges in decentralised land management. Firstly, conditions for success include institutional support to formalise local committees; the integration of scientific monitoring with local and traditional ecological knowledge (TEK); financial incentives that make conservation economically rewarding; and capacity building tailored to close technical skill gaps in rural populations. Secondly, specific replicable models include the VCF cadre, which can be adapted to states such as Arunachal Pradesh and Nagaland; community-led resource assessments (forest inventory, bamboo assessment) that provide practical blueprints for managing privately and community-owned resources; and the dual reclamation–livelihood model – such as aromatic grass cultivation – that combines ecological restoration with income generation and serves as a template for areas affected by mining or industrial degradation worldwide.

### 20.4.1 Call to Action: Securing the Green Future

For the Government of Meghalaya and its partners, the call to action is to finalise the transition from project-based success to a permanently funded governance norm.

Firstly, secure financial continuity by fully leveraging carbon market revenue streams and sustaining the current dedicated state budget allocations to PES and VCF/NRMC structures.

Secondly, resolve land tenure conflicts by aggressively addressing the non-legal status of community mapping data – developing state-level frameworks that formally recognise community-led boundary delineations to mitigate ownership disputes and ensure long-term conservation commitments.

Thirdly, deepen technical resilience by expanding the Apprenticeship Training Programme to provide continuous capacity building in emerging areas (e.g., AI-driven GIS analysis) and integrating this trained cadre into permanent positions within local governance structures.

Meghalaya's experience is a compelling argument that global climate action cannot be delivered by centralised mandates, but by trusting the tenure of the earth to its most committed and knowledgeable stewards. When local knowledge is amplified by modern technology, and stewardship is valued by the market, communities become the unstoppable engine of sustainable development. The challenge is no longer *if* decentralised conservation can work, but how quickly the rest of the world can implement the lessons learned in the Khasi, Jaintia, and Garo Hills.

# INDEX

For Product Safety Concerns and Information please contact our EU
representative  GPSR@taylorandfrancis.com
Taylor & Francis Verlag GmbH, Kaufingerstraße 24, 80331 München, Germany